Laser & Magnetism

Laser & Magnetism

Editor:

Olga Moreira, Ph.D.

ArclerPress

www.arclerpress.com

Laser & Magnetism

Editor: Olga Moreira, Ph.D.

© 2017 Arcler Press LLC
708 3rd Avenue, 6th Floor
New York
NY 10017
United States of America
www.arclerpress.com

ISBN: 978-1-68094-431-0

Library of Congress Control Number: 2016951123

About the Editor

Olga Moreira, Ph.D.

Olga Moreira obtained her Ph.D. in Astrophysics from the University of Liege (Belgium) in 2010, her BSc. in Physics and Applied Mathematics from the University of Porto (Portugal). Her post-graduate travels and international collaborations with the European Space Agency (ESA) and European Southern Observatory (ESO) led to great personal and professional growth as a scientist. Currently, she is working as an independent researcher, technical writer, and editor in the fields of Mathematics, Physics, Astronomy and Astrophysics.

Contents

List of Contributors

C. S.Gonçalves
Departamento de Física e Astronomia and IFIMUP-IN, Faculdade de Ciências, Universidade do Porto, Rua do Campo Alegre 687, 4169-007 Porto, Portugal.

A. S. Silva
Departamento de Física e Astronomia and IFIMUP-IN, Faculdade de Ciências, Universidade do Porto, Rua do Campo Alegre 687, 4169-007 Porto, Portugal.

D. Navas
Departamento de Física e Astronomia and IFIMUP-IN, Faculdade de Ciências, Universidade do Porto, Rua do Campo Alegre 687, 4169-007 Porto, Portugal.

M. Miranda
Department of Physics, Lund University, P.O. Box 118, SE-221 00 Lund, Sweden.

F. Silva
Departamento de Física e Astronomia and IFIMUP-IN, Faculdade de Ciências, Universidade do Porto, Rua do Campo Alegre 687, 4169-007 Porto, Portugal.

H. Crespo
Departamento de Física e Astronomia and IFIMUP-IN, Faculdade de Ciências, Universidade do Porto, Rua do Campo Alegre 687, 4169-007 Porto, Portugal.

D. S. Schmool
Departamento de Física e Astronomia and IFIMUP-IN, Faculdade de Ciências, Universidade do Porto, Rua do Campo Alegre 687, 4169-007 Porto, Portugal
Laboratoire PROMES CNRS (UPR 8521), Université de Perpignan Via Domitia, Perpignan, France.
Correspondence and requests for materials should be addressed to D.N.

Haitian Xu
Department of Physics and Astronomy, University of Victoria, Victoria V8P 5C2, Canada

Ghazal Hajisalem
Department of Electrical and Computer Engineering, University of Victoria, Victoria V8P 5C2, Canada. Correspondence and requests for materials should be addressed to B.C.C.

Geoffrey M. Steeves
Department of Physics and Astronomy, University of Victoria, Victoria V8P 5C2, Canada

Reuven Gordon
Department of Electrical and Computer Engineering, University of Victoria, Victoria V8P 5C2, Canada. Correspondence and requests for materials should be addressed to B.C.C.

Byoung C. Choi
Department of Physics and Astronomy, University of Victoria, Victoria V8P 5C2, Canada.

Xijun Wang
Changchun Institute of Optics, Fine Mechanics and Physics, State Key Laboratory of Applied Optics, Chinese Academy of Sciences, Changchun 130033, China

T. Taniuchi
Institute for Solid State Physics, University of Tokyo, Chiba 277-8581, Japan.
CREST, Japan Science and Technology Agency, Tokyo 102-0075, Japan.

Y. Motoyui
Institute for Solid State Physics, University of Tokyo, Chiba 277-8581, Japan.

K. Morozumi
Institute for Solid State Physics, University of Tokyo, Chiba 277-8581, Japan.

T.C. Ro¨del
CSNSM, CNRS/IN2P3 and Universite´ Paris-Sud, Baˆtiments 104 et 108, Orsay 91405, France.
Synchrotron SOLEIL, L'Orme des Merisiers, Saint-AubinBP48, Gif-sur-Yvette 91192, France. Correspondence and requests for materials should be addressed to T.T.

F. Fortuna
CSNSM, CNRS/IN2P3 and Universite´ Paris-Sud, Baˆtiments 104 et 108, Orsay 91405, France.

A.F. Santander-Syro
CSNSM, CNRS/IN2P3 and Universite´ Paris-Sud, Ba^timents 104 et 108, Orsay 91405, France.

S. Shin
Institute for Solid State Physics, University of Tokyo, Chiba 277-8581, Japan.
CREST, Japan Science and Technology Agency, Tokyo 102-0075, Japan.

D. Bossini
Institute for Molecules and Materials, Spectroscopy of Solids and Interfaces, Radboud University, Heyendaalseweg 135, 6525 AJ Nijmegen, The Netherlands

S. Dal Conte
Dipartimento di Fisica, Politecnico di Milano, Piazza Leonardo da Vinci 32, I-20133 Milano, Italy
Istituto di Fotonica e Nanotecnologie, Consiglio Nazionale delle Ricerche, Piazza Leonardo da Vinci 32, I-20133 Milano, Italy

Y. Hashimoto
[1] Institute for Molecules and Materials, Spectroscopy of Solids and Interfaces, Radboud University, Heyendaalseweg 135, 6525 AJ Nijmegen, The Netherlands

A. Secchi
Institute for Molecules and Materials, Spectroscopy of Solids and Interfaces, Radboud University, Heyendaalseweg 135, 6525 AJ Nijmegen, The Netherlands

R.V. Pisarev
Ioffe Physical-Technical Institute, Ferroics Physics Laboratory, Russian Academy of Sciences, St Petersburg 194021, Russia

Th. Rasing
Institute for Molecules and Materials, Spectroscopy of Solids and Interfaces, Radboud University, Heyendaalseweg 135, 6525 AJ Nijmegen, The Netherlands

G. Cerullo
Dipartimento di Fisica, Politecnico di Milano, Piazza Leonardo da Vinci 32, I-20133 Milano, Italy
Istituto di Fotonica e Nanotecnologie, Consiglio Nazionale delle Ricerche, Piazza Leonardo da Vinci 32, I-20133 Milano, Italy

A.V. Kimel
Institute for Molecules and Materials, Spectroscopy of Solids and Interfaces, Radboud University, Heyendaalseweg 135, 6525 AJ Nijmegen, The Netherlands

Mohammed-Amin Alubaidy
Department of Mechanical Engineering, Ryerson University, 350 Victoria Street, Toronto, ON, M3N 2H8, Canada

Krishnan Venkatakrishnan
Department of Mechanical Engineering, Ryerson University, 350 Victoria Street, Toronto, ON, M3N 2H8, Canada

Bo Tan
Department of Aerospace Engineering, Ryerson University, 350 Victoria Street, Toronto, ON, M3N 2H8, Canada

Oleg R. Pavlyniuk
Department of Physics, Taras Shevchenko National University of Kyiv, 64 Volodymyrska Str., 01601 Kyiv, Ukraine

Vitaly V. Datsyuk
Department of Physics, Taras Shevchenko National University of Kyiv, 64 Volodymyrska Str., 01601 Kyiv, Ukraine

J J Santos
Univ. Bordeaux, CNRS, CEA, CELIA (Centre Lasers Intenses et Applications), UMR 5107, F-33405 Talence, France

M Bailly-Grandvaux
Univ. Bordeaux, CNRS, CEA, CELIA (Centre Lasers Intenses et Applications), UMR 5107, F-33405 Talence, France

L Giuffrida
Univ. Bordeaux, CNRS, CEA, CELIA (Centre Lasers Intenses et Applications), UMR 5107, F-33405 Talence, France
Current address: ELI-Beamline Project, Institute of Physics, ASCR, PALS Center, Prague, Czech Republic.

P Forestier-Colleoni
Univ. Bordeaux, CNRS, CEA, CELIA (Centre Lasers Intenses et Applications), UMR 5107, F-33405 Talence, France

S Fujioka
Institute of Laser Engineering, Osaka University, Osaka 565-0871, Japan

Z Zhang
Institute of Laser Engineering, Osaka University, Osaka 565-0871, Japan

P Korneev
Univ. Bordeaux, CNRS, CEA, CELIA (Centre Lasers Intenses et Applications), UMR 5107, F-33405 Talence, France
National Research Nuclear University MEPhI, 115409, Moscow, Russian Federation, Russia

R Bouillaud
Laboratoire pour l'Utilisation des Lasers Intenses, Ecole Polytechnique, CNRS, CEA, UMR 7605, F-91128 Palaiseau, France

S Dorard
Laboratoire pour l'Utilisation des Lasers Intenses, Ecole Polytechnique, CNRS, CEA, UMR 7605, F-91128 Palaiseau, France

D Batani
Univ. Bordeaux, CNRS, CEA, CELIA (Centre Lasers Intenses et Applications), UMR 5107, F-33405 Talence, France

M Chevrot
Laboratoire pour l'Utilisation des Lasers Intenses, Ecole Polytechnique, CNRS, CEA, UMR 7605, F-91128 Palaiseau, France

J E Cross
Department of Physics, University of Oxford, Parks Road, Oxford OX1 3PU, UK

R Crowston
Department of Physics, Heslington, University of York, YO10 5DD, UK

J-L Dubois
CEA/DAM/CESTA, BP 12, F-33405 Le Barp, France

J Gazave
CEA/DAM/CESTA, BP 12, F-33405 Le Barp, France

G Gregori
Department of Physics, University of Oxford, Parks Road, Oxford OX1 3PU, UK

E d'Humières
Univ. Bordeaux, CNRS, CEA, CELIA (Centre Lasers Intenses et Applications), UMR 5107, F-33405 Talence, France

S Hulin
Univ. Bordeaux, CNRS, CEA, CELIA (Centre Lasers Intenses et Applications), UMR 5107, F-33405 Talence, France

K Ishihara
Institute of Laser Engineering, Osaka University, Osaka 565-0871, Japan

S Kojima
Institute of Laser Engineering, Osaka University, Osaka 565-0871, Japan

E Loyez
Laboratoire pour l'Utilisation des Lasers Intenses, Ecole Polytechnique, CNRS, CEA, UMR 7605, F-91128 Palaiseau, France

J-R Marquès
Laboratoire pour l'Utilisation des Lasers Intenses, Ecole Polytechnique, CNRS, CEA, UMR 7605, F-91128 Palaiseau, France

A Morace
Institute of Laser Engineering, Osaka University, Osaka 565-0871, Japan

P Nicolaï
Univ. Bordeaux, CNRS, CEA, CELIA (Centre Lasers Intenses et Applications), UMR 5107, F-33405 Talence, France

O Peyrusse
Univ. Bordeaux, CNRS, CEA, CELIA (Centre Lasers Intenses et Applications), UMR 5107, F-33405 Talence, France

A Poyé
Univ. Bordeaux, CNRS, CEA, CELIA (Centre Lasers Intenses et Applications), UMR 5107, F-33405 Talence, France

D Raffestin
CEA/DAM/CESTA, BP 12, F-33405 Le Barp, France

J Ribolzi
CEA/DAM/CESTA, BP 12, F-33405 Le Barp, France

M Roth
Institut für Kernphysik, Tech. Univ. Darmstadt, Germany

G Schaumann
Institut für Kernphysik, Tech. Univ. Darmstadt, Germany

F Serres
Laboratoire pour l'Utilisation des Lasers Intenses, Ecole Polytechnique, CNRS, CEA, UMR 7605, F-91128 Palaiseau, France

V T Tikhonchuk
Univ. Bordeaux, CNRS, CEA, CELIA (Centre Lasers Intenses et Applications), UMR 5107, F-33405 Talence, France

P Vacar
Laboratoire pour l'Utilisation des Lasers Intenses, Ecole Polytechnique, CNRS, CEA, UMR 7605, F-91128 Palaiseau, France

N Woolsey
Department of Physics, Heslington, University of York, YO10 5DD, UK

N. Šantić
Department of Physics, University of Zagreb, Bijenička c. 32, 10000 Zagreb, Croatia.
Institute of Physics, Bijenička c. 46, 10000 Zagreb, Croatia. Correspondence and requests for materials should be addressed to H.B.

T. Dubček
Department of Physics, University of Zagreb, Bijenička c. 32, 10000 Zagreb, Croatia.

D. Aumiler
Department of Physics, University of Zagreb, Bijenička c. 32, 10000 Zagreb, Croatia.
Institute of Physics, Bijenička c. 46, 10000 Zagreb, Croatia. Correspondence and requests for materials should be addressed to H.B.

H. Buljan
Department of Physics, University of Zagreb, Bijenička c. 32, 10000 Zagreb, Croatia.
Institute of Physics, Bijenička c. 46, 10000 Zagreb, Croatia. Correspondence and requests for materials should be addressed to H.B.

T. Ban
Department of Physics, University of Zagreb, Bijenička c. 32, 10000 Zagreb, Croatia.

Jun Lin
Yunnan Observatories, Chinese Academy of Sciences, P.O. Box 110, Kunming, Yunnan 650216, China

Nicholas A. Murphy
Harvard-Smithsonian Center for Astrophysics, 60 Garden Street, Cambridge, MA 02138, USA

Chengcai Shen
Harvard-Smithsonian Center for Astrophysics, 60 Garden Street, Cambridge, MA 02138, USA

John C. Raymond
Harvard-Smithsonian Center for Astrophysics, 60 Garden Street, Cambridge, MA 02138, USA

Katharine K. Reeves
Harvard-Smithsonian Center for Astrophysics, 60 Garden Street, Cambridge, MA 02138, USA

Jiayong Zhong
Key Laboratory of Optical Astronomy, National Astronomical Observatories of China, Chinese Academy of Sciences, 20A Datun Road, Chaoyang District, Beijing 100012, China

Ning Wu
School of Tourism and Geography, Yunnan Normal University, 293 Street 121, Kunming, Yunnan 650092, China

Yan Li
Yunnan Observatories, Chinese Academy of Sciences, P.O. Box 110, Kunming, Yunnan 650216, China

J.Y. Zhong
Key Laboratory of Optical Astronomy, National Astronomical Observatories, Chinese Academy of Sciences, Beijing 100012, China

Y.T. Li
Beijing National Laboratory for Condensed Matter Physics, Institute of Physics, Chinese Academy of Sciences, Beijing 100190, China

X.G. Wang
State Key Laboratory of Nuclear Physics & Technology, School of Physics, Peking University, Beijing 100871, China

J.Q. Wang
State Key Laboratory of Nuclear Physics & Technology, School of Physics, Peking University, Beijing 100871, China

Q.L. Dong
Beijing National Laboratory for Condensed Matter Physics, Institute of Physics, Chinese Academy of Sciences, Beijing 100190, China

X. Liu
Beijing National Laboratory for Condensed Matter Physics, Institute of Physics, Chinese Academy of Sciences, Beijing 100190, China

X.X. Lin
Beijing National Laboratory for Condensed Matter Physics, Institute of Physics, Chinese Academy of Sciences, Beijing 100190, China

D.W. Yuan
Beijing National Laboratory for Condensed Matter Physics, Institute of Physics, Chinese Academy of Sciences, Beijing 100190, China

F. Du
Beijing National Laboratory for Condensed Matter Physics, Institute of Physics, Chinese Academy of Sciences, Beijing 100190, China

S.J. Wang
Beijing National Laboratory for Condensed Matter Physics, Institute of Physics, Chinese Academy of Sciences, Beijing 100190, China

L. Zhang
Beijing National Laboratory for Condensed Matter Physics, Institute of Physics, Chinese Academy of Sciences, Beijing 100190, China

L. An
Beijing National Laboratory for Condensed Matter Physics, Institute of Physics, Chinese Academy of Sciences, Beijing 100190, China

C.J. Xiao
State Key Laboratory of Nuclear Physics & Technology, School of Physics, Peking University, Beijing 100871, China

H.G. Wei
Key Laboratory of Optical Astronomy, National Astronomical Observatories, Chinese Academy of Sciences, Beijing 100012, China

K. Zhang
Key Laboratory of Optical Astronomy, National Astronomical Observatories, Chinese Academy of Sciences, Beijing 100012, China

F.L. Wang
Key Laboratory of Optical Astronomy, National Astronomical Observatories, Chinese Academy of Sciences, Beijing 100012, China

S.E. Jiang
Research Center for Laser Fusion, China Academy of Engineering Physics, Mianyang 621900, China

Y.K. Ding
Research Center for Laser Fusion, China Academy of Engineering Physics, Mianyang 621900, China

Z.R. Cao
Research Center for Laser Fusion, China Academy of Engineering Physics, Mianyang 621900, China

Z. Yuan
Research Center for Laser Fusion, China Academy of Engineering Physics, Mianyang 621900, China

H.Y. Zhang
Research Center for Laser Fusion, China Academy of Engineering Physics, Mianyang 621900, China

Z.W. Yang
Research Center for Laser Fusion, China Academy of Engineering Physics, Mianyang 621900, China

J.Q. Zhu
National Laboratory on High Power Lasers and Physics, Shanghai 201800, China

X.T. He
Center for Applied Physics and Technology, Peking University, Beijing 100871, China
Institute for Fusion and simulation, Zhejiang University, Hangzhou 310027, China
Institute of Applied Physics and Computational Mathematics, Beijing 100094, China

H.B. Cai
Institute for Fusion and simulation, Zhejiang University, Hangzhou 310027, China

G. Zhao
Key Laboratory of Optical Astronomy, National Astronomical Observatories, Chinese Academy of Sciences, Beijing 100012, China

J. Zhang
Departmem of Physics, Shanghai Jiao Tong University, Shanghai 200240, China

Marketa Ryvolova
Central European Institute of Technology, Brno University of Technology, Technicka 3058/10, CZ-616 00 Brno, Czech Republic
Department of Chemistry and Biochemistry, Faculty of Agronomy, Mendel University in Brno, Zemedelska 1, CZ-613 00 Brno, Czech Republic

Jana Chomoucka
Central European Institute of Technology, Brno University of Technology, Technicka 3058/10, CZ-616 00 Brno, Czech Republic

Jana Drbohlavova
Central European Institute of Technology, Brno University of Technology, Technicka 3058/10, CZ-616 00 Brno, Czech Republic

Pavel Kopel
Central European Institute of Technology, Brno University of Technology, Technicka 3058/10, CZ-616 00 Brno, Czech Republic
Department of Chemistry and Biochemistry, Faculty of Agronomy, Mendel University in Brno, Zemedelska 1, CZ-613 00 Brno, Czech Republic

Petr Babula
Central European Institute of Technology, Brno University of Technology, Technicka 3058/10, CZ-616 00 Brno, Czech Republic

David Hynek
Central European Institute of Technology, Brno University of Technology, Technicka 3058/10, CZ-616 00 Brno, Czech Republic
Department of Chemistry and Biochemistry, Faculty of Agronomy, Mendel University in Brno, Zemedelska 1, CZ-613 00 Brno, Czech Republic

Vojtech Adam
Central European Institute of Technology, Brno University of Technology, Technicka 3058/10, CZ-616 00 Brno, Czech Republic
Department of Chemistry and Biochemistry, Faculty of Agronomy, Mendel University in Brno, Zemedelska 1, CZ-613 00 Brno, Czech Republic

Tomas Eckschlager
Department of Paediatric Haematology and Oncology, 2nd Faculty of Medicine Charles University in Prague and University Hospital Motol, Prague, Czech Republic

Jaromir Hubalek
Central European Institute of Technology, Brno University of Technology, Technicka 3058/10, CZ-616 00 Brno, Czech Republic

Marie Stiborova
Department of Biochemistry, Faculty of Science, Charles University, Albertov 2030, CZ 128 40 Prague 2, Czech Republic

Jozef Kaiser
Central European Institute of Technology, Brno University of Technology, Technicka 3058/10, CZ-616 00 Brno, Czech Republic

Rene Kizek
Central European Institute of Technology, Brno University of Technology, Technicka 3058/10, CZ-616 00 Brno, Czech Republic
Department of Chemistry and Biochemistry, Faculty of Agronomy, Mendel University in Brno, Zemedelska 1, CZ-613 00 Brno, Czech Republic

Vidya Sagar
Center for Personalized Nanomedicine/Institute of Neuroimmune Pharmacology, Department of Immunology, Herbert Wertheim College of Medicine, Florida International University, Miami, Florida 33199, USA.

V. S. R.Atluri
Center for Personalized Nanomedicine/Institute of Neuroimmune Pharmacology, Department of Immunology, Herbert Wertheim College of Medicine, Florida International University, Miami, Florida 33199, USA.

A.Tomitaka
Center for Personalized Nanomedicine/Institute of Neuroimmune Pharmacology, Department of Immunology, Herbert Wertheim College of Medicine, Florida International University, Miami, Florida 33199, USA.

P. Shah
Department of Biomedical engineering, College of Engineering and Computing, Florida International University, Miami, 33174 Florida, USA.

A. Nagasetti
Department of Biomedical engineering, College of Engineering and Computing, Florida International University, Miami, 33174 Florida, USA.

S. Pilakka-Kanthikeel
Center for Personalized Nanomedicine/Institute of Neuroimmune Pharmacology, Department of Immunology, Herbert Wertheim College of Medicine, Florida International University, Miami, Florida 33199, USA.

N. El-Hage
Center for Personalized Nanomedicine/Institute of Neuroimmune Pharmacology, Department of Immunology, Herbert Wertheim College of Medicine, Florida International University, Miami, Florida 33199, USA.

A. McGoron
Department of Biomedical engineering, College of Engineering and Computing, Florida International University, Miami, 33174 Florida, USA.

Y.Takemura
Department of Electrical and Computer Engineering, Yokohama National University, Yokohama 240-8501, Japan. Correspondence and requests for materials should be addressed to M.N.

M. Nair
Center for Personalized Nanomedicine/Institute of Neuroimmune Pharmacology, Department of Immunology, Herbert Wertheim College of Medicine, Florida International University, Miami, Florida 33199, USA.

Sarah E. Norred
Biomedical Engineering Department, University of Tennessee, Knoxville, TN 37996, USA

Jacqueline Anne Johnson
Biomedical Engineering Department, University of Tennessee, Knoxville, TN 37996, USA
University of Tennessee Space Institute, Tullahoma, TN 37388, USA

G. A. Pablo Cirrone
INFN-LNS, Via S. Sofia 62, Catania 95123, Italy

Giacomo Cuttone
INFN-LNS, Via S. Sofia 62, Catania 95123, Italy

Francesco Romano
INFN-LNS, Via S. Sofia 62, Catania 95123, Italy
Medical Physics School, University of Catania, Via S. Sofia 64, Catania 95123, Italy

Francesco Schillaci
INFN-LNS, Via S. Sofia 62, Catania 95123, Italy
Medical Physics School, University of Catania, Via S. Sofia 64, Catania 95123, Italy

Valentina Scuderi
INFN-LNS, Via S. Sofia 62, Catania 95123, Italy
Institute of Physics ASCR, v.v.i. (FZU), ELI-Beamlines Project, Na Slovance 2, Prague 18221, Czech Republic

Antonino Amato
INFN-LNS, Via S. Sofia 62, Catania 95123, Italy

Giacomo Candiano
INFN-LNS, Via S. Sofia 62, Catania 95123, Italy
Medical Physics School, University of Catania, Via S. Sofia 64, Catania 95123, Italy

Michele Costa
INFN-LNS, Via S. Sofia 62, Catania 95123, Italy

Giuseppe Gallo
INFN-LNS, Via S. Sofia 62, Catania 95123, Italy

Giuseppina Larosa
INFN-LNS, Via S. Sofia 62, Catania 95123, Italy

Georg Korn
Institute of Physics ASCR, v.v.i. (FZU), ELI-Beamlines Project, Na Slovance 2, Prague 18221, Czech Republic

Renata Leanza
INFN-LNS, Via S. Sofia 62, Catania 95123, Italy
Dipartimento di Fisica e Astronomia, Universitá degli Studi di Catania, Via S. Sofia 64, Catania 95123, Italy

Rosanna Manna
INFN-LNS, Via S. Sofia 62, Catania 95123, Italy

Mario Maggiore
INFN-LNL, Viale dell'Universitá 2, Legnaro (PD) 35020, Italy

Valentina Marchese
INFN-LNS, Via S. Sofia 62, Catania 95123, Italy

Daniele Margarone
Institute of Physics ASCR, v.v.i. (FZU), ELI-Beamlines Project, Na Slovance 2, Prague 18221, Czech Republic

Giuliana Milluzzo
INFN-LNS, Via S. Sofia 62, Catania 95123, Italy
Dipartimento di Fisica e Astronomia, Universitá degli Studi di Catania, Via S. Sofia 64, Catania 95123, Italy

Giada Petringa
INFN-LNS, Via S. Sofia 62, Catania 95123, Italy

Antonella Tramontana
INFN-LNS, Via S. Sofia 62, Catania 95123, Italy
Dipartimento di Fisica e Astronomia, Universitá degli Studi di Catania, Via S. Sofia 64, Catania 95123, Italy

M. Ghoranneviss
Plasma Physics Research Center, Tehran Science and Research Branch, Islamic Azad University, Tehran, Iran

A. Salar Elahi
Plasma Physics Research Center, Tehran Science and Research Branch, Islamic Azad University, Tehran, Iran

Hirokuni Oda
Institute of Geology and Geoinformation, Geological Survey of Japan, AIST, Central 7, 1-1-1 Higashi, Tsukuba 305-8567, Japan.

Chuang Xuan
School of Ocean and Earth Science, National Oceanography Centre Southampton, University of Southampton, Waterfront Campus, European Way, Southampton SO14 3ZH, UK.

Yuhji Yamamoto
Center for Advanced Marine Core Research, Kochi University, B200 Monobe, Nankoku, Kochi 783-8502, Japan.

Preface

The main purpose of this book is to offer an overview of the current research work that combines laser and magnetism. The first part will focus mainly on experiments designed to observe and measure how short laser pulses that can affect dynamical processes in magnetic systems. Chapter 1 and 2 are dedicated to recent advances in laser-induced ultrafast demagnetization techniques using pump-probe spectroscopy. Chapters 3, 4, 5, 6 and 7 investigate the behaviour of magnetic nanoparticles and nanostructures using pulsed-laser methods, dual scanning laser speckle interferometry and laser-based ultrahigh resolution photoemission electron microscopy (laser-PEEM). Potential applications of these techniques in the development of future nanotechnologies are various and include heat-assisted magnetic recording (HAMR) technology and growing nanocrystalline magnetic nanofibers. Chapter 8 and 9 are focused on laser-induced techniques for the generation of synthetic magnetic fields.

The second part will focus on the applications of the laser-induced methods in astrophysical, biomedical, paleomagnetic and fusion energy studies. Chapter 10 and 11 show how astrophysical phenomena such as solar flares can now be studied in laboratory by using laser-induced magnetic fields. Chapters 12 to 15 reviews the current use of nanoparticles, laser breakdown spectroscopy and magnetic resonance imaging in medical applications. This includes non-invasive treatment strategies based on magnetic nanoparticles and magnetic resonance-guided laser-induced thermotherapy (MRgLITT). It concludes with an overview of ELIMED (ELI-Beamlines MEDical applications) and description of the detectors dedicated to diagnostics and dosimetry as well as some preliminary results obtained both with an accelerator-driven and laser-driven beam.

The last two chapters review the recent developments in laser-driven inertial fusion and the use laser interferometry for paleomagnetic measurements.

Editor
Olga Moreira, Ph.D.

INTRODUCTION

A laser is a coherent and focused beam of photons; coherent, in this context, means that it is all one wavelength, unlike ordinary light which showers on us in many wavelengths. The acronym laser stands for "light amplification by stimulated emission of radiation." Lasers work as a result of resonant effects. The output of a laser is a coherent electromagnetic field. In a coherent beam of electromagnetic energy, all the waves have the same frequency and phase. In a basic laser, a chamber called a cavity is designed to internally reflect infrared (IR), visible-light, or ultraviolet (UV) waves so they reinforce each other. The cavity can contain gases, liquids, or solids. The choice of cavity material determines the wavelength of the output. At each end of the cavity, there is a mirror. One mirror is totally reflective, allowing none of the energy to pass through; the other mirror is partially reflective, allowing approximately 5 percent of the energy to pass through. Energy is introduced into the cavity from an external source; this is called pumping.

As a result of pumping, an electromagnetic field appears inside the laser cavity at the natural (resonant) frequency of the atoms of the material that fills the cavity. The waves reflect back and forth between the mirrors. The length of the cavity is such that the reflected and re-reflected wave fronts reinforce each other in phase at the natural frequency of the cavity substance. Electromagnetic waves at this resonant frequency emerge from the end of the cavity having the partially-reflective mirror. The output may appear as a continuous beam, or as a series of brief, intense pulses. The ruby laser, a simple and common type, has a rod-shaped cavity made of a mixture of solid aluminum oxide and chromium. The output is in pulses that last approximately 500 microseconds each. Pumping is done by means of a helical flash tube wrapped around the rod. The output is in the red visible range. A blue laser has a shorter wavelength than the red laser, and the ability to store and read two to four times the amount of data.

The helium-neon laser is another popular type, favored by electronics hobbyists because of its moderate cost. As its name implies, it has a cavity filled with helium and neon gases. The output of the device is bright crimson. Other gases can be used instead of helium and neon, producing beams of different wavelengths. Argon produces a laser with blue visible output. A mixture of nitrogen, carbon dioxide, and helium produces IR output. Lasers are one of the most significant inventions developed during the20th century. They have found a tremendous variety of uses in electronics, computer hardware, medicine, and experimental science.

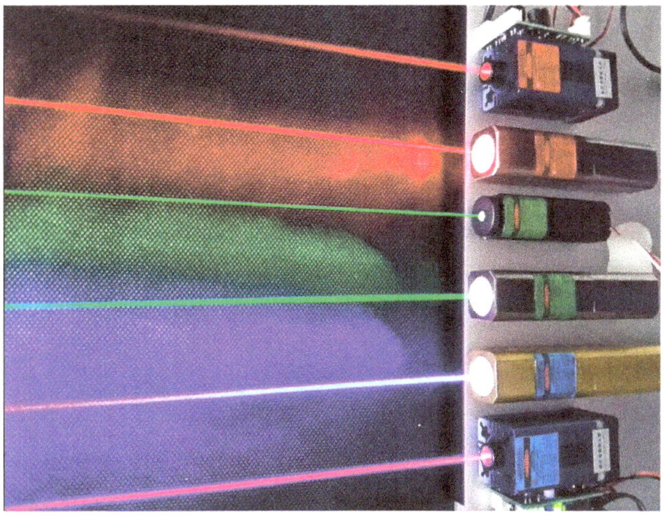

Fundamentals of Laser

Modern telescopes use laser technologies to compensate for the blurring effect of the Earth's atmosphere. Lasers are distinguished from other light sources by their coherence. Spatial coherence is typically expressed through the output being a narrow beam, which is diffraction-limited. Laser beams can be focused to very tiny spots, achieving a very high irradiance, or they can have very low divergence in order to concentrate their power at a great distance. Temporal (or longitudinal) coherence implies a polarized wave at a single frequency whose phase is correlated over a relatively great distance (the coherence length) along the beam. A beam produced by a thermal or other incoherent light source has an instantaneous amplitude and phase that vary randomly with respect to time and position, thus having a short coherence length. Lasers are characterized according to their wavelength in a vacuum. Most "single wavelength" lasers actually produce radiation in several modes having slightly differing frequencies (wavelengths), often not in a single polarization. Although temporal coherence implies monochromaticity, there are lasers that emit a broad spectrum of light or emit different wavelengths of light simultaneously. There are some lasers that are not single spatial mode and consequently have light beams that diverge more than is required by the diffraction limit. However, all such devices are classified as "lasers" based on their method of producing light, i.e., stimulated emission. Lasers are employed in applications where light of the required spatial or temporal coherence could not be produced using simpler technologies.

Magnetism

Magnetism is a class of physical phenomena that are mediated by magnetic fields. Electric currents and the magnetic moments of elementary particles give rise to a magnetic field, which acts on other currents and magnetic moments. Every material is influenced to some extent by a magnetic field. The most familiar effect is on permanent magnets, which have persistent magnetic moments caused by ferromagnetism. The prefix Ferro refers to iron, because permanent magnetism was first observed in a form of natural iron ore called magnetite, Fe_3O_4. Most materials do not have permanent moments. Some are attracted to a magnetic field (paramagnetism); others are repulsed by a magnetic field (diamagnetism); others have a more complex relationship with an applied magnetic field (spin glass behavior and antiferromagnetism). Substances that are negligibly affected by magnetic fields are known as non-magnetic substances. These include copper, aluminum, gases, and plastic. Pure oxygen exhibits magnetic properties when cooled to a liquid state.

Fundamentals of Magnetism

Basic to magnetism are magnetic fields and their effects on matter, as, for instance, the deflection of moving charges and torques on other magnetic objects. Evidence for the presence of a magnetic field is the magnetic force on charges moving in that field; the force is at right angles to both the field and the velocity of the charge. This force deflects the particles without changing their speed. The deflection can be observed in the torque on a compass needle that acts to align the needle with the magnetic field of Earth. The needle is a thin piece of iron that has been magnetized—i.e., a small bar magnet. One end of the magnet is called a north pole and the other end a south pole. The force between a north and a south pole is attractive, whereas the force between like poles is repulsive. The magnetic field is sometimes referred to as magnetic induction or magnetic flux density; it is always symbolized by B. Magnetic fields are measured in units of tesla (T). (Another unit of measure commonly used for B is the gauss, though it is no longer considered a standard unit. One gauss equals 10^{-4} tesla.)

A fundamental property of a magnetic field is that its flux through any closed surface vanishes. (A closed surface is one that completely surrounds a volume.) This is expressed mathematically by div B = 0 and can be understood physically in terms of the field lines representing B. These lines always close on themselves, so that if they enter a certain volume at some point, they must also leave that volume. In this respect, a magnetic field is quite different from an electric field. Electric field lines can begin and end on a charge, but no equivalent magnetic charge has been found in spite of many searches for so-called magnetic monopoles.

The most common source of magnetic fields is the electric current loop. It may be an electric current in a circular conductor or the motion of an orbiting electron in an atom. Associated with both these types of current loops is a magnetic dipole moment, the value of which is iA, the product of the current i and the area of the loop A. In addition, electrons, protons, and neutrons in atoms have a magnetic dipole moment associated with their intrinsic spin; such magnetic dipole moments represent another important source of magnetic fields. A particle with a magnetic dipole moment is often referred to as a magnetic dipole. (A magnetic dipole may be thought of as a tiny bar magnet. It has the same magnetic field as such a magnet and behaves the same way in external magnetic fields.) When placed in an external magnetic field, a magnetic dipole can be subjected to a torque that tends to align it with the field; if the external field is not uniform, the dipole also can be subjected to a force.

All matter exhibits magnetic properties to some degree. When placed in an inhomogeneous field, matter is either attracted or repelled in the direction of the gradient of the field. This property is described by the magnetic susceptibility of the matter and depends on the degree of magnetization of the matter in the field. Magnetization depends on the size of the dipole moments of the atoms in a substance and the degree to which the dipole moments are aligned with respect to each other. Certain materials, such as iron, exhibit very strong magnetic properties because of the alignment of the magnetic moments of their atoms within certain small regions called domains. Under normal conditions, the various domains have fields that cancel, but they can be aligned with each other to produce extremely large magnetic fields. Various alloys, like NdFeB (an alloy of neodymium, iron, and boron), keep their domains aligned and are used to make permanent magnets. The strong magnetic field produced by a typical three-millimetre-thick magnet of this material is comparable to an electromagnet made of a copper loop carrying a current of several thousand amperes. In comparison, the current in a typical light bulb is 0.5 ampere. Since aligning the domains of a material produces a magnet, disorganizing the orderly alignment destroys the magnetic properties of the material. Thermal agitation that results from heating a magnet to a high temperature destroys its magnetic properties.

Ferromagnetism

When a ferromagnetic material is placed near a magnet, it will be attracted toward the region of greater magnetic field. This is what we are most familiar with when our magnet picks up a bunch of paperclips. Iron, cobalt, nickel, gadolinium, dysprosium and alloys containing these elements exhibit ferromagnetism because of the way the electron spins within one atom interact with those of nearby atoms. They will align themselves, creating magnetic domains forming a temporary magnet. If a piece of iron is placed within a strong magnetic field, the domains in line with the field will grow in size as the domains perpendicular to the field will shrink in size.

Diamagnetism

When a diamagnetic material is placed near a magnet, it will be repelled from the region of greater magnetic field, just opposite to a ferromagnetic material. It is exhibited by all common materials, but is very weak. People and frogs are diamagnetic. Metals such as bismuth, copper, gold, silver and lead, as well as many nonmetals such as graphite, water and most organic compounds are diamagnetic.

What is a magnetic field?

Suppose you put a bar magnet (shaped like a rectangle, sometimes with the north and south poles painted different colors) or a horseshoe magnet (bent round into a tight U-shape) onto a table and place an iron nail nearby. If you push the magnet slowly toward the nail, there will come a point when the nail jumps across and sticks to the magnet. That's what we mean by magnets having an invisible magnetic field that extends all around them. Another way of describing this is to say that a magnet can "act at a distance": it can cause a pushing or pulling force on other objects it isn't actually touching).

Magnetic fields can penetrate through all kinds of materials, not just air. You probably have little notes stuck to the door of your refrigerator with brightly colored magnets—so you can see that magnetic fields cut through paper. You may have done the trick where you use a magnet to pick up a long chain of paperclips, with each clip magnetizing the next one along. That little experiment tells us that a magnetic field can penetrate through magnetic materials such as iron.

What is Magnetomotive Force?

When physicists discuss the force of a magnetic field, they refer to it as magneto motive force (mmf) or magnetic potential. Magnetomotive force is the work that carries a measurable unit of strength through a magnetic circuit. This unit of strength is measured in ampere-turns (AT).

Magnetism in a circuit flows from the north to the South Pole. Following a specific path, the force of the magnetism is similar to the force in an electrical circuit. While electricity follows a certain path from a point of origin to a receiving point and back, magnetism always flows from one pole to the other. As magnetism flows through the circuit, it produces lines of force, or flux lines, and generates a magnetic field.

In simplest terms, anything that produces magnetism can be described as exerting mmf. Understanding the basic meaning of the word itself makes it easy to understand its application. Magnetomotive force translates, quite literally, to "magnetic motion causing." The motion that the force generates is always exerted perpendicular to the magnetic circuit's resistance.

Although magnetism exerts a force all its own, it does exhibit a reluctance to travel across large spans of air. This reluctance is similar to the resistance of an electrical circuit. Magnetism travels much more successfully through iron. A magnetic circuit composed exclusively of iron will have much lower reluctance than a magnetic circuit crosses an air gap.

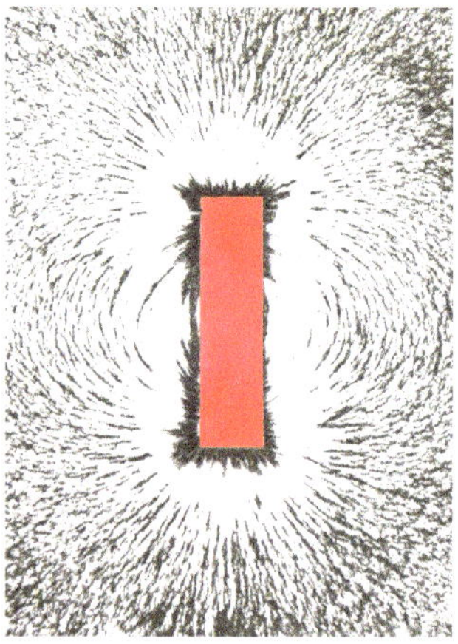

Either permanent magnets or coiled wire conducting electricity may produce a magnetomotive force. When a coil of wire produces a magnetomotive force, the number of turns of the wire is directly related to the value of the force generated. For instance, the magnetic motion force of a coil with 50 turns would be 25 times greater than the magnetomotive force of a coil with only two turns. The magnetic flux of a magnetic circuit is, then, equal to the magnet motion force divided by the circuit's reluctance. This relationship may be better understood if one takes time to consider that magnetic flux represents the journey of a substance's magnetism through a magnetic circuit. Each turn in a coil concentrates the force of the magnetic. Overcoming the reluctance of the circuit to travel through each of the coils, the natural strength of the magnet is demonstrated as sheer force.

How do Magnets Work?

Magnets have magnetic fields that attract items containing iron. For example, some of the substances attracted to magnets include iron, nickel, and steel. Magnets are used for a variety of purposes, ranging from attaching items to refrigerator doors and creating compasses to providing for fast roller coaster rides and converting mechanical energy into electrical energy. They are even used in some toys.

Understanding how magnets work means figuring out the dynamics of a magnetic field. Consider the space that surrounds a magnet. This space is occupied by a magnetic force and is called a magnetic field. If a magnet is placed within this field, it will be acted upon by magnetic forces.

A magnetic field is created as the result of moving charges. A good example of this is electric current that flows through a wire. When this occurs, there are negatively charged, subatomic particles, called electrons, moving through the wire. As these charges move, a magnetic field forms around the wire. Likewise, the magnetic field of a magnet is created by the movement of electrons.

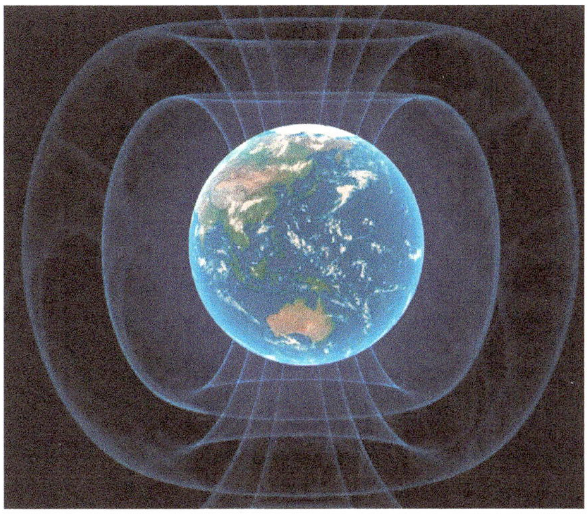

A magnet's magnetic field either attracts or repels certain metals, as well as other magnets. A magnet has two ends that are referred to as poles. One pole is called north and the other one is referred to as south. To attract magnets to each other, you have to place opposite ends of two magnets near each other. Placing like ends of two magnets near each other causes the opposite to occur; the two magnets repel each other.

Interestingly, the Earth has a natural, magnetic field at is core. If you take a look at a compass, you will notice that one side is marked "N" and faces toward the Earth's northern magnetic pole. As such, this part of the compass is referred to as the North-seeking pole. However, the Earth's north magnetic pole should not be confused with its North Pole. The North Pole is located hundreds of miles away from the north-seeking pole.

Though most people are familiar with simple, refrigerator magnets, there are many other types of magnets. They are categorized as permanent, temporary, and electromagnetic. Permanent magnets maintain their magnetic properties for an extended period of time while temporary magnets lose their magnetism more quickly. An electromagnet, on the other hand, is one that is created using electricity. Its strength can be altered based on the electrical current that is allowed to move through it.

CHAPTER 1

SPECTROSCOPY OF ULTRAFAST MAGNETIZATION DYNAMICS

C. S.Gonçalves, A. S. Silva, D. Navas, M. Miranda, F. Silva, H. Crespo & D. S. Schmool

INTRODUCTION

Laser-induced femtosecond magnetism or femtomagnetism[1] is a new research field in modern magnetism, whose origin is usually dated to 1996, when Beaurepaire et al.[2] first showed that ultrafast 60 fs laser pulses could be used to demagnetize a Ni thin film and to study the evolution of the magnetization dynamics within the first few picoseconds after excitation. Since this important discovery, a wealth of research has been dedicated to the study of several crucial questions in this field, such as: how fast can the magnetization of a system be changed, how can the corresponding change be measured, how can it be controlled on an ultrafast time scale and what are the shortest excitation pulses that can affect ultrafast dynamical processes in magnetic systems.

Based on the premise that the photo-excitation of a magnetic system with ultrashort laser pulses significantly alters the thermodynamic equilibrium among the constituent degrees of freedom of the system (charge carriers, spins, and lattice), a variety of dynamical processes have been observed on an extremely fast timescale, which are inaccessible via thermal equilibrium transitions[1,2,3,4,5,6,7,8]. These provide essential insight into the understanding of fundamental transient magnetic phenomena, and have clear implications for future high-speed, multifunctional magneto-optical device technology[3,4]. Therefore, the experimental observation of magnetization dynamics, induced by pump-probe methods, has gained significant impetus with the use of femtosecond laser sources, permitting the manipulation of the magnetization of ferro- and ferrimagnetic materials[2,5,6,7] and to study their related properties on timescales ranging from tens of femtoseconds to hundreds of picoseconds[8,9,10,11].

However, the exact way in which laser pulses can change the collective spin ordering in the coherent regime, or induce a complete spin reversal or a magnetic phase transition on ultrafast times, is not completely understood. In this ultrashort temporal regime, the highly non-equilibrium processes and photo-excited coherences are the most interesting and least understood issues, since the relevant characteristic timescales are comparable to (or even shorter than) one single oscillation cycle of phonons and magnons, spin-dependent scattering times, and dephasing times[6]. This raises some fundamental questions with respect to the cornerstones of our current understanding of magnetism and phase transitions, including the microscopic origin of angular momentum conservation and the validity of the thermodynamic description of magnetism at these extremely short timescales[5].

Although rapid advances in ultrafast optical methods have allowed the temporal resolution to be gradually improved, the most recent measurements have been routinely performed with a temporal resolution that is typically limited to few tens of fs (e.g., 50 fs in reference 5). This prevents the excitation of ultrafast processes at shorter timescales, and a lack of information is expected on timescales below that of the pulse width.

Moreover, the need to achieve shorter temporal resolutions has been recently asserted by Bossini et al.[12] where it was demonstrated that a successful excitation of high-frequency magnons demands laser pulses shorter than 50 fs. They showed that the femtosecond spin dynamics triggered by the impulsive excitation of the two-magnon (2M) mode in an anriferromagnet such as $KNiF_3$, is only accessible with using experimental pump-probe system with sub-20 femtosecond resolution.

In this paper we present a new system for nondegenerate pump-probe measurements of magneto-dynamic processes, hence minimizing potential contamination from the pump beam while achieving the highest temporal resolution for a system of this type. The hollow-core fibre (HCF) compressor behind our design is extremely broadband, from 450 to 1050 nm, with a Fourier-limit below 3 fs[13]. Via dichroic beamsplitting and careful dispersion compensation, we generate dual-colour pump and probe pulses with sub-10 fs durations. Our system has been fully optically characterized using autocorrelation

Citation: C. S.Gonçalves, A. S. Silval, D. Navas, M. Miranda, F. Silva, H. Crespo & D. S. Schmool, Spectroscopy of Ultrafast Magnetization Dynamics, doi:10.1038/srep22872.

and d-scan techniques[14]. To demonstrate the capability of our time-resolved magneto-optic Kerr effect (TR-MOKE) system, we measured the dynamical behaviour of a GdFeCo thin film, where we observed both ultrafast demagnetization ("fast") and its related precessional dynamics ("slow"), in agreement with literature[3,4,6,15,16,17]. This work and results enabled us to attain the following two key objectives:

- Demonstration of unprecedented sub-10-fs temporal resolution from a novel (dual-colour) TR-MOKE system, which opens the way to assessing information on dynamical processes taking place at timescales well below 50 fs.
- Demonstration of sub-10-fs excitation: although the interaction of femtosecond laser pulses with spins, using Gaussian pulses of 10 fs width, has been theoretically studied by Vonesch et al.[18], we have experimentally confirmed for the first time that an intense sub-10-fs pulse is able to efficiently excite a ferrimagnetic system such as a GdFeCo thin film.

EXPERIMENTAL DUAL-COLOUR PUMP-PROBE SYSTEM

Previous to the present work, two main designs of pump-probe TR-MOKE systems have been employed: the first is based on a degenerate interaction, where both pump and probe beams have the same spectral bandwidth (or frequency range), even though they can have different polarizations; the second corresponds to a nondegenerate setup, where the pump and probe beams exhibit significantly different spectral bandwidths. One of the major disadvantages of a degenerate system lies in the fact that the strong pump beam can contaminate the signal obtained from the probe beam. This rise to a rapidly oscillating signal at the detector when the pump and probe pulses overlap in both space and time at the sample surface. This phenomenon is commonly referred to as a "coherent artifact"[19,20].

In conventional dual-color systems, one or more optical parametric amplifiers (OPAs)[21] are usually employed as a secondary source of spectrally tunable femtosecond pulses, with spectra typically ranging from the visible to the near-infrared. The temporal resolution of these techniques is determined by the available laser pulse width. OPAs are commercial systems with temporal regimes typically from the picosecond to tens of femtoseconds. In this latter case, OPAs are usually pumped by titanium:sapphire lasers with a central wavelength of 800 nm (or by their second-harmonic at 400 nm), with pulse energies ranging from 0.5 to 1 mJ and compressed pulse durations from 20 to 100 fs[22]. Here we present a new methodology for the development of a dual-colour pump-probe setup that eliminates contamination of the pump beam and provides laser pulses with spectral width larger than previous techniques and allowing the generation of sub-10-fs pump and probe pulses (described in detail in the following sections).

ARCHITECTURE OF THE NOVEL DUAL-COLOUR SYSTEM

Our methodology consists in obtaining the two-colour pump and probe pulses from the same ultra-broadband hollow fibre source. Unlike OPA systems, this approach has several advantages, namely the fact that both pump and probe pulses have very broad bandwidths capable of supporting durations below 10 fs (Fig. 1).

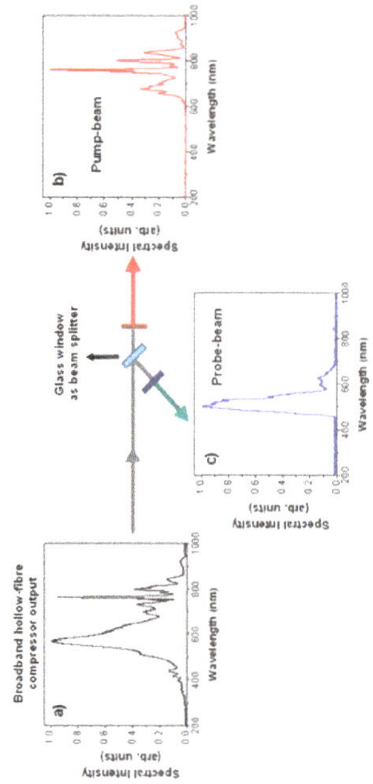

Figure 1: The spectral ranges of (**b**) the pump (transmitted) and (**c**) probe (reflected) beams after a fused silica window and colors filters, as well as (**a**) the full spectral range after the hollow fibre compressor and before beam splitting.

In our experimental set-up, we have used a thin window that transmits 96% of the pulse energy to be used as the pump pulse, while the remaining 4% is reflected and used as the probe beam. The spectral overlap between the pump and probe spectra pulses is then reduced using optical bandpass filters. An advantage of this configuration is that by switching the two filters, the roles of the pulses can be exchanged when studying specific samples that may benefit from a particular combination of central wavelengths for the pump and the probe pulses. Moreover, the full spectrum of the broadband source can be used for both pump and probe pulses with only small changes in the setup (namely by removing the spectral filters), which is expected to increase the temporal resolution to below 5 fs. In this case, the setup will correspond to a degenerate system, albeit one with extremely high resolution.

EXPERIMENTAL APPARATUS

Our pump-probe system (see Fig. 2) is based on a commercial titanium:sapphire laser amplifier (Femtolasers Compact Pro CE-phase) delivering sub-30-fs laser pulses (approximately 40 nm bandwidth centered at 800 nm) with 1 mJ of energy at a repetition rate of 1 kHz and with carrier-envelope phase (CEP) stabilization. These pulses are further post-compressed in a home-built state-of-the-art HCF and chirped-mirror compressor that uses the dispersion-scan (d-scan) technique for the simultaneous measurement of the pulses (as described in sections C and D)[13,14,23]. We make use of a refractive telescope, consisting of two lenses with focal lengths of 100 and 400 mm, to reduce the diameter of the beam prior to sending the pulses through a second chirped-mirror set (Fig. 2). These chirped mirrors effectively compensate the dispersion induced by the various optical elements along the beam paths. They have a working spectral bandwidth between 500 and 950 nm and are capable of compensating, up to high-order, the material dispersion associated with the telescope, the air paths and all remaining optical components (beamsplitter, filters, lenses, wave-plates and polarizers) traversed by the pulsed beams on their way to the sample under study.

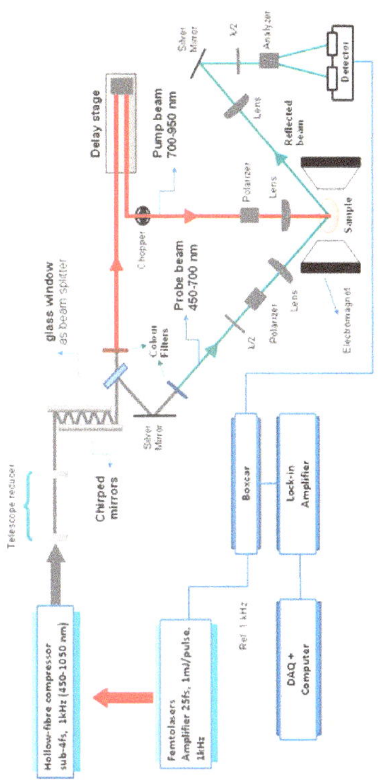

Figure 2: Schematic of the dual-colour pump-probe TR-MOKE apparatus.

After the chirped mirrors, a fused silica window is used as a broadband beamsplitter that reflects 4% of the energy to be used as the probe pulse (Fig. 1). The rest of the beam is transmitted and used as the pump pulse. To limit the spectral bandwidth of the pump, we use a lowpass filter with transmission ranging from 700 to 950 nm, and it is sent through a mechanical delay line (with a minimum step size of 0.3 fs). Furthermore, in the optical path of the pump beam, a polarizer cube has also been added. The spectral range of the probe beam is selected using a color bandpass filter that transmits in the range of 450–650 nm. In order to obtain a more homogeneous polarization of the probe beam, a combination of a polarizer cube and half-wave plate is also introduced allowing for the control of both probe beam polarization angle and intensity. Therefore, the setup permits the control of polarization and energy of both pulses, allowing us to determine the polarization rotation and ellipticity.

An optical chopper operating at 500 Hz and synchronized with the laser pulse train provides the reference for the detection electronics, which include a boxcar integrator and a lock-in amplifier coupled to an analog-to-digital converter. In both pump and probe beams, we place lenses with 200 and 100 mm focal length which produce focused spots with diameters of about 350 and 180 μm, respectively. In any optical pumping experiment, these spots must be accurately focused and overlapped on the sample surface, which has been performed with the help of a CCD camera. The presently installed magnet and associated bipolar power supply can generate a magnetic field of up to 1 T at the sample plane.

The detection of the magnetic state of the sample is performed via the measurement of the magneto-optic Kerr effect (MOKE) signal obtained from the probe beam after reflection from the sample surface. In the experiments discussed in section III, and although other MOKE configurations are possible upon small adjustments of the sample geometry, we have used the longitudinal Kerr effect, where the rotation angle of the polarization is directly proportional to the in-plane magnetization contribution. While the pump beam was set at normal incidence, the probe beam was set to 45° in order to maximize the magneto-optical response. The reflected beam is then directed through a broadband half-wave-plate and an analyzer to separate (and balance) the s and p polarized components prior to measuring their corresponding intensities using a balanced amplified photodetector. The difference in the signals allows us to evaluate the relative magnetization state of the sample as a function of the time delay introduced between pump and probe beams.

HOLLOW CORE FIBRE AND CHIRPED-MIRROR COMPRESSOR

The generation of ultrashort high-energy pulses with femtosecond durations is normally achieved via two main techniques: chirped-pulse amplification (CPA), usually based on titanium:sapphire technology, and optical parametric chirped-pulse amplification (OPCPA)[24].

By using such techniques, it is possible to obtain pulses with durations of the order of 20–100 fs. Although few-cycle 10 fs pulses have been generated from a CPA system by using specialized optics to minimize the gain narrowing effects in the laser amplification process[25], these techniques are not commonplace and the final durations are still far from the single-cycle regime. On the other hand, several few-cycle OPCPA systems have been demonstrated, also making use of Ti:Sapph technology[24]. These systems are mostly designed for intense laser-matter interaction and high-harmonic generation, having high repetition rates of the order of 1 MHz and a relatively low energy per pulse, usually in the microjoule range[26]. They also tend to be complex in their construction compared to CPA systems.

In order to generate intense radiation, with pulse energies of hundreds of microjoules and durations in the sub-4-fs regime, as envisaged for versatile high resolution pump-probe experiments, a high-energy pulse compression technique based on pulses initially created by CPA, is employed. This technique is composed of two steps: i) nonlinear spectral broadening and ii) broadband dispersion compensation[13,27,28](Fig. 3). In the first step, the optical pulse spectrum is broadened inside a gas-filled hollow core fibre by self-phase modulation (SPM). The hollow-core fiber (HCF) compressor used in this setup comprises a standard 1 meter long hollow core fiber with a diameter of 250 micrometers and placed in a static pressure chamber with 1 atm of Argon. This is a quite standard configuration of HCF compressor in terms of fiber parameters and nominal gas pressure. The choice of Argon gas was motivated by its higher Kerr-nonlinearity compared to lighter noble gases such as Neon. We found that Argon gas was more effective for generating broadband, octave-spanning spectra with a well-behaved spectral phase when pumped with the sub-mJ and sub-30-fs pulses of our laser amplifier[13]. For higher energy and/or shorter pulse laser systems, other gases, such as Neon, can in principle be used too.

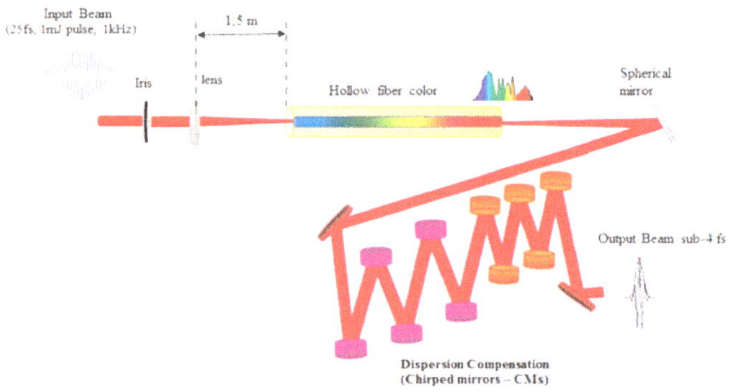

Figure 3: Simplified set-up of our hollow fibre and chirped mirror compressor.

Although the chirped mirrors are identical, we used different colours to note that they should be placed at two different angles (5° and 19°).

However, as the pulse spectrum broadens during nonlinear propagation along the hollow fibre, the pulse itself acquires a chirp due to SPM as well as due to dispersion in the gas and optical windows of the chamber containing the hollow fibre. This chirp needs to be compensated in order to obtain pulses with the shortest possible temporal duration. Dispersion compensation can be achieved in many ways; among the most common are prism pairs or chirped mirrors[29]. In our case, the 1 mJ, 27 fs pulses from our CPA amplifier are coupled into a hollow fibre through a lens with a focal length of 1.5 m (Fig. 3). The hollow-fibre is placed inside a chamber filled with Argon gas at a pressure of 1 bar. For input pulse energy of 500–700 µJ, the fibre transmission is approximately 40%. Dispersion compensation is performed using a set of ultra-broadband chirped mirrors (CMs). The CMs (Ultrafast Innovations GmbH) are designed in such a way that when two reflections are combined, with incidence angles of 5° and 19°, respectively, the residual group delay oscillations are minimized[13,23]. This compressor consistently delivers sub-4-fs pulses, with 200–300 µJ energy and a stabilized carrier-envelope phase. Its output is extremely broadband, spanning from 450 to 1050 nm, and with a Fourier-limit below 3 fs[13].

PULSE CHARACTERIZATION METHOD: THE D-SCAN TECHNIQUE

The time domain waveform and pulse width are key parameters in any study involving ultrashort optical pulses, since they determine the timescale of the excitation, and hence the interaction regime, as well as the achievable temporal resolution. Therefore, a precise knowledge of the optical pulse duration is fundamental. There are several techniques for ultrashort pulse characterization, from autocorrelation (AC)[30], which can only provide an estimate of the pulse duration, to methods capable of retrieving the complete field (amplitude and phase) of the pulses, such as frequency-resolved optical gating (FROG)[31] and spectral phase interferometry for direct electric-field reconstruction (SPIDER)32. In this work, our pulses were characterized by the new technique known as dispersion-scan, or d-scan[14,33], which provides complete information about the ultrashort optical pulses[13].

THE D-SCAN TECHNIQUE

The d-scan technique has been described in detail in references[13,14,33], where it was used to successfully characterize the ultrashort laser pulses produced by the oscillator of our CPA system, as well as the output of the broadband hollow-core fibre compressor. In this work, d-scan was used to measure both the pump and probe pulses of our TR-MOKE system at the plane of the sample.

The experimental implementation of d-scan consists in measuring the second-harmonic generation (SHG) spectrum of the pulse (or the spectrum of another nonlinear process) while the dispersion is varied via the insertion of one of two glass wedges around the point of maximum compression (minimum pulse duration). As a result, a spectrally resolved SHG trace as a function of wedge insertion (dispersion) is obtained (Fig. 4 a). Retrieval of the complete electric field of the pulses is then performed with an iterative numerical algorithm14(Fig. 4 b). In the present set-up, we focus the pump (probe) pulses with a lens with focal length of 200 mm (100 mm) on a nonlinear crystal (Beta Barium Borate BBO, 20 µm thick, cut for type I SHG at 800 nm) placed in the position where the pump and probe pulses are overlapped on the sample. The SHG signal is collimated with a lens, and a blue filter is used to remove the remaining fundamental frequency signal before detection with an optical spectrometer (HR4000, Ocean Optics Inc.).

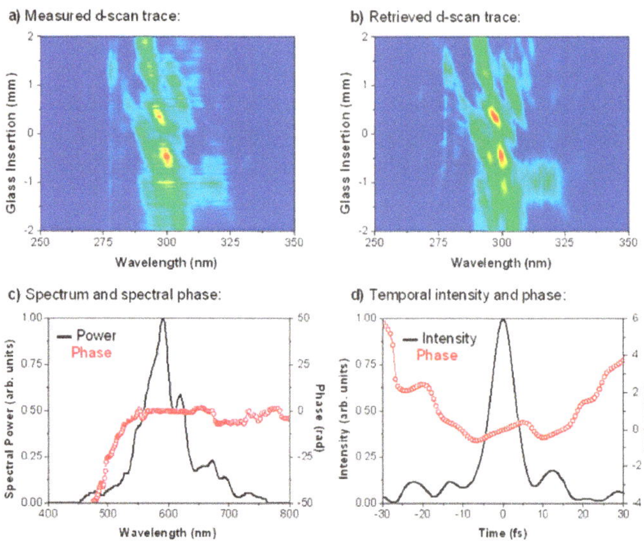

Figure 4 (**a**) Measured and (**b**) retrieved d-scan traces for the probe pulse (central wavelength of approx. 600 nm); re-trieved probe pulse in the spectral (**c**) and temporal (**d**) do-mains; probe pulse width is 7.2 fs (FWHM).

Figures 4 and 5 show the measurements obtained for the probe and pump beams, with temporal durations of 7.2 and 7.3 fs, respectively. These demonstrate that dual-colour ultrashort pump and probe pulses with sub-10 fs duration have been successfully obtained at the plane of the sample, which ensures an ultra-high temporal resolution.

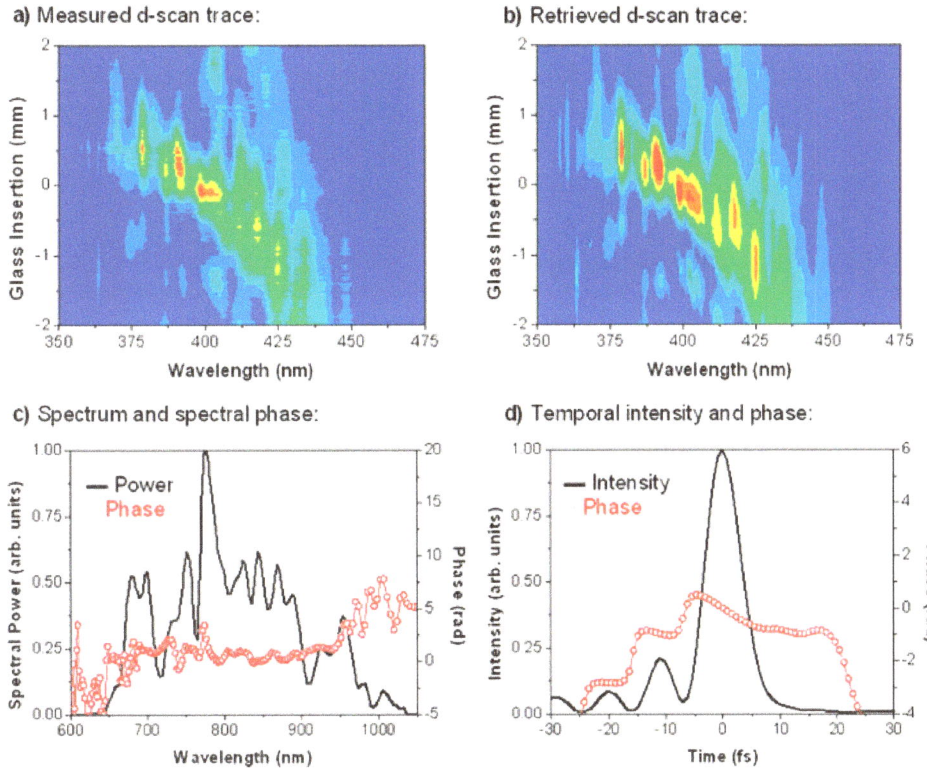

Figure 5: (**a**) Measured and (**b**) retrieved d-scan traces for the pump pulse (central wavelength of approx. 800 nm); retrieved pump pulse in the spectral (**c**) and temporal (**d**) domains; pump pulse width is 7.3 fs (FWHM).

RESULTS AND DISCUSSION

In order to demonstrate the capabilities of our apparatus, we have studied the magnetic behavior of a ferrimagnetic rare earth–transition metal (RE-TM) thin film, in the form of a GdFeCo alloy. In this alloy, the Co and Fe sublattice magnetizations are coupled ferromagnetically with each other and the total magnetization of Fe and Co is coupled antiferromagnetically with that of Gd, yielding a ferrimagnetic order. This RE-TM alloy has been extensively studied due to its strong magneto-optical response[6,34,35].

In particular, we carried out our measurements on a 20 nm thick $Gd_{23}Fe_{68}Co_{9}$ thin film which was grown by magnetron sputtering in the following multilayer structure: glass/AlTi (10 nm)/SiN(5 nm)/GdFeCo(20 nm)/SiN(60 nm) (see References 16 and 17 for more details). The AlTi layer serves as a heat sink and the SiN as buffer and capping layers. The latter also serves as an antireflection coating. This film shows out-of-plane magnetic anisotropy and the Curie temperature (T_C) is about $\approx$ 500 K[16,17,36].

The laser-induced dynamics was firstly confirmed by measuring the in-plane hysteresis loops of the ferromagnetic GdFeCo thin film with and without applying the pump beam (Fig. 6). While in the latter case we obtain the standard static MOKE hysteresis loop, the curve, obtained in the presence of a pump fluence of 2.7 mJ/cm^2 with a pulse width lower than 8 fs and for the time delay when the observed demagnetization was maximum, shows that the sample magnetization is reduced to about 70% of its normal saturation value.

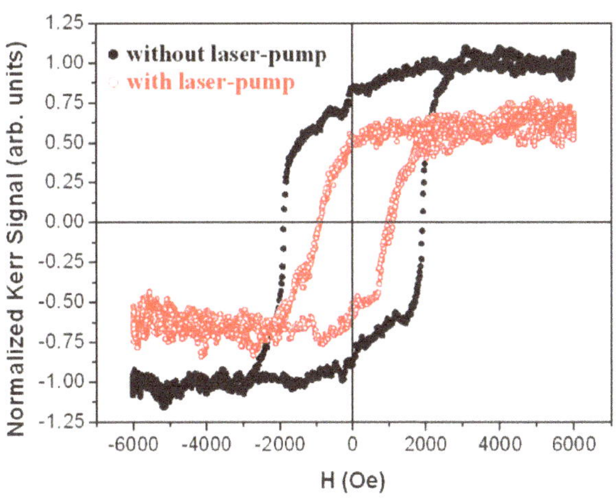

Figure 6: In-plane hysteresis loops of the GdFeCo thin film without (•) and with pump pulse (○) for a pump fluence of 2.7 mJ/cm^2.

Once we verified that the ferrimagnetic system could be excited using sub-10-fs pulses, we studied its time-resolved magnetization dynamics. However, we should note that with our set-up configuration, which uses linearly polarized laser pulses with a fluence of 2.7 mJ/cm^2 and an in-plane external magnetic field, magnetization switching of the sample is not allowed and we can focus our attention on the ultrafast demagnetization and precessional processes[3,6].

As was previously observed[16,37], the magnetization shows a sudden drop within the first picosecond after the pump pulse, a fast recovery (remagnetization) within a few picoseconds, followed by a clear oscillation, or precession, during a slower magnetization recovery, which is characterized by the ferromagnetic resonance frequency $\omega_{FMR}$and the Gilbert damping parameter α. Our TR-MOKE results are summarized in Fig. 7 where different in-plane external magnetic fields were applied to the sample.

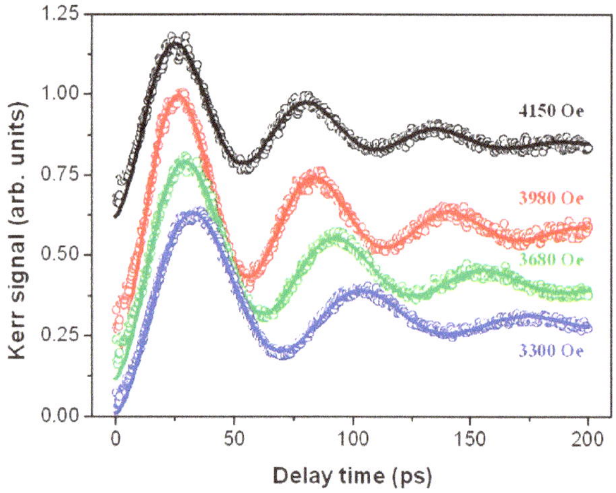

Theoretical curves (solid curves) are fit to the experimental data (open symbols) using Equation 1.

Figure 7: Time-resolved magneto-optical Kerr effect (TRMOKE) signals measured at a pump laser fluence of 2.7 mJ/cm^2 for a 20-nm-thick GdFeCo film under different in-plane external applied fields.

We begin our analysis by focusing on the oscillatory behavior of the magnetization, as induced by the ultrafast-photo-excitation from the pump pulse. In our experiment, we applied an in-plane magnetic field strong enough to keep the sample magnetization in plane and parallel to the effective anisotropy field. The effective anisotropy field depends on several factors, such as the external applied magnetic field (strength and direction), the sample magnetocrystalline anisotropy energy and the shape anisotropy energy.

Initially, the pump beam energy is transferred to the electrons of the system and they are excited to above the Fermi level. Then, the electrons thermalize by electron-electron interactions, and transfer heat to the lattice by electron-phonon scattering and to the spin system by electron-spin scattering1·2. Therefore, the laser increases the spin temperature, modifying the strength of the magnetization, as well as the magnetocrystalline and shape anisotropy energies, and consequently inducing a change in the total effective anisotropy field. As this change is fast, the magnetization vector will not be able to follow the net internal field and will start to precess around its new equilibrium4. The magnetization recovery is described by a damped precessional motion around the effective field direction on a timescale of hundreds of picoseconds, until the original state is regained, and with a resonance frequency, ω_{FMR}. Time-domain measurements on the excited precession give quantitative information on anisotropy, switching, and damping phenomena. In order to obtain a correct fit of the precessional processes, without any contamination from the ultrafast demagnetization, the TR-MOKE measurements were fitted, after a delay time of 10 ps, to a damped-harmonic function superposed with an exponential decay background[10,38]:

$$\theta = \theta_0 + A * \exp(-t/t_0) + B * \mathrm{Sin}(2\pi\omega_{FMR}t + \varphi) * \exp(-t/\tau) \quad (1)(1)$$

where θ_0 and A are the background magnitudes and t_0 is the background recovery time. The final term represents the precessional motion with the resonance frequency $\omega_{FMR,}$ relaxation time τ and initial phase φ. The solid curves in Fig. 7 are the fits using Equation 1, which show that the model agrees very well with the experimental data.

Therefore, the resonance frequency of the precessional motion (ω_{FMR}) can be determined[16] and its field-dependence is shown in Fig. 8. Assuming that both the external magnetic field and the sample magnetization are in the sample plane, we obtained the field dependence of the resonance frequency, ω_{FMR}(H), from the Kittel approximation:

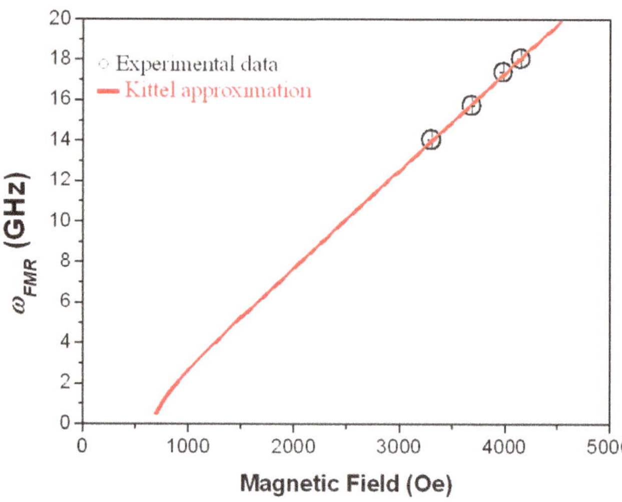

Field dependence of the resonance frequency (○) and the theoretical fit using the Kittel approximation (▬).

Figure 8

$$\omega_{FMR} = \frac{\gamma}{2\pi}\sqrt{H \times (H - H_u)} \quad (2) \quad (2)$$

where γ is the gyromagnetic ratio and H_u is the effective anisotropy field of the sample. This analysis confirms that the GdFeCo thin film shows an effective out-of-plane anisotropy energy with an effective anisotropy field of $H_u = (0.7 \pm 0.2)$ kOe and an effective g-factor of $g_{eff} = (3.4 \pm 0.2)$, in agreement with the values obtained by Kato et al.[15]

Moreover, the variation of (1/τ) versus H were also obtained from the fits and compared with the theoretical calculations (Fig. 9 a)[10]:

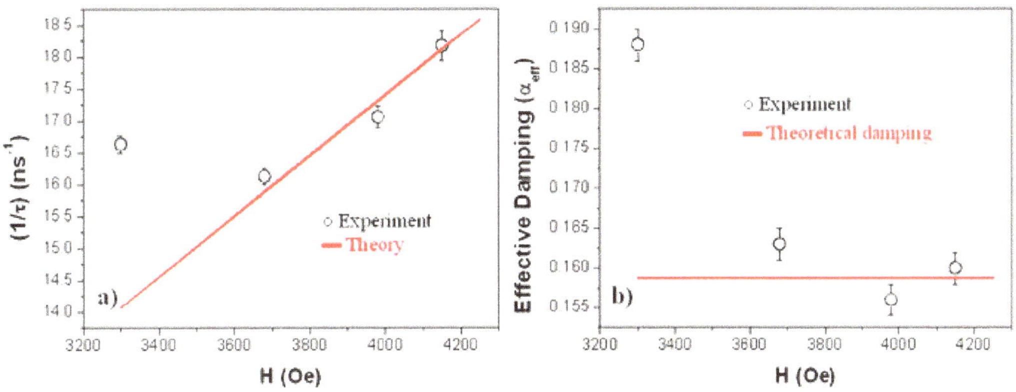

Figure 9 (**a**) Field dependence of (1/τ): Experimental data (○) and theoretical fit (▬) using equation (3). (**b**) Field dependence of (α_{eff}): Experimental data (○) and theoretical damping coefficient (α) (▬) calculated from Equation (3).

$$(1/\tau) = \pi\alpha\gamma(2H - H_u) \qquad (3)$$

(3)

where α is the theoretical damping parameter, H the external applied field, H_u the effective anisotropy field and γ the gyromagnetic ratio. We obtained that the damping parameter is α = (0.159 ± 0.001), which is also in agreement with the values reported in other works[15,16].

However, and although (1/τ) should be roughly proportional to ω_{FMR}, a significant deviation from the theoretical curve has been observed at H = 3300 Oe. An alternative approach to evaluate α is based on defining the effective Gilbert damping parameter, α_{eff}[10]:

$$\alpha_{eff} = \frac{1}{2\pi\omega_{FMR}\tau} \qquad (4)$$

(4)

where ω_{FMR} and τ values at each applied H were estimated from the experiment (Fig. 9 b). The α_{eff} value is identical to or larger than α, and corresponds to an upper-bound value of α. As observed in Fig. 9 b, α_{eff} is constant for large applied magnetic fields, independent of H and close to α = (0.159 ± 0.001). However, at lower fields, α_{eff} is larger than α.

We now focus our attention on the ultrafast demagnetization and the fast remagnetization processes within the first few picoseconds after excitation by the pump pulse. For this purpose, we have performed time-resolved pump-probe experiments using a pump pulse fluence of 2.7 mJ/cm², with a pulse width of ~7.3 fs, an external applied field of 3300 Oe and a time delay from −1 to 2 ps. Under these experimental conditions, we previously determined that the sample was demagnetized by up to 70% of its normal saturation value (see Fig. 6). The normalized demagnetization signal is shown in Fig. 10 a. An ultrafast laser-induced demagnetization process is observed, with a minimum in the magnetization at a delay time of ≈800 fs, followed by its recovery, which is consistent with previous studies[17,39].

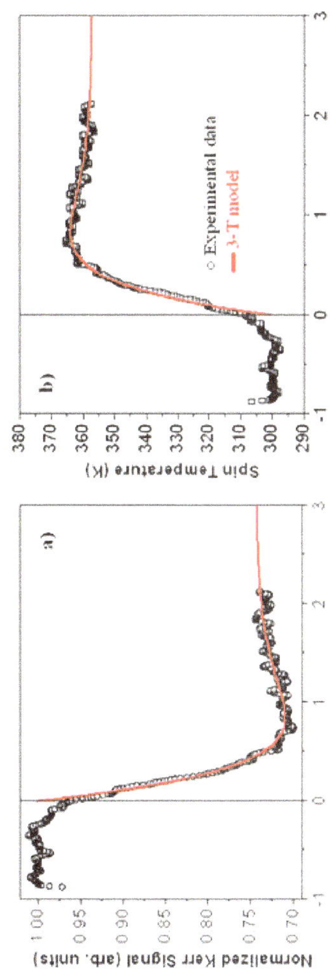

Figure 10 (**a**) Experimental demagnetization data of GdFeCo and (**b**) expected temporal evolution of the spin temperature using Equation 6 (open black symbols). Both curves were fitted within the 3T model (▬).

Our analysis is based on the 3-Temperature model[1,2] where it has been assumed that the energy supplied to the sample by the photons can be analyzed on different timescales via three different channels: the energy of the electrons (e), the phonon or lattice energy (l) and the spin energy (s), which is related with the magnetization change. Therefore, within the 3T model, the electron temperature, $T_e(t)$, is coupled to the lattice temperature, $T_l(t)$, and the spin temperature, $T_s(t)$, in the form of the following three coupled differential equations[1,2]:

$$\begin{aligned} C_e(T_e)\frac{\partial T_e}{\partial t} &= -G_{el}(T_e - T_l) - G_{es}(T_e - T_s) + P(t) \\ C_l(T_l)\frac{\partial T_l}{\partial t} &= -G_{el}(T_l - T_e) - G_{sl}(T_l - T_s) \\ C_s(T_s)\frac{\partial T_s}{\partial t} &= -G_{es}(T_s - T_e) - G_{sl}(T_s - T_l) \end{aligned} \quad (5)$$

where $C_e(T_e)$, $C_l(T_l)$ and $C_s(T_s)$ are the electronic, lattice and spin specific heats, respectively. G_{el}, G_{es} and G_{sl} denote the electron-lattice, the electron-spin and the spin-lattice coupling constants, respectively, and determine the rate of the energy exchange between the electrons, the lattice and the spin baths. P(t) is the laser power density delivered to the electrons, which is described by a Gaussian function with a 8 fs pulse width (FWHM) and a fluence of 2.7 mJ/cm^2.

In order to study the evolution of the magnetization, we first need to extract the experimental spin temperature (Fig. 10 b) which is achieved using the Bloch $T^{3/2}$ law17:

$$M(T_s)/M(T_{room}) = [1 - (T_s/T_C)^{3/2}] \quad (6)$$

where T_C is the Curie Temperature ($T_C = 500$ K for GdCoFe)[17,36]. T_s reaches a maximum of around 365 K at t ≈ 800 fs after the pump pulse. Considering that the specific heat of the electrons is proportional to the electronic temperature[17,40], $C_e(T_e) = cT_e$ with $c = 714$ J/m^{3}K^2, and solving the three coupled differential equations (Eq. 5), the experimental spin temperature has been tuned until a reasonable agreement was obtained between the calculation and the experiment (see Fig. 10b). Our best fit was achieved for $C_l(T_l) = (2.33 \times 10^6)$ J/m^{3}K, $C_s(T_s) = (0.34 \times 10^4)$ J/m^{3}K, $G_{el} = (8.19 \times 10^{17})$ W/m^{3}K, $G_{es} = (0.2 \times 10^{16})$ W/m^{3}K and $G_{sl} = (0.55 \times 10^{16})$ W/m^{3}K, which are reasonable values when compared with the typical values for metals2 and for a GdFeCo alloy with similar chemical composition17. Finally, and using Equation 6, the evolution of the magnetization is presented by the solid red line in Fig. 10 a.

In summary, we have developed a new dual-colour pump-probe MOKE spectrometer dedicated to the study of ultrafast demagnetization processes in a range of magnetic materials and structures. This versatile set-up uses state-of-the-art ultrafast optical methods to deliver ultrashort laser pump and probe pulses, at the sample position, permitting the observation of ultrafast magnetization dynamics at unprecedented temporal resolutions for a system of its type, well within the sub-10-fs range. This few-cycle regime is also highly promising for the direct excitation and observation of coherent ultrafast magnetodynamic behavior, which is expected to help us to gain a deeper understanding of both the conservation of angular momentum associated with the demagnetization process, and the relation between the spin thermalization and the corresponding quasi-equilibrium magnetization5. Furthermore, CEP-stabilized sub-5-fs pulses open the way for exploring the possible dependence of the induced magnetization dynamics on the electric field of the excitation pulses.

To demonstrate the essential capability of our dual-colour set-up, we have measured the ultrafast demagnetization and precessional dynamics in a ferrimagnetic GdFeCo thin film using a pump pulse fluence of 2.7 mJ/cm^2, with sub-8-fs pulse width and ranging the time delay from −1 to hundreds of ps. We observed that the magnetization shows a sudden drop within the first picosecond after the pump pulse and a fast recovery (remagnetization) within a few picoseconds. This is followed by a clear oscillation or precession during a slower magnetization recovery. Regarding the precessional behaviour, we have determined both the ferromagnetic resonance frequency ω_{FMR} and the effective Gilbert damping parameter α_{eff}, which are in good agreement with the current literature.

Furthermore, the ultrafast demagnetization and the fast remagnetization processes within the initial picoseconds after excitation by the pump pulse has been analyzed using an external applied field of 3300 Oe and a time delay ranging from −1 to 2 ps. Under these experimental conditions, measurements of the laser-induced demagnetization revealed a minimum at t ≈ 800 fs. Within the 3T model, the electronic, lattice and spin specific heats, as well as the electron-lattice, the electron-spin and the spin-lattice coupling constants were calculated.

Therefore, we have experimentally confirmed that our sub-8-fs pump pulses are able to efficiently induce demagnetization in a ferrimagnetic material and so we have established that the minimum pulse duration required for the demagnetization process is clearly in the sub-10-fs range. We expect that the present pump-probe system, when operating in a degenerate configuration with sub-5-fs CEP stabilized pulses, will enable us to further explore this limit, which we believe is an important aim for future developments in femtomagnetism.

ACKNOWLEDGEMENTS

The authors thank Dr. A.V. Kimel for supplying the GdFeCo thin film used in this study. This work was partially supported by the Portuguese funding agency Fundação para a Ciência e Tecnologia (FCT) through grants PTDC/FIS/115561/2009 and PTDC/FIS/122511/2010, co-funded by COMPETE and FEDER, and the project NORTE-07-0124-FEDER-000070. C.S.G. would like to acknowledge financial support from FCT via a PhD grant SFRH/BD/61466/2009 and CSF-CAPES via a PhD grant BEX 9461/13-2. D.N. thanks FCT (Contract No. IF/01191/2013) for financial support.

REFERENCES

1. Zhang, G. P., Hübner, W., Beaurepaire, E. & Bigot, J.-Y. In Topics of applied physics: spin dynamics in confined magnetic structures I(eds Hillebrands, B. & Ounadjela, K.). Laser-induced ultrafast demagnetization: Femtomagnetism, a new frontier?. Vol. 83, 245–289 (Springer-Verlag Berlin Heidelberg 2002).
2. Beaurepaire, E., Merle, J.-C., Daunois, A. & Bigot, J.-Y. Ultrafast spin dynamics in ferromagnetic nickel. Phys. Rev. Lett. 76, 4250–4253 (1996).
3. Stanciu, C. D. et al. All-optical magnetic recording with circularly polarized light. Phys. Rev. Lett. 99, 047601 (2007).
4. Kirilyuk, A., Kimel, A. V. & Rasing, T. Ultrafast optical manipulation of magnetic order. Rev. Mod. Phys. 82, 2731–2784 (2010).
5. Bigot, J.-Y., Vomir, M. & Beaurepaire, E. Coherent ultrafast magnetism induced by femtosecond laser pulses. Nature Phys. 5, 515–520 (2009).
6. Kirilyuk, A., Kimel, A. V. & Rasing, T. Laser-induced magnetization dynamics and reversal in ferrimagnetic alloys. Rep. Prog. Phys. 72, 026501 (2013).

7. Zhang, G. P. & Hübner, W. Laser-induced ultrafast demagnetization in ferromagnetic metals. Phys. Rev. Lett. 85, 3025–3028 (2000).
8. Hicken, R. et al. Magnetooptical studies of magnetism on pico and femtosecond timescales. J. Magn. Magn. Mater. 242–245, 559–564 (2002).
9. Faahnle, M., Seib, J. & Illg, C. Relating gilbert damping and ultrafast laser-induced demagnetization. Phys. Rev. B 82, 144405 (2010).
10. Mizukami, S. et al. Fast magnetization precession observed in L10-FePt epitaxial thin film. Appl. Phys. Lett. 98, 052501 (2011).
11. Zhang, Z. et al. Ultrafast laser-induced magnetization precession dynamics in FePt/CoFe exchange-coupled films. Appl. Phys. Lett.97, 172508 (2010).
12. Bossini, D. et al. Macrospin dynamics in antiferromagnets triggered by Sub-20 femtoseconds injection of nanomagnons. arXiv:1507.03377v1 [cond-mat.mes-hall].
13. Silva, F. et al. Simultaneous compression, characterization and phase stabilization of GW-level 1.4 cycle VIS-NIR femtosecond pulses using a single dispersion-scan setup. Opt. Express 22, 10181–10191 (2014).
14. Miranda, M., Fordell, T., Arnold, C., L'Huillier, A. & Crespo, H.Simultaneous compression and characterization of ultrashort laser pulses using chirped mirrors and glass wedges. Opt. Express20, 688–697 (2012).
15. Kato, T. et al. Compositional Dependence of g-Factor and Damping Constant of GdFeCo Amorphous Alloy Films. IEEE Trans. Magn. 44, 3380–3383 (2008).
16. Stanciu, C. D. et al. Ultrafast spin dynamics across compensation points in ferrimagnetic GdFeCo: The role of angular momentum compensation. Phys. Rev. B 73, 220402(R) (2006).
17. Mekonnen, A. et al. Role of the inter-sublattice exchange coupling in short-laser-pulse-induced demagnetization dynamics of GdCo and GdCoFe alloys. Phys. Rev. B 87, 180406(R) (2013).
18. Vonesch, H. & Bigot, J.-Y. Ultrafast spin-photon interaction investigated with coherent magneto-optics. Phys. Rev. B 85, 180407(R) (2012).
19. Luo, C. W., Wang, Y. T., Chen, F. W., Shih, H. C. & Kobayashi, T.Eliminate coherence spike in reflection-type pump-probe measurements. Opt. Express 17, 11321–11327 (2009).
20. Lebedev, M. V., Misochko, O. V., Dekorsy, T. & Georgiev, N. On the nature of coherent artifact. J. Exp. Theor. Phys. 100, 272–282 (2005).
21. Cerullo, G. & De Silvestri, S. Ultrafast optical parametric amplifiers. Rev. Sci. Instrum. 74, 1 (2003)
22. Steinmeyer, G., Sutter, D. H., Gallmann, L., Matuschek, N. & Keller, U. Frontiers in ultrashort pulse generation: Pushing the limits in linear and nonlinear optics. Science 286, 1507–1512 (1999).
23. Alonso, B. et al. Characterization of sub-two-cycle pulses from a hollow-core fiber compressor in the spatiotemporal and spatio-spectral domains. Appl. Phys. B 112, 105–114 (2013).
24. Witte, S. & Eikema, K. S. E. Ultrafast optical parametric chirped-pulse amplification. IEEE J. Sel. Top. Quantum Electron. 18, 296–307 (2012).
25. Eilanlou, A. A., Nabekawa, Y., Ishikawa, K. L., Takahashi, H. &Midorikawa, K. Direct amplification of terawatt sub-10-fs pulses in a CPA system of Ti:sapphire laser. Opt. Express 16, 13431–13438 (2008).
26. Rothhardt, J., Demmler, S., Hädrich, S., Limpert, J. & Tünnermann, A. Octave-spanning OPCPA system delivering CEP-stable few-cycle pulses and 22 W of average power at 1 MHz repetition rate.Opt. Express 20, 10870–10878 (2012).
27. Nisoli, M. et al. Compression of high-energy laser pulses below 5 fs. Opt. Lett. 22, 522–524 (1997).
28. Schenkel, B. et al. Generation of 3.8-fs pulses from adaptive compression of a cascaded hollow fiber supercontinuum. Opt. Lett. 28, 1987–1989 (2003).
29. Szipöcs, R., Ferencz, K., Spielmann, C. & Krausz, F. Chirped multilayer coatings for broadband dispersion control in femtosecond lasers. Opt. Lett. 19, 201–203 (1994).
30. Kärtner, F. X. In MITopencourseware, Chapter 10: Pulse Characterization Ultrafast Optics, 333–370 (Spring Term 2005).
31. Trebino, R. In Frequency-Resolved Optical Gating: The Measurement of Ultrashort Laser Pulses (Springer Science + Business Media, LLC, 2000).
32. C. Iaconis, C. & Walmsley, I. A. Spectral phase interferometry for direct electric-field reconstruction of ultrashort optical pulses.Opt. Lett. 23, 792–794 (1998).
33. Miranda, M. et al. Characterization of broadband few-cycle laser pulses with the d-scan technique. Opt. Express 20, 18732–18743 (2012).
34. Mansuripur, M. In The Physical Principles of Magneto-Optical Recording (Cambridge: Cambridge University Press, 1995).
35. Awano, H. et al. Magnetic domain expansion readout for amplification of an ultra high density magneto-optical recording signal. Appl. Phys. Lett. 69, 4257 (1996).
36. Ostler, T. A. et al. Crystallographically amorphous ferrimagnetic alloys: Comparing a localized atomistic spin model with experiments. Phys. Rev. B 84, 024407 (2011).
37. Hohlfeld, J. et al. Fast magnetization reversal of GdFeCo induced by femtosecond laser pulses. Phys. Rev. B 65, 012413 (2001).
38. Song, H.-S. et al. Observation of the intrinsic Gilbert damping constant in Co/Ni multilayers independent of the stack number with perpendicular anisotropy. Appl. Phys. Lett. 102, 102401 (2013).
39. Stanciu, C. D. et al. Subpicosecond Magnetization Reversal across Ferrimagnetic Compensation Points. Phys. Rev. Lett. 99, 217204 (2007).
40. Kittel, C. In Introduction to Solid State Physics (John Wiley & Sons, New York, 1996).

CHAPTER 2

LASER-INDUCED ULTRAFAST DEMAGNETIZATION

Haitian Xu, Ghazal Hajisalem, Geoffrey M. Steeves, Reuven Gordon & Byoung C. Choi

INTRODUCTION

A ferromagnetic (FM) thin film excited by femtosecond pulsed laser will undergo ultrafast loss of magnetic order (demagnetization) within several hundred femtoseconds, due to rapid energy transfer from thermalized electrons to the spin system. This is followed by a slower recovery (remagnetization) over the picosecond timescale as electrons equilibrate with the phonons, and eventually to complete cooling via nanosecond lattice diffusion[1,2]. The timescale of laser-induced ultrafast demagnetization is orders of magnitude below the limit imposed by conventional switching of magnetic order via magnetic field pulse (~100 ps[3]), and has been the focus of considerable research since its discovery in 1996[4,5,6,7].

Orders-of-magnitude enhancement of electric fields, light absorption and scattering can occur as a result of the resonant excitation of conduction electrons – also known as localized surface plasmons (LSPs), in a metal nanoparticle (NP) under illumination[8]. The resonant frequency is highly sensitive to NP geometry and environment[9], and is tunable from visible to near infrared. With recent advances in the synthesis of highly monodisperse colloidal noble metal NPs[10], plasmon NPs have been the subject of investigation in a wide range of plasmonic applications, including cancer diagnosis and therapy, remote delivery and photothermal imaging[11,12]. Furthermore, an increasing number of 'magnetoplasmonic' studies have emerged, focusing on combining magnetic and plasmonic functionalities[13,14]. Of particular relevance to information storage is the on-going development of heat-assisted magnetic recording (HAMR) technology, which uses a near-field plasmon NP transducer to resonantly focus laser light onto a magnetic recording medium, temporarily reducing its coercivity through heat generation, thus lowering the write-field requirement[15,16,17,18]. A primary issue associated with current HAMR prototypes is the accelerated thermal degradation of the transducer due to thermal cycling, which limits their long-term reliability[19,20]. Several solutions have been proposed, including two-stage heating of the recording medium[21] as well as the ongoing search for alternative, low-loss plasmon materials[22]. Nevertheless, thermal degradation of the near-field transducer remains a major concern.

In this work, we investigate the improvement in laser-induced demagnetization signal through LSP-enhanced localized heating. We show that, by coating FM thin films with gold nanorods (AuNRs) dimensionally-tuned to support LSPs near the peak laser frequency, the demagnetization can be increased by over 50% for ~4% AuNR coverage, without increasing the laser fluence. The AuNR coating functions in a similar capacity to NP transducers in HAMR recording heads, and the arrangement offers a possible solution to the thermal degradation problem in current HAMR prototypes by uniformly distributing or patterning the plasmon NPs across the surface of the recording medium, thus subjecting individual AuNRs to a small fraction of the overall thermal cycling.

RESULTS

Colloidal 10 nm × 41 nm AuNR was drop-evaporated onto 20 nm ferromagnetic $Ni_{80}Fe_{20}$ (permalloy) thin films to achieve an average density of 96 AuNR/μm² (a total AuNR cross-sectional area of 39360 nm² per μm², or 3.94% areal coverage), with a standard deviation of 41 AuNR/μm², as shown in the SEM image in Fig. 1(a). Due to capillary pressure, there is a highly non-uniform aggregation of AuNR at the edge of the evaporated spot (Fig. 1(a), lower inset), and care was taken during subsequent experiments to avoid this edge formation. To independently verify the plasmon resonance effect of the AuNR coating, dark-field scattering spectra of the thin film samples were obtained with a white light source at 70° incidence using a setup described elsewhere[23,24]. Molar extinction coefficient of the colloidal solution was measured using the Beer-Lambert law[25]. A stroboscopic pump-probe magneto-optic Kerr effect setup in the polar configuration, schematically shown in Fig. 1(b), was used to measure time-

Citation: Haitian Xu, Ghazal Hajisalem, Geoffrey M. Steeves, Reuven Gordon & Byoung C. Choi, Laser-Induced Ultrafast Demagnetization, doi:10.1038/srep15933.

resolved out-of-plane magnetization, M_z under the illumination of a femtosecond Ti:sapphire pulsed laser, tuned to the plasmon frequency of the AuNRs[26].

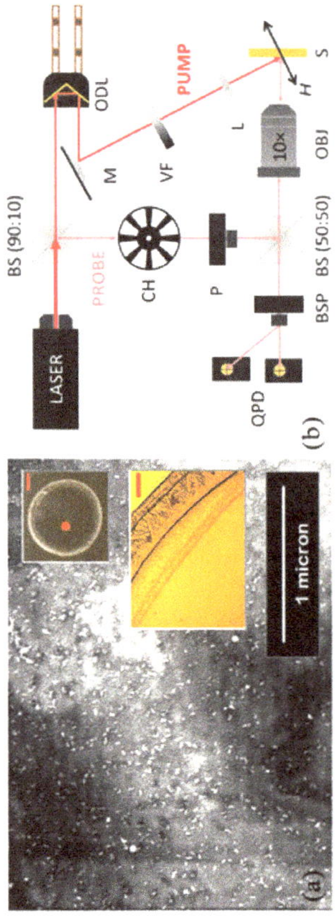

Figure 1 (**a**) SEM image showing drop-evaporated AuNR distribution on film surface. Top inset: drop-evaporated spot. Red dot indicates the approximate position where the SEM image was recorded. Scale bar = 1 mm. Bottom inset: bright-field image taken at the edge of the drop-evaporated spot under 20× objective, showing details of the highly non-uniform AuNR aggregation. Scale bar = 50 µm. (**b**) Pump-probe setup for measuring transient magnetization. Abbreviations: BS: beam splitter, BSP: polarizing beam-splitter, CH: chopper,H: external field, L: lens, M: mirror, ODL: optical delay line, OBJ: objective, P: polarizer, QPD: quadrant photodiode, S: sample, VF: variable filter.

Figure 2(a) shows the dark-field scattering spectra of the AuNR-coated sample (red) normalized against the regular permalloy film (blue), as well as the molar extinction coefficient spectra of the colloidal AuNR solution (green) and the Ti:sapphire laser (magenta). The AuNR-coated thin film shows a distinct spectral peak slightly below 800 nm, close to the central wavelength of the Ti:sapphire laser. The uncoated sample by comparison does not exhibit any discernable peaks over the entire (500–1000 nm) frequency range. The spectral peak of the AuNR-coated sample exhibits a slight blue-shift and significant broadening compared to the colloidal solution, which may be attributed to the modification of LSP due to interaction between nanorods and the close proximity of the permalloy substrate[27,28]. Simulations using a commercial finite-difference time-domain (FDTD) package (Lumerical Solutions, www.lumerical.com) reveals a qualitative agreement with experimental observations (Fig. 2(b)).

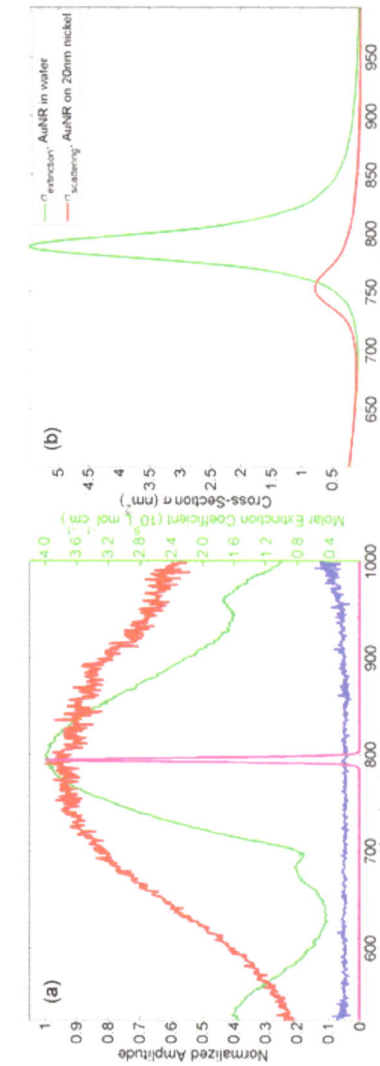

Figure 2 (**a**) Dark-field scattering spectra of AuNR-coated (red) and uncoated (blue) permalloy films (arbitrary units). Magenta: Ti:sapphire laser spectrum (arbitrary units). Green: molar extinction coefficient spectrum of the colloidal AuNR solution (unit: 10^9 L mol^{-1} cm^{-1}, right y-axis). (**b**) Simulated scattering spectra of 10 × 40 nm AuNR in water (red) and on 20 nm nickel (green). Material data taken from ref. 30,31.

Laser-induced ultrafast demagnetization signals in M_z were recorded for a range of pump fluences, $F \approx 0.08 - 0.26$ mJ/cm^2 (315–1007 mW). Selected results are shown in Fig. 3 for uncoated (blue) and AuNR-coated (red) permalloy samples for different pump fluences. The characteristic sub-picosecond demagnetization, the subsequent picosecond remagnetization as well as the increase in demagnetization signal with fluence are in general agreement with existing literature[29,30,31,32,33]. A long scan range (~100 ps with 500 fs time-steps) was used in order to capture the slow remagnetization process. This temporal regime is orders of magnitude longer than the subpicosecond pulse duration and thermalization time[34], therefore the remagnetization process can be adequately described by a single exponential expression[35], $M(t) = C_1\exp(-t/\tau) + C_2$, where the characteristic remagnetization time τ is quantitatively defined as the time at which magnetization recovers by 63%.

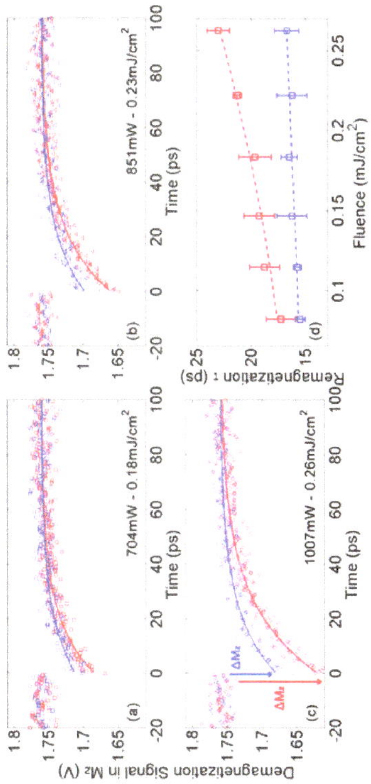

Figure 3 Laser pulse-induced ultrafast demagnetization measured with (red) and without (blue) AuNR coating at pump fluences of (**a**) 0.18 mJ, (**b**) 0.23 mJ and (**c**) 0.26 mJ. Time-step used = 500 fs. Solid lines are single exponential fits. (**d**) Characteristic remagnetization time τ versus laser fluence for samples with (red) and without (blue) AuNR. Lines serve as visual guides.

Most interestingly, the results in Fig. 3 reveal an increase in ΔM_z in the AuNR-coated sample compared to the uncoated sample at the same pump fluence. Plots of characteristic remagnetization times, τ against pump fluence are shown in Fig. 3(d). τ increases with increasing pump fluence, due to the increasing loss of magnetic correlation with increasing temperature[36]. For the AuNR-coated sample, τ is longer at all pump fluences, which complements the demagnetization data and indicates a higher 'effective temperature' in the LSP-enhanced thin films.

In order to better interpret these results, we performed first-order calculations of the localized AuNR field enhancement, and the subsequent heat generation and demagnetization in a FM thin film with 4% AuNR coverage. Using experimentally measured laser fluences F, the averaged temperature increase in the regular FM substrate after pulse absorption ΔT_{FM} was estimated using

$$\Delta T_{FM} = \frac{(1 - R_{py})F}{lC_{py}\rho_{py}} \qquad (1)$$

(1)

where R_{py} is the experimentally measured reflectance (scattering is ignored for the film sample), laser penetration depth l = 15 nm and C_{py} and ρ_{py} are the heat capacity and density of permalloy-80, taken from ref. [37]. Given the scan range and temporal resolution (100 ps/500 fs) of the current investigation, thermalization (<1 ps34·35) was considered instantaneous. Intricacies resulting from the interplay between the various demagnetization and thermalization timescales are left for a future study. For FM film coated with AuNR, enhancement in heating originates from the enhanced electric field |***E***| inside the FM substrate at plasmon resonance and can be calculated as follows[38]:

$$Q = \frac{\omega\varepsilon_0\,\mathrm{Im}(\varepsilon)}{2}\int_{\mathrm{substrate}} |\boldsymbol{E}(\boldsymbol{r})|^2 \mathrm{d}r, \qquad (2)$$

(2)

where Q is the heat generation, ω is the angular frequency of the laser, and ε_0 and ε are the vacuum and relative permittivity of the FM film, respectively. For a given incident electric field amplitude and polarization, E(r), and hence Q and ΔT_{FM}, can be computed using Lumerical Solutions. To take into account random orientations of the AuNRs with respect to the linearly polarized pump pulse in our experiment, an extra factor of 0.5 was applied to the Q calculated for the case where AuNR is aligned in parallel to electric field polarization (Fig. 4(a)), in order to better estimate an 'average' temperature increase. The capping layer around the colloidal AuNR (cetyltrimethylammonium bromide (CTAB)) was modeled as a 1 nm spacer layer with refractive index equal to that of air, which has been previously shown to be a reasonable assumption for thin CTAB layers in the near-field[39]. As a result of the field enhancement inside the AuNR, they act as sources of heat[40] and must be taken into account. COMSOL Multiphysics (version 4.3.0.233) was used to investigate conductive heat transfer from AuNRs to the FM layer using initial AuNR temperatures calculated from equation (2).

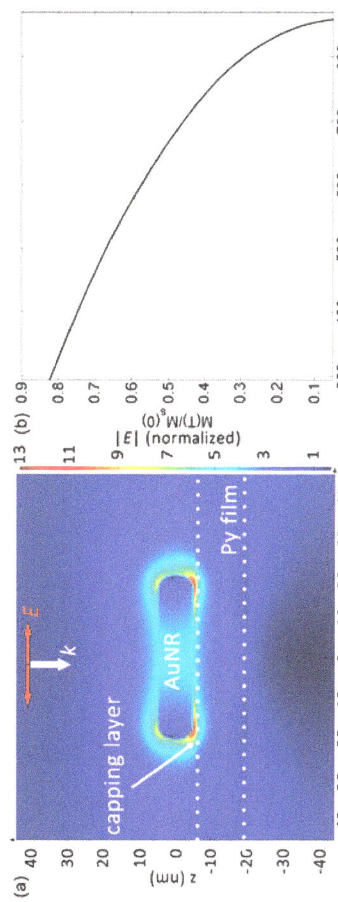

Figure 4 (**a**) Cross-sectional electric field |***E***| profile for 10 nm × 40 nm AuNR on 15 nm permalloy with 1 nm spacer layer at plasmon resonance, calculated using Lumerical Solutions. Simulation mesh size = 0.5 nm in the vicinity of the AuNR. Dotted lines represent the boundaries of the permalloy (Py) film. (**b**) Temperature-dependence of magnetization for permalloy, calculated using $M(T) \sim (1 - T/T_C)^{\beta}$.

The additional temperature increase in the FM layer in this case was found to be negligible over the demagnetization timescale (~ps), due to the presence of the CTAB ligand and the small area of contact between the AuNR and the FM film. Using calculated values of ΔT_{FM}, ΔM_z at different fluences were found for both regular and AuNR-coated permalloy substrates using the temperature-dependence of magnetization, $M(T) \sim (1 - T/T_C)^{\beta}$ (Fig. 4(b)), where Curie temperature$T_C = 860\,K$ for permalloy[41] and critical exponent $\beta = 0.5$ from mean-field theory[42].

The fluence dependence of demagnetization from both experiment and modeling are presented in Fig. 5. There is a consistent enhancement in ΔM_z for AuNR-coated samples over the entire experimental fluence range. At higher pump fluences, this enhancement becomes more significant, and ΔM_z for the AuNR-coated sample exhibits divergent behavior due to the increased curvature M(T) at temperatures approachT_C (Fig. 4(b)). At the highest experimental pump fluence ($F \sim 0.26$ mJ/cm^2), ΔM_z in the AuNR-coated sample is increased by ~60% compared to the uncoated sample. By extrapolating the linear fit in Fig. 5(a), it was found that the uncoated sample would require a pump fluence $F \sim 0.42$ mJ/cm^2 in order to achieve the same ΔM_z as the AuNR-coated sample at $F \sim 0.26$ mJ/cm^2. Despite its simplicity, our model calculations were able to qualitatively reproduce the experimental results (Fig. 5(b)). We note that the simulation does not take into account the variation in density and orientation of the drop-evaporated nanorods, or thermal diffusion to the surroundings, leading to an over-estimation in demagnetization.

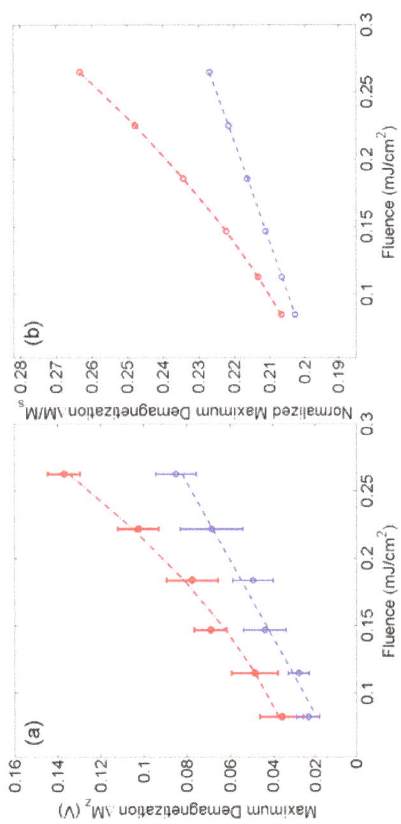

(**a**) experimental results; (**b**) results from model calculations. Lines serve as visual guides.

Figure 5: Fluence dependence of demagnetization for AuNR-coated (red) and uncoated (blue) samples.

In conclusion, we demonstrated the LSP-enhancement of laser-induced ultrafast demagnetization on permalloy thin films coated with drop-evaporated colloidal AuNRs supporting LSP near the peak laser frequency. Compared to the uncoated sample, an increase in demagnetization signal of >50% was observed in the AuNR-coated samples with 4% AuNR coverage, which increases with increasing laser fluence. Despite the low areal coverage of AuNR, LSP-enhanced electric-field intensity in the permalloy substrate in the vicinity of AuNR contributes disproportionately to the overall demagnetization signal. We were able to qualitatively reproduce the experimental results with FDTD modeling and mean-field theory.

METHODS

Sample Preparation

20 nm films of ferromagnetic $Ni_{80}Fe_{20}$ (permalloy) with 5 nm titanium adhesion were deposited using electron-beam evaporation in 1.0×10^{-6} Torr vacuum at 1 Å/s on native-oxide coated (110) Si wafers. The film thickness was verified post-deposition with a spectroscopic ellipsometer (J.A. Woollam Alpha-SE), and found to be 20.348 nm, with an average roughness of 0.327 nm. 10 μL of commercial monodisperse 10 nm × 41 nm colloidal AuNR solution (Nanopartz, 7.24×10^{11}/mL, CTAB-capped) with longitudinal plasmon resonance wavelength $\lambda = 808$ nm was drop-cast onto the film surface and allowed to evaporate under closed ambient conditions to minimize density variation and band formation[43]. AuNR density and variability were measured by visually inspecting 3 × 3 μm^2 areas of the SEM images of the sample.

Experimental Methods

In order to trigger ultrafast demagnetization, a 76 MHz Ti:sapphire laser (Coherent Mira 900) pump beam with 795 nm central wavelength and 135 fs pulse duration was focused on a $r \approx 40$ μm spot, for a maximum pulse fluence $F \approx 0.33$ mJ/cm^2 (~16 nJ/pulse). This fluence is sufficiently large to induce measurable demagnetization, but does not melt the AuNRs or cause a significant, permanent temperature increase during measurement[44]. The ratio of pump to probe intensity at sample surface was >20:1. The probe beam was modulated at 300 Hz using a chopper and was focused on an overlapping area with $r \approx 10$ μm with the help of a CCD camera. When measuring AuNR-coated samples, care was taken to position the probe beam over areas on the sample with close to average AuNR density. The delay between pump and probe was mechanically controlled via a 220 mm (1466 ps) optical delay line (Thorlabs ODL220-FS), with a delay precision of ±0.67 fs. An external field H = 140 kA/m inclined at 30° to the normal was applied to create a canted, out-of-plane saturation magnetization. Transient changes in M_z in terms of Kerr rotation θ_K, averaged over the 10 μm probe spot in the Py film, were measured using a balanced photodiode optical bridge detector, which isolates the Kerr and the reflectivity signals45. The detector system consists of a polarizing beam splitter (Wollaston prism) and a pair of silicon quadrant photodiodes (Thorlabs PDQ80A) with lock-in amplification referenced at the modulation frequency of the probe beam. The difference between two photodiode signals is proportional to the Kerr rotation. The experiment was carried out at room temperature.

REFERENCES

1. Kirilyuk, A., Kimel, A. V. & Rasing, T. Ultrafast optical manipulation of magnetic order. Rev. Mod. Phys. 82, 2731–2784 (2010).
2. Mathias, S. et al. Ultrafast element-specific magnetization dynamics of complex magnetic materials on a table-top. J. Electron Spectros. Relat. Phenomena 189, 164–170 (2013).
3. Tudosa, I. et al. The ultimate speed of magnetic switching in granular recording media. Nature 428, 831–833 (2004).
4. Beaurepaire, E., Merle, J.-C., Daunois, A. & Bigot, J.-Y. Ultrafast Spin dynamics in Ferromagnetic Nickel. Phys. Rev. Lett. 76, 4250–4253 (1996).
5. Bigot, J. Y., Guidoni, L., Beaurepaire, E. & Saeta, P. N. Femtosecond spectrotemporal magneto-optics. Phys. Rev. Lett. 93, 077401 (2004).
6. Stamm, C. et al. Femtosecond modification of electron localization and transfer of angular momentum in nickel. Nat. Mater. 6, 740–743 (2007).
7. Koopmans, B. et al. Explaining the paradoxical diversity of ultrafast laser-induced demagnetization. Nat. Mater. 9, 259–265 (2010).
8. Baffou, G. & Quidant, R. Thermo-plasmonics: using metallic nanostructures as nano-sources of heat. Laser Photon. Rev. 7, 171–187 (2012).
9. Atay, T., Song, J. H. & Nurmikko, A. V. Strongly interacting plasmon nanoparticle pairs: from dipole-dipole interaction to conductively coupled regime. Nano Lett. 4, 1627–1631 (2004).
10. Richardson, H. H., Carlson, M. T., Tandler, P. J., Hernandez, P. &Govorov, A. O. Experimental and theoretical studies of light-to-heat conversion and collective heating effects in metal nanoparticle solution. Nano Lett. 9, 1139–1146 (2009).
11. Lee, S. E., Liu, G. L., Kim, F. & Lee, L. P. Remote optical switch for localized and selective control of gene interference. Nano Lett. 9, 562–570 (2009).
12. Boyer, D., Tamarat, P., Maali, A., Lounis, B. & Orrit, M.Photothermal imaging of nanometer-sized metal particles among scatterers. Science 297, 1160–1163 (2002).
13. Jain, P. K., Xiao, Y., Walsworth, R. & Cohen, A. E. Surface plasmon resonance enhanced magneto-optics (SuPREMO): Faraday rotation enhancement in gold-coated iron oxide nanocrystals.Nano Lett. 9, 1644–1650 (2009).
14. Armelles, G., Cebollada, A., García-Martín, A. & González, M. U.Magnetoplasmonics: magnetoplasmonics: combining magnetic and plasmonic functionalities (Advanced Optical Materials 1/2013). Adv. Opt. Mater. 1, 2–2 (2013).
15. Kryder, M. H. et al. Heat assisted magnetic recording. Proc. IEEE96, 1810–1835 (2008).
16. Challener, W. A. et al. Heat-assisted magnetic recording by a near-field transducer with efficient optical energy transfer. Nat. Photonics 3, 220–224 (2009).
17. Stipe, B. C. et al. Magnetic recording at 1.5 Pb m−2 using an integrated plasmonic antenna. Nat. Photonics 4, 484–488 (2010).
18. Zhou, N. et al. Plasmonic near-field transducer for heat-assisted magnetic recording. Nanophotonics 3, 141–155 (2014).
19. Link, S., Burda, C., Nikoobakht, B. & El-Sayed, M. A. Laser-induced shape changes of colloidal gold nanorods using femtosecond and nanosecond laser pulses. J. Phys. Chem. B 104, 6152–6163 (2000).
20. Xu, B. et al. Relationship between near field optical transducer laser absorption and its efficiency. in IEEE Trans. Magn. 48, 1789–1793 (2012).
21. Xiong, S., Kim, J., Wang, Y., Zhang, X. & Bogy, D. A two-stage heating scheme for heat assisted magnetic recording. J. Appl. Phys. 115, 17B702 (2014).
22. Naik, G. V, Kim, J. & Boltasseva, A. Oxides and nitrides as alternative plasmonic materials in the optical range. Opt. Mater. Express 1, 1090–1099 (2011).
23. Mock, J. J., Hill, R. T., Tsai, Y. J., Chilkoti, A. & Smith, D. R. Probing dynamically tunable localized surface plasmon resonances of film-coupled nanoparticles by evanescent wave excitation. Nano Lett. 12, 1757–1764 (2012).
24. Hajisalem, G., Min, Q., Gelfand, R. & Gordon, R. Effect of surface roughness on self-assembled monolayer plasmonic ruler in nonlocal regime. Opt Express 22, 9604–9610 (2014).
25. Park, K., Biswas, S., Kanel, S., Nepal, D. & Vaia, R. A. Engineering the optical properties of gold nanorods: independent tuning of surface plasmon energy, extinction coefficient, and scattering cross section. J. Phys. Chem. C 118, 5918–5926 (2014).
26. Choi, B. C. & Freeman, M. R. Ultrathin Magnetic Structures III. New York (Springer, 2005).
27. Noguez, C. Surface plasmons on metal nanoparticles: the influence of shape and physical environment. J. Phys. Chem. C 111, 3806–3819 (2007).
28. Mock, J. J. et al. Distance-dependent plasmon resonant coupling between a gold nanoparticle and gold film. Nano Lett. 8, 2245–2252 (2008).
29. Johnson, P. B. & Christy, R. W. Optical constants of the noble metals. Phys. Rev. B 6, 4370–4379 (1972).
30. Palik, E. D. Handbook of optical constants. Proc. Natl. Acad. Sci. USA 2, 1096 (1991).
31. Atxitia, U. & Chubykalo-Fesenko, O. Ultrafast magnetization dynamics rates within the Landau-Lifshitz-Bloch model. Phys. Rev. B - Condens. Matter Mater. Phys. 84, 144414 (2011).
32. Jiaqi, Z., Boyin, C., Zongzhi, Z., Bin, M. & Qingyuan, J. Laser-induced magnetization dynamics for L10-FePt thin films with perpendicular anisotropy. J. Korean Phys. Soc. 56, 1269–1273 (2010).
33. Roth, T. et al. Temperature dependence of laser-induced demagnetization in Ni: A key for identifying the underlying mechanism. Phys. Rev. X 2, 021006 (2012).
34. Vodungbo, B. et al. Laser-induced ultrafast

demagnetization in the presence of a nanoscale magnetic domain network. Nat. Commun. 3, 999 (2012).

35. Moisan, N. et al. Investigating the role of superdiffusive currents in laser induced demagnetization of ferromagnets with nanoscale magnetic domains. Sci. Rep. 4, 4658 (2014).

36. Atxitia, U. et al. Micromagnetic modeling of laser-induced magnetization dynamics using the Landau-Lifshitz-Bloch equation. Appl. Phys. Lett. 91, 232507 (2007).

37. Fangohr, H., Chernyshenko, D. S., Franchin, M., Fischbacher, T. &Meier, G. Joule heating in nanowires. Phys. Rev. B - Condens. Matter Mater. Phys. 84, 054437 (2011).

38. Baffou, G., Quidant, R. & Girard, C. Heat generation in plasmonic nanostructures: Influence of morphology. Appl. Phys. Lett. 94, 153109 (2009).

39. Abate, Y., Schwartzberg, A., Strasser, D. & Leone, S. R.Nanometer-scale size dependent imaging of cetyl trimethyl ammonium bromide (CTAB) capped and uncapped gold nanoparticles by apertureless near-field optical microscopy.Chem. Phys. Lett. 474, 146–152 (2009).

40. Baffou, G. & Rigneault, H. Femtosecond-pulsed optical heating of gold nanoparticles. Phys. Rev. B - Condens. Matter Mater. Phys. 84, 035415 (2011).

41. Yu, P. et al. Curie temperatures of fcc and bcc nickel and permalloy: Supercell and Green's function methods. Phys. Rev. B - Condens. Matter Mater. Phys. 77, 054431 (2008).

42. Stanley, H. E. Introduction to phase transitions and critical phenomena. Am. J. Phys. 40, 927 (1972).

43. Sharma, V., Park, K. & Srinivasarao, M. Colloidal dispersion of gold nanorods: historical background, optical properties, seed-mediated synthesis, shape separation and self-assembly. Mater. Sci. Eng. R Reports 65, 1–38 (2009).

44. Ekici, O. et al. Thermal analysis of gold nanorods heated with femtosecond laser pulses. J. Phys. D. Appl. Phys. 41, 185501 (2008).

45. Kasiraj, P., Shelby, R. M., Best, J. S. & Horne, D. E. Magnetic domain imaging with a scanning kerr effect microscope. IEEE Trans. Magn. 22, 837–842 (1986).

CHAPTER 3 MAGNETIC NANOPARTICLES

Xijun Wang

INTRODUCTION

A magnetic fluid is a colloidal suspension of the paramagnetic nanoparticles in a liquid carrier which can be easily actuated by applying a magnetic field [1]. Therefore the application areas have been widened into the nanotechnologies such as nanogenerators, nanopumps, and nanomotors, and many other nanodevices with a variety of magnetic fluids have been coming of age [2–4].

The dynamic property of magnetic fluids in combination with high absorption coefficient to visible light has been found useful in optical device applications such as information displays and light modulators [5–8].

Because the performance of magnetic-fluid-based optical devices will depend on the movement of magnetic fluid responding to the applied magnetic field, it is significantly crucial to characterize the behavior of the magnetic particles in a magnetic field. However, the dynamics of magnetic particles in magnetic fluid, unlike the nanoparticles fixed in a solid bulk material, is difficult to characterize by adopting the conventional methods such as AFM and SEM. Instead, it is possible to use noncontact and nondestructive method based on the laser speckle technique [9–12].

Although the light scattering method has discovered applications in the characterizations of the dynamic laser speckle from a magnet fluibuiltd [13, 14], the contrast of the speckle or the scattering intensity may open only a general observation to the bulk magnet solution. The magnetic nanoparticles were also covered in the fluid that could not be observed. The average signal of the PMT (photomultiplier tube) or photodiode is too deficient to meet the requirement for the nanoparticle movement in the solution when the exerted magnetic field drives the magnetic particles rotating, aligning, or piling up. So the speckle images captured by CCD image sensor will record an array of optoelectronic data of the speckle in each equal interval.

How to coordinate the speckle movement from a variation of the surface profile resulted in the solution of the moving magnetic particles which could be driven by an electromagnetic tesla coil array. The optical interferometry that the crucial component works in an AFM is the first choice to build up the coordinates in the speckle images. In order to make stationary interferometry strips, the two questions should be answered that the higher-order interferometer speckle decreases the coherence of the scattering light so significantly that the strips could not be found; while the transverse scanning find a uniform view field, the nonlinear additional phase should be compensated by the phase delay to make the strip "still."

While a dual scanning laser speckle interferometry to observe the dynamic behavior of the magnetic fluid actuated by a magnetic field was built up, the experimental results were discussed to characterize the movement of the granules in the magnetic fluid.

EXPERIMENTAL SETUP

Magnetic Fluid and the Device

In the experiment, magnetite Fe_3O_4 nanoparticles were from the F^{2+} and F^{3+} precipitation with the F^{2+}/F^{3+} molar rate 1 : 1.8, acting temperature 80°C [15], and keeping pH 9~10 by gradually dropping the concentrated ammonia into the agitation solution. The magnetite Fe_3O_4 particles were separated by a magnet directly. The average particle size was around 40 nm by AFM and the saturation magnetization was around 70 emu/g. The magnetite particles were coated with sodium oleate keeping pH 5~6 and dispersed in n-nonane (C_9H_{20}), which was one clear ingredient of commercial kerosene.

The operation of a magnetic fluid display device relied on the modulation of the thickness of a magnetic fluid layer which was placed on the substrate, and the characterization was focused on a magnetic fluid actuator to form a pixel of a display device. The actuator consisted of two units each with an

Citation: Xijun Wang, "Magnetic Nanoparticle, vol. 2014, Article ID 287813, 7 pages, 2014. doi:10.1155/2014/287813.

area of $1.5 \times 0.5\,mm^2$ formed by a thin layer of ferromagnetic thin film enclosing a spiral conductor electrode which could be manipulated easily by controlling the current going through the spiral coils in Figure 1(a) [8]. The channel stored the magnet fluid and formed the effective path to the mesa. The length of the channel was 1.5 mm, and its width was the same as the gap in the mesa, so the width was 15 micron in the design.

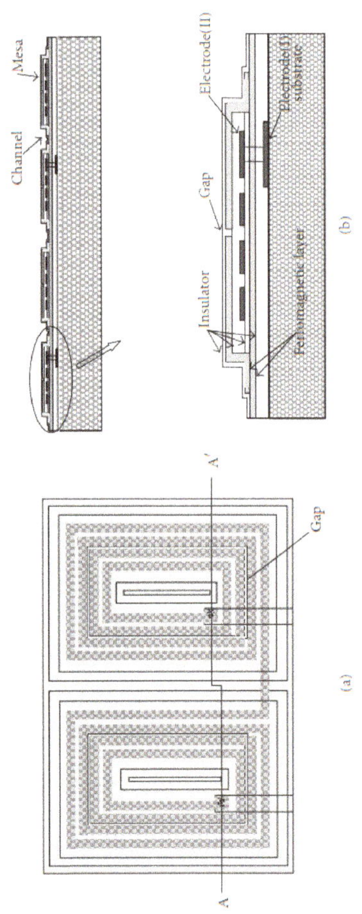

Figure 1: (a) Top view of the device with the spiral electrode coils and the rectangular magnetic gaps. (b) Cross-sectional view of the device with a magnified view to the vertical layer structure.

Most of the magnetic fluid would initially stay in the channel around the mesa due to the capillarity when no electric field was applied, and only a minute thin solution layer of the liquid would reside on top of the mesa area in Figure 1(b).

The thickness of the magnetic fluid was modulated by the step magnetic field which was generated by the electric current pulse through the circuit underneath the actuator mesa. The magnetic force exerted by the electric field drove the Ferro fluids into the region of strong magnetic intensity and made the display pixel darker. When the current was off, on the other hand, the surface tension of the fluid restored the force of the liquid back to the original state and the pixel became brighter, in which the light and dark states of the pixel could be thereby produced correspondingly.

Experimental Design

The scanning interferometer used a He-Ne laser with the beam collimating, expanding, splitting, and combining to generate the reference and object light to create interferometer patterns practicably. Two nearly parallel light beams, one as the object beam was incident to the magnetic unit containing magnetic fluid and scattered by the magnetic particles to form a speckle field and the other as the reference beam was incident to the speckle field, which generated interferometer fringes in the speckle field.

The transverse scanning device that was a spinning optical parallel plate keeping the incident angle invariable aligned the object light to the pixel mesa by transverse scanning precisely with an additional phase. The phase delay was used to compensate the variation of the additional phase caused by the view field requirement of the object light. The experimental setup was shown in Figure 2.

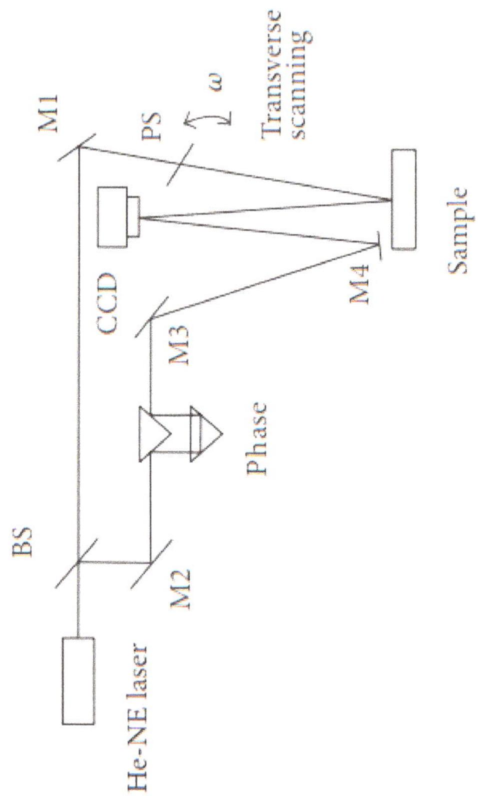

Figure 2: Schematic of a dual-scan laser speckle interferometer experiment.

Adjusting the angle of the two light beams could change the interferometer fringe spacing and the fringe contrast in the laser speckle field. Assuming that the object light and reference light were in the same incident angle i and symmetrically incident to CCD, the spacing of the interferometer fringe was given by

$$\Delta x = \frac{\lambda}{2 \sin i}. \tag{1}$$

Phase Compensation

An additional phase shift was into the object light path while using a transverse scanning plate as shown in Figure 3, where α was the incidence angle of the plate and β was the refraction angle in Snell's law: $n \cdot \sin \beta = \sin \alpha$, where n was the refractive index of the plate and the air refractive index was 1.0.

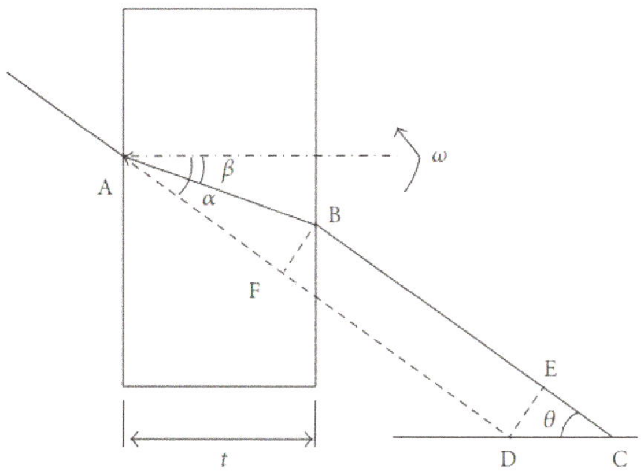

Figure 3: Transverse scanning plate to subjoin phase delay.

The optical path length L from the incidence point A that was the spinning center on the transverse scanning plate to the point C on the sample surface was given by

$$\begin{aligned} L &= n\,\mathrm{AB} + \mathrm{BC} = n\,\mathrm{AB} + (\mathrm{AD} - \mathrm{AF} + \mathrm{CE}), \\ \mathrm{AB} &= \frac{t}{\cos\beta}, \\ \mathrm{AF} &= \mathrm{AB}\cdot\cos(\alpha-\beta), \\ \mathrm{CE} &= \mathrm{AB}\cdot\sin(\alpha-\beta)\cot\theta, \end{aligned} \tag{2}$$

where θ was the incident light angle of the Ferro fluids sample surface. Thus, the optical path length could be written as

$$L = \mathrm{AD} + \{n - \cos(\alpha-\beta) + \sin(\alpha-\beta)\cot\theta\}\cdot\frac{t}{\cos\beta}, \tag{3}$$

where AD and θ were invariants in the experimental setup. The derivative of L versus α which gave the changing rate of the optical path length varied with the spinning angle of the transverse scanning plate. Consider

$$\begin{aligned} \frac{dL}{d\alpha} &= \frac{t\sin\beta\cos\alpha}{n\cos^3\beta}\left[n - \cos(\alpha-\beta) + \sin(\alpha-\beta)\cot\theta\right] \\ &\quad + \frac{t(n\cos\beta - \cos\alpha)}{n\cos^2\beta}\left[\sin(\alpha-\beta) + \cos(\alpha-\beta)\cot\theta\right]. \end{aligned} \tag{4}$$

The rotation of the transverse scanning plate was driven by a stepper motor with a miniature step angle 0.009°. OPD (optical path difference) from transverse scanning could be obtained by formulas (3) and (4). The OPD curve was showed in Figures 4(a) and 4(b), while the thickness of the parallel plate is 0.5 mm.

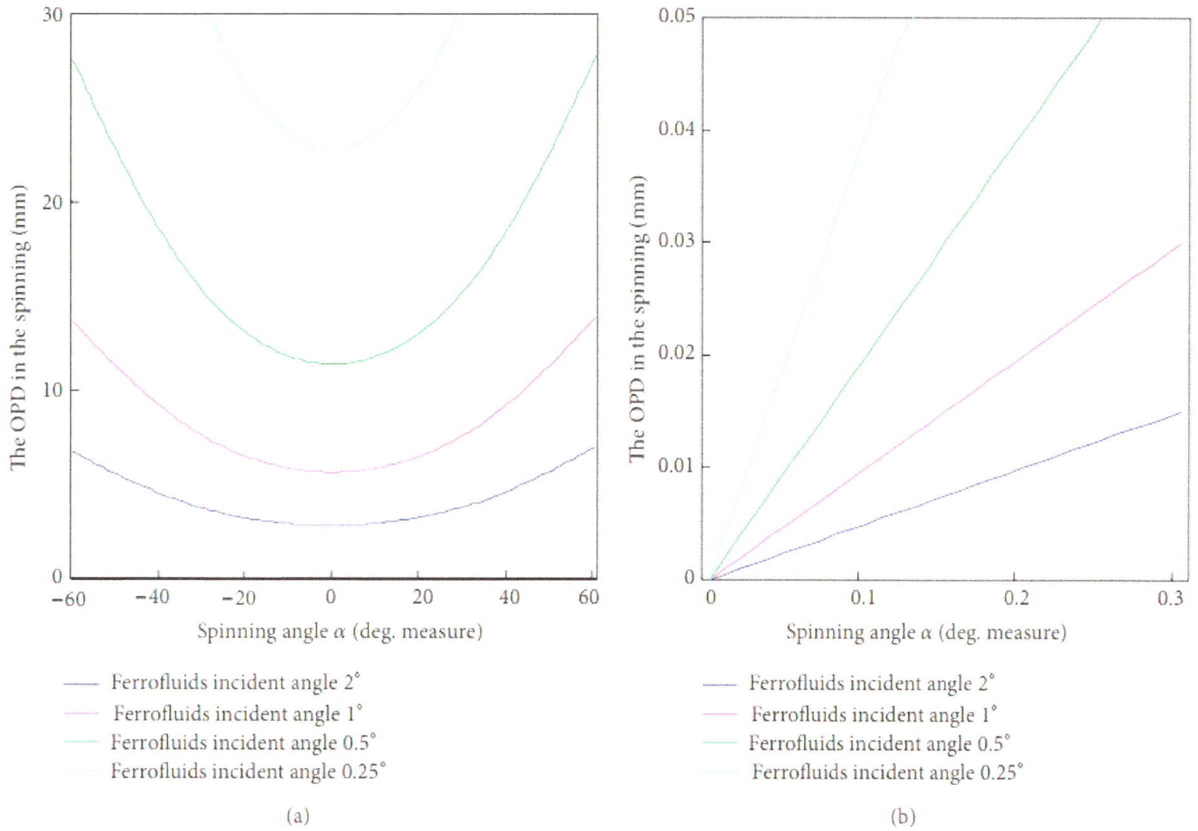

Figure 4: The OPD induced from the transverse scanning in all angle range; (b) was the OPD linear in small angle.

The transverse scanning could be controlled in half wavelength in the setup and Figure 4(a) showed that the tendency of the OPD varied with the spinning angle depending on formula (4) which is symmetrical with the spinning angles and nonlinear in the angle range. The ferrofluids incident angle θ became smaller and the OPD varied faster. So the incident angle was around 2° in the experiment.

Nearby the spinning angle zero location, the curve became linear. But some light might have reflected back the laser source and made some noise in the zero angle, so in the experiment, the transverse scanning would deviate a little more to get the higher resolution in tens nanometers.

The optical path of the reference light was adjusted using a 180° prism phase compensation device driven by another stepper motor in a straight line reciprocating motion. The OPD created by the phase delay was twice the movement distance.

For a given incidence angle α of the transverse scanning plate, a certain amount of phase compensation was to be determined so as to determine the rotation angle of the phase delay step motor and also to synchronize the phase delay with the transverse scanning by creating the driving parameters for both of the step motors controlled by microcomputer units.

EXPERIMENTAL RESULTS

He-Ne laser with an output power of 3 mW at 632.8 nm was used as the light source, and a magnetic fluid film with a thickness of about 20 μm was used as the device for test. The magnetic fluid display cell was driven by step rectangular pulses with a duty ratio of 1 : 2 and a pulse width of 100 ms. The response time of the device was measured to be around 400 ms with a driving current of 70 mA. The device was placed horizontally in the optical path.

To get an unlimited localization of the speckle image, the unmounted lens CCD and the free space propagation model were adopted. The CCD was placed at 30 cm vertically above the sample surface.

The energy ratio between the object and the reference light was adjusted to balance each other using polarizers in order to produce visible interferometer fringes.

The speckle patterns were recorded continuously by the CCD camera with a frame rate of 25 fps and an image size of 640 × 480 pixels, and the CCD operation was synchronized with the electric current pulse driving the magnetic fluid. While the incident angle in which the light beams shined upon the CCD was one degree, the fringe spacing was estimated to be 19 μm by formula (1). The incident angle increased twice; on the contrary, the fringe space decreased twice nearly.

Because the speckle was the highest order coherent of the laser beam, the interferometry fringe would not be easy to capture even if the laser had an excellent coherence. Making the two beams equal was the first step to utilize the polarization method.

After the speckle had appeared in the object, a beam shined on the magnet fluid device and it was necessary to find a zero order fringe by the detailed adjusting phase delay with miniature step driving. In spite of the probably appearing fringes, interferometer strip is difficult to observe in naked sight because of the lesser strip contrast within background noises.

Partly, the unstable fringes were captured with the CCD, until the transverse scanning both stabilized with the magnet particle movement and synchronized with the phase delay. A set of laser speckle interferometer fringes in the same intervals recorded by the CCD was showed from Figures 4(a) to 4(b). The thin and crowed strips with an arc were the interferometry fringes reflected by the surfaces of the multiple lenses and became main noises. The intensity relation correlation coefficient in neighbor named "neighbor coefficient" that means the coefficient calculation as the pairwise correlation. From Figures 5(a) to 5(f), the neighbor coefficient was, respectively, 0.5197, 0.4329, 0.5340, 0.8125, and 0.2175, in Figures 5(a)~5(f) sequence.

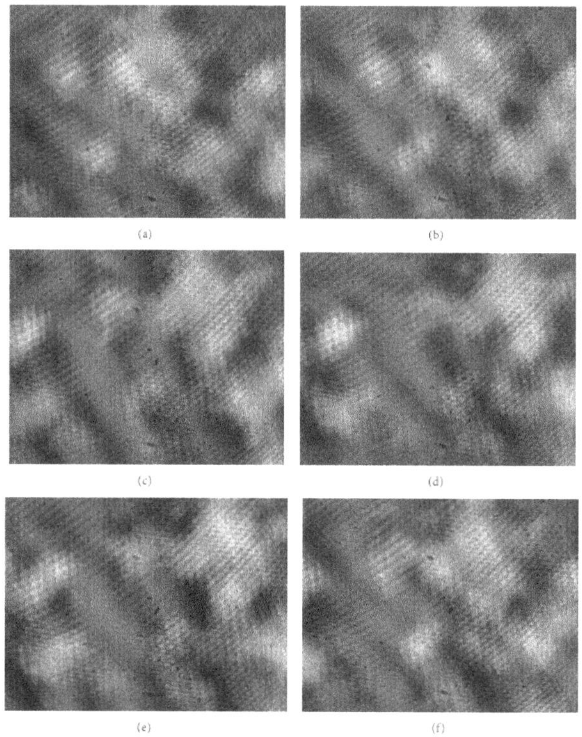

Figure 5: The sequent patterns of the dual-scan laser speckle interferometry.

The coefficient showed that Figure 5(b) compared with Figure 5(a) began to change as the coefficient decreased; Figure 5(c) compared with Figure 5(b) changed intensively as the coefficient became less; Figure 5(d) compared with Figure 5(c) began to change slowly as the coefficient recovered; Figure 5(e) compared with Figure 5(d)tended to stop the movement as the coefficient became more; Figure 5(f) compared with Figure 5(e) stopped the movement as the coefficient jumped when the driving current of the coil was off.

To study the speckle profile movement, the centerlines and laser speckle interferometer bright fringes were extracted simply by the wavelet denoising, the contrast enhancement, and the binarization algorithm in a MATLAB software. The calculation from Figure 5 is converted to Figure 6.

With the center line of the interferometer, bright strips were as a calibration, and the lower left corner of the center line from article 1 interferometer fringe calibration to 1, to the upper right corner to the center line of the interferometer fringes in turn calibration for .2,3,...

A speckle pattern was identified and circled as a reference for tracking as shown in Figures 6(a)~6(f), where the number next to the circle represents the relative position of the speckle with respect to an interferometer fringe. The black mark in the middle of the reference speckle was the centroid of the speckle. The origin of the graph corresponds to Figure 6(a). The marked number 3.9 in Figure 6(a) was shown as the speckle and strip tracking coordinates.

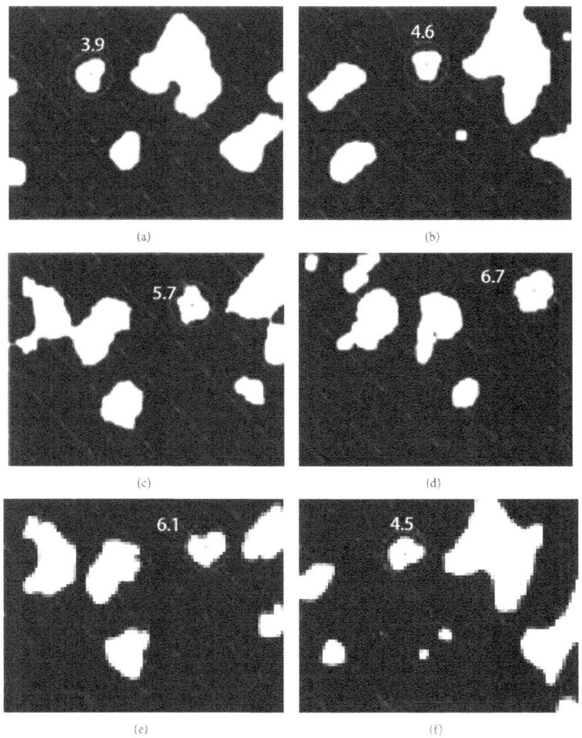

Figure 6: Image binarization and tracking of the speckle pattern movement.

To calculate the speckle movement from Figures 6(a) to 6(b), the speckle movement was 0.7 strip, and the others were 1.1, 1.0, −0.6, and −1.6 strips. The speckle moved forward or backward in the interferometry strips because the nanoparticles were driven by the coil current in, on, or off state.

The speckle pattern moving amount in the fringes showed that the nanoparticles moved quicker in the piling up than in the quitting in the same time intervals. Therefore, the electromagnetic force was less than the gravity and capillarity to keep laminar flow of the nanoparticle fluid but not the turbulence flow of the solution.

Every strip reflected light expressed twice OPD of one wavelength 632.8 nm, so one strip is related to $632.8/2 = 316.4$ nm variation of the magnetic nanoparticles in the interval 1/25 second. Furthermore, the nanoparticle piling rate is around 7.6 micron per second. The mesa area was $1.5 \times 0.5\ mm^2$ and the effective "channel" area was $1.5 \times 0.015\ mm^2$, so the relative speed between nanoparticles in the channel was 0.3 mm/s. In other words, one step pulse made the nanoparticle move ~0.3/25 mm (12 micron), in which the distance was far less than the path length of the solution. Furthermore, the nanoparticle's relative speed variation also proved that the nanoparticle's movement in the experiment was a laminar flow.

CONCLUSIONS

Although the laser speckle was a higher order coherent phenomenon and the speckle led to the distinguished annihilation of the light coherence, the speckle of the magnetic nanoparticle fluid movement would be recorded by CCD within the interferometry strips. The development of the magnetic nanoparticle fluid characterization focused on a dynamic interaction of the magnet nanoparticles. It was obvious that the scanning interferometry speckle could capture the nanoparticle and nanostructure's movements, which was different from the light scatting studies which focused on measuring the scattering intensity with a statistical averaged calculation in contrast with the patterns, or with a phase microscope method in a static micron dimension of the speckle patterns [13, 14].

The experiment showed that the nanoparticles fluid dynamic properties could synergistically appear in the fringe speckles. The analysis of the nanoparticle relative speed and the speckle pattern moving amount in the fringes proved the nanoparticles' movement in a laminar flow in the experiment.

CONFLICT OF INTERESTS

The author declares that there is no conflict of interests regarding the publication of this paper.

ACKNOWLEDGMENTS

The author is fully grateful for the support of the study by the National Science Fund of China (NSFC) (no. 60748026) and Korea Science and Engineering Foundation (KOSEF) (Grant no. F01-2005-000-10169-0).

REFERENCES

1. B. M. Berkovsky, V. F. Medvedev, and M. S. Krakov, Magnetic Fluids Engineering and Applications, Oxford University Press, Oxford, UK, 1993.
2. M. Zahn, "Magnetic fluid and nanoparticle applications to nanotechnology," Journal of Nanoparticle Research, vol. 3, no. 1, pp. 73–78, 2001.
3. K. Nakatsuka, "Trends of magnetic fluid applications in Japan," Journal of Magnetism and Magnetic Materials, vol. 122, no. 1–3, pp. 387–394, 1993.
4. K. Raj, B. Moskowitz, and R. Casciari, "Advances in ferrofluid technology," Journal of Magnetism and Magnetic Materials, vol. 149, no. 1-2, pp. 174–180, 1995.
5. J.-W. Seo and S. J. Park, "An experimental study of light modulator using magnetic fluid for display applications," Journal of Magnetism and Magnetic Materials, vol. 192, no. 3, pp. 499–504, 1999.
6. J.-W. Seo, S.-M. Jeon, S. J. Park, and H.-S. Lee, "An experimental and numerical investigation of flat panel display cell using magnetic fluid," Journal of Magnetism and Magnetic Materials, vol. 252, pp. 353–355, 2002.
7. J.-W. Seo, H. Kim, and S. Sung, "Design and fabrication of a magnetic microfluidic light modulator using magnetic fluid," Journal of Magnetism and Magnetic Materials, vol. 272–276, pp. E1787–E1789, 2004.
8. J.-W. Seo and X.-J. Wang, "Magnetic-fluid microelectrornechanical light modulator," Optics and Precision Engineering, vol. 13, no. 5, pp. 542–547, 2005.
9. C.A. Thompson, K. J. Webb, and A. M. Weiner, "Diffusive media characterization with laser speckle,"Applied Optics, vol. 36, no. 16, pp. 3726–3734, 1997.
10. R.-S. Lu, G.-Y. Tian, D. Gledhill, and S. Ward, "Grinding surface roughness measurement based on the co-occurrence matrix of speckle pattern texture," Applied Optics, vol. 45, no. 35, pp. 8839–8847, 2006.

11. W.-W. Feng, M.-J. Liu, X.-Q. Wang, F.-R. Shi, J. Zhang, and R.-X. Jiang, "Feature extraction and recognition of laser speckle for special material surface," Infrared and Laser Engineering, vol. 36, no. 2, pp. 186–188, 2007 (Chinese).

12. X.-F. Li, J. Xu, J.-J. Luo, L. Cao, and S. Zhang, "Noise analyzing and denoising of intensity image for laser active imaging system," Infrared and Laser Engineering, vol. 40, no. 2, pp. 332–337, 2011 (Chinese).

13. J. Philip and J. M. Laskar, "Optical properties and applications of ferrofluids—a review," Journal of Nanofluids, vol. 1, no. 1, pp. 3–20, 2012.

14. S. Brojabasi and J. Philip, "Magnetic field dependant backscattering of light in water based ferrofluid containing polymer covered Fe_3O_4 nanoparticles," Journal of Applied Physics, vol. 113, Article ID 054902, 2013.

15. S. Wu, A.-Z. Sun, F.-Q. Zhai et al., "Fe_3O_4 magnetic nanoparticles synthesis from tailings by ultrasonic chemical co-precipitation," Materials Letters, vol. 65, no. 12, pp. 1882–1884, 2011.

CHAPTER 4 FERROMAGNETIC NANO-DOMAINS AT THE SURFACE OF A NON-MAGNETIC OXIDE

T. Taniuchi, Y. Motoyui, K. Morozumi, T.C. Ro¨del, F. Fortuna, A.F. Santander-Syro & S. Shin

INTRODUCTION

Thanks to their wide range of physical properties in the bulk, and their epitaxial compatibility, transition-metal oxide perovskites provide an ideal arena to explore the competition, interaction and creation of novel ground states at their interfaces[1,2,3,4,5,6,7,8,9,10,11,12]. The two-dimensional electron gas (2DEG) at the interface between $SrTiO_3$ (STO) and $LaAlO_3$ (LAO), two wide-gap insulating perovskites in the bulk, is a paradigmatic example of such emerging behaviour. The ground state at the LAO/STO interface can be tuned by external bias gating to show metal-to-insulator transitions[2], superconductivity below 300 mK (refs3, 4) or magnetoresistance[5]. Moreover, it was recently demonstrated that the LAO/STO heterostructures also show sub-micrometre patches of ferromagnetism[6,7,8], although neither LAO nor STO are magnetic in the bulk. However, it is difficult to verify if such intrinsic magnetism exists only at the interface. Most of the previous studies have used macroscopic probes, such as magnetoresistance, net magnetization or torque magnetometry[5,6,7,8]. In the case of LAO/STO heterostructures, for instance, scanning superconducting quantum interference device microscopy experiments found micrometre-sized ferromagnetic patches in the plane parallel to the interface[7,13,14], but the specific depth where the moments reside cannot be determined from this probe. Thus, it is essential to determine the microscopic nature of the observed ferromagnetism: is it intrinsic to the LAO/STO interface? Can it be also found (and under what conditions) at the bare STO surface? Or is it an effect extended over the bulk of STO? More generally, the observation of low-dimensional ferromagnetism in non-magnetic Ti-based oxides would be extremely important both from a fundamental perspective and for its potential in applications in oxide-based electronics[10,11,12,15,16.] Besides the case of the $SrTiO_3$-based heterostructures, it was also found that the bare $SrTiO_3$ surface sustains a 2DEG arising from oxygen vacancies[10,11,15], whereas recent experiments indicated the possible existence of magnetism at the surface of $SrTiO_3$ (ref. 12). In principle, magnetism in $SrTiO_3$ could arise from anion defects in the bulk[17,18,19,20], or could be specific to the surface. Such a surface ferromagnetism would be a leading example of truly emergent phenomena at transition-metal oxide surfaces. In the present work, we investigated the magnetism at the oxygen-deficient surface of $SrTiO_3$ using laser-based ultrahigh resolution photoemission electron microscopy (laser-PEEM)[21], a recently developed technique allowing a direct magnetic imaging of a sample surface, even in bulk insulators, with a high spatial resolution of 2.6 nm. Our data show a high concentration of ferromagnetic nano-domains developing below ~900 K, of sizes around 40 nm at room temperature, at the oxygen-deficient surface of $SrTiO_3$. The existence of such nano-domain magnetism up to temperatures well above room temperature is appealing for applications using oxide-based electronics. More generally, given the variety and rich phase diagrams of complex oxides, our results highlight the use of oxide surface states as model systems towards new physics and functional devices.

RESULTS

Laser-PEEM magnetic imaging

Our magnetic microscopy studies were carried out using magnetic circular dichroism (MCD) imaging of PEEM. Although MCD conventionally requires synchrotron X-rays with circular polarizations[22], and has been used recently to study the origin of ferromagnetism is LAO/STO heterostructures[23,24], we applied instead the threshold photoemission technique using ultraviolet laser sources[21]. In fact, threshold photoemission laser-PEEM is a powerful tool for observing magnetic domain structures, particularly for metallic magnetic layers on top of bulk insulators. The photoelectron yield in the case of threshold photoemission depends on the work function of the material. For metallic states at the surface of $SrTiO_3$, the photoemission threshold for the insulating bulk is higher than that for the

Citation: T. Taniuchi, Y. Motoyui, K. Morozumi, T. C. Rödel, F. Fortuna, A. F. Santander-Syro & S. Shin, Ferromagnetic nano-domains at the surface of a non-magnetic oxide, doi:10.1038/ncomms11781.

conductive surface. In our case, the photon energy of our laser source (4.66 eV) is not enough to make photoelectrons from the insulating bulk transcend the vacuum level at the surface, which is of at least 7.4 eV (4.2 eV of work function +3.2 eV of band gap). As a result, these photons excite photoelectrons coming only from metallic surface layers (Fig. 1a). Thus, threshold photoemission is selectively sensitive to the metallic states at the surface. For magnetic imaging, we use right- and left-handed circular polarizations, exploiting the MCD near the Fermi level[25,26]. Therefore, PEEM imaging with the threshold photoemission MCD technique is considered to be suitable to investigate ferromagnetism on a thin metallic layer with a small magnetization. Using this technique, we can also determine the direction of magnetization. For instance, by illuminating samples with perpendicular incidence, as indicated in Fig. 1b, the MCD signal exhibits a clear asymmetry, directly related to the magnetization components perpendicular to the surface[21]. The MCD asymmetry is defined here by $A=(I^{left}-I^{right})/(I^{left}+I^{right})$, where I^{left} andI^{right} are the photoelectron intensities at a lateral position in the PEEM images taken with left- and right-circularly polarized lights, respectively.

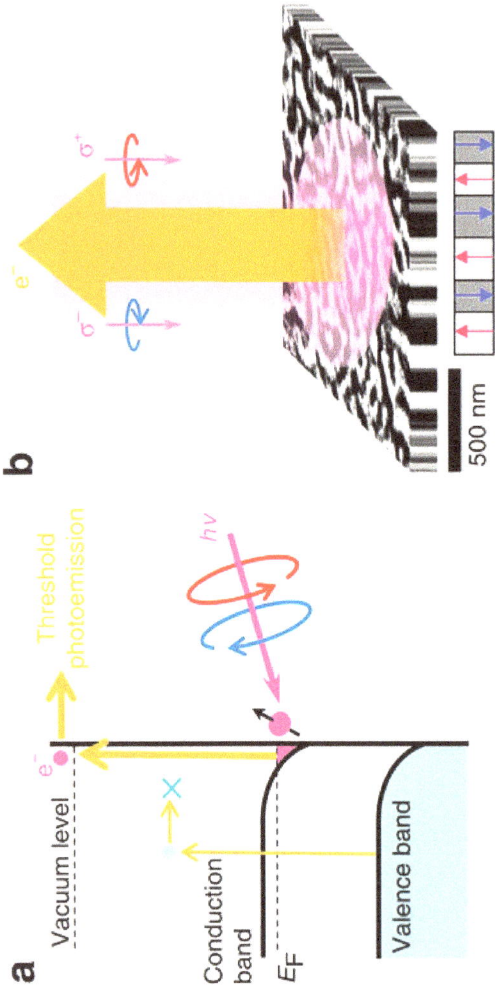

(**a**) Excitation process of threshold photoemission at the surface $SrTiO_3$. The photon energy of our laser is 4.66 eV. The band gap of bulk $SrTiO_3$ is 3.2 eV. The laser selectively excites photoelectrons derived from the metallic surface. Circularly polarized photons are irradiated to the sample surface, and the spin polarization of the ejected photoelectrons is detected. (**b**) Magnetic imaging technique using ultrahigh resolution laser-PEEM. The magnetic image in the illustration shows a magnetic domain structure in a FePt epitaxial thin film with perpendicular magnetic anisotropy as a typical example. The field of view is 2 μm. The difference between two images taken with left-handed and right-handed circular polarizations (σ^+ and σ^-, respectively) shows magnetic contrast. The magnetic contrast originates from magnetization components along the incident direction of photons. The best lateral resolution of our microscope is ~2.6 nm (ref. 15).

Figure 1: Schematic illustration and demonstration of magnetic imaging.

Samples

For our experiments, we used non-doped TiO_2-terminated $SrTiO_3$ (001) single crystals. The crystals were then flash-annealed at 1,000 K for 30–60 s in ultrahigh vacuum (better than 1×10^{-6} Pa) to produce oxygen vacancies at the surface[15,27]. After annealing, the bulk crystals were still transparent and showed an atomically flat surface with step-and-terrace structure (Fig. 2a). The sample-preparation procedure and characterization are further detailed in the Methods section.

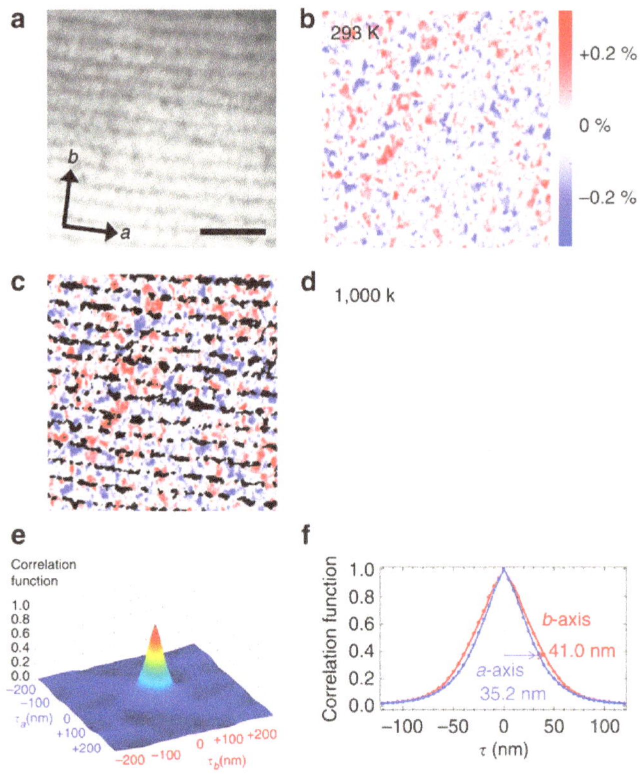

(**a**) Raw laser-PEEM image after annealing at 1,000 K. The black lines in the image correspond to atomic steps. The scale bar at the bottom right of the image corresponds to 500 nm. We confirmed that the terrace width did not change during the annealing by comparison with atomic force microscopy and laser-PEEM images before the annealing, which means that step bunching did not occur. After annealing at 1,000 K, the sample with metallic states at the surface is still transparent after the experiments. (**b**) Magnetic image at room temperature at the same position as the raw image. Red and blue colours indicate positive and negative MCD signals, respectively, which correspond to perpendicular magnetization with spin-up and spin-down. The MCD asymmetry is defined here as $A=(I^{left}-I^{right})/(I^{left}+I^{right})$, where I^{left} and I^{right} are the photoelectron intensities at a lateral position in the PEEM images taken with left- and right-circularly polarized photons, respectively.

In these measurements, the light incidence is perpendicular to the sample's surface. With this geometry, the MCD measures the perpendicular components of the magnetization vector. (**c**) Magnetic image overlaid on the step structure from the previous two images. The magnetic domains are partially defined at terraces. (**d**) Magnetic image at 1,000 K. The colour scale of magnetization is the same as in **b**. The magnetic domains have disappeared at 1,000 K. (**e**) Auto-correlation function estimated from the magnetic image of **b**. $_a$ and $_b$are the displacement lengths along the *a*- and *b*-axes, as shown in the inset of **a**. (**f**) Auto-correlation function parallel (blue curve) and perpendicular (red curve) to the step edges. Their respective correlation lengths are 35 and 41 nm. The correlation length is estimated as the distance between the origin to the point at 1/*e* height of the correlation function.

Figure 2: Direct evidence for ferromagnetic nano-domains at the $SrTiO_3$ surface.

Laser-PEEM imaging of ferromagnetic nano-domains at the $SrTiO_3$surface

The MCD imaging was performed in the geometry with perpendicular incidence, which detects magnetization components perpendicular to the surface as described above. Figure 2b shows the MCD laser-PEEM image observed at the same position as Fig. 2a. The measurement was performed at room temperature. The image presents clear magnetic-domain structures of opposite MCD contrast, shown by the red and blue coloured areas, indicating perpendicular components of the magnetization with opposite orientations. In fact, the contrast in MCD signal in Fig. 2b should come from the surface, as essentially all photoelectrons come from the metallic surface layer as mentioned above. Moreover, we observe that the magnetic domains are homogeneously distributed over a whole field of view, unlike the previous studies in the $LaAlO_3/SrTiO_3$ heterostructures using scanning superconducting quantum interference device microscopy7, which showed only very localized magnetic signals separated by several micrometres or tens of micrometres. This can also be seen from the magnetic image overlaid on the step pattern extracted from the direct surface image, as shown in Fig. 2c. Note that the magnetic domains are not specifically localized around surface atomic steps, although, as we will see later, the size of the domains is slightly different along and perpendicular to the step edges. All these observations directly demonstrate the existence of intrinsic room-temperature ferromagnetism at the oxygen-deficient surface of $SrTiO_3$. As shown inFig. 2d, the contrast in MCD signal vanishes when measured at higher temperature (~1,000 K), indicating that it arises from ferromagnetic ordering occurring below 1,000 K. These results were reproduced in samples with different annealing temperatures above 900 K. For samples annealed at lower temperatures, no ferromagnetism was found (see Supplementary Note 1 with Supplementary Fig. 1). One possible reason is that a critical concentration of oxygen vacancies at the surface, obtained only after annealing above 900 K, is necessary to create a magnetic order.

The size of the magnetic domains was estimated using an autocorrelation technique. The autocorrelation function of the magnetic image along a given direction *x*, noted ϕ(*x*), was obtained from the cross section of the MCD asymmetry along *x*, *A*(*x*), through the autocorrelation relation:

$$\varphi(x) = \int_{min}^{max} A(x) \cdot A(x - \tau)d\tau \qquad (1)$$

Here is a displacement length of the magnetic image along *x*, and the integration limits are given by the size of the magnetic image. Figure 2eshows the two-dimensional autocorrelation function calculated from the magnetic image shown in Fig. 2b. $_a$ and $_b$ are displacement lengths along the *a*- and *b*-axes defined in Fig. 2a: the directions parallel and perpendicular to the step edges, respectively. Thus, the average domain size is defined as the value of the autocorrelation length at 1/*e* height of the maximum correlation function, and is estimated from Fig. 2e to be 38 nm. This value is much smaller than the terrace widths of the sample surface (125 nm).

Using this approach, we found that the size of the magnetic domains depends on the atomic step direction at the sample surface. Figure 2fshows the cross-sections of the two-dimensional autocorrelation function along the *a*- and *b*-axes, blue and red lines, respectively. We see that the average correlation lengths are ~35 nm along the step edges, and 41 nm perpendicular to the step edges. This difference indicates that the observed magnetism originates from the sample surface, as it is correlated to the topography of the topmost atomic layer (see Supplementary Note 1 and Supplementary Fig. 2 for further discussions).

Temperature evolution of the ferromagnetic nano-domains

To determine the Curie temperature of this system, we performed the magnetic imaging at different temperatures. As mentioned previously, the samples were annealed at 1,000 K and subsequently measured at different temperatures with perpendicular light incidence. The results are shown in Fig. 3. The asymmetry in the perpendicular MCD signal, extracted from the MCD laser-PEEM images, is plotted as a function of temperature in Fig. 3a. Figures 3b–i are MCD laser-PEEM images for the indicated temperatures. The red and blue contrasts show the magnetization components perpendicular to the surface. Based on these results, the estimated Curie temperature for ferromagnetic ordering is around 900 K. These measurements were reproducible over different samples.

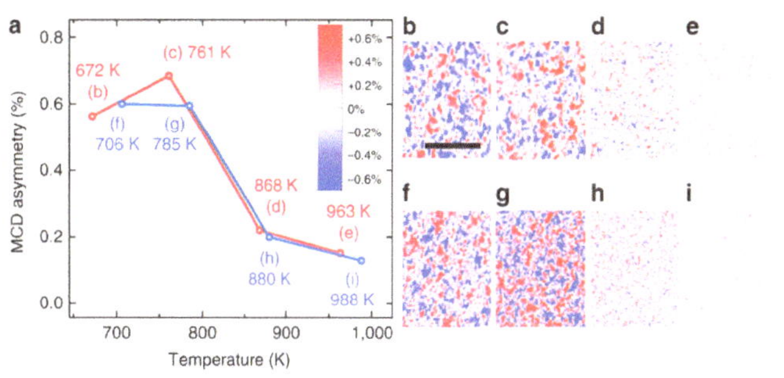

(**a**) The asymmetry in the out-of-plane MCD signal, extracted from the acquired MCD images, is plotted as a function of measurement temperature. The samples were first annealed at 1,000 K. (**b**–**e**) and (**f**–**i**) Perpendicular magnetic domain structure images for the indicated temperatures in **a**, which were measured by the Laser-PEEM with threshold photoemission MCD. The scale bars correspond to

1,000 nm. From the evolution of the MCD contrast, the Curie temperature is estimated to be around 900 K. The red and blue lines in panel **a**, and their associated MCD images in panels **b–e** and **f–i**, respectively, correspond to two independent sets of measurements.

Figure 3: Temperature dependence of the magnetic contrast.

Response to an external perpendicular magnetic field

To confirm the magnetic origin of the MCD domains, we measured the change of the MCD contrast after application of an external magnetic field. The experimental procedure is shown in Fig. 4a: we performed the MCD imaging in an as-annealed sample (Fig. 4b). Then the sample was heated up to 700 K in ultrahigh vacuum while applying a perpendicular magnetic field of ~1,000 Oe from a commercial samarium-cobalt magnet held in front of the sample. After cooling to room temperature in magnetic field, we carried out again an MCD imaging at room temperature and zero field, as shown in Fig. 4c. The increase of the MCD contrast after the magnetic field was applied is evident by comparing Fig. 4b,c. Such a response to the magnetic field clearly shows that the image contrast between different domains comes from their magnetism. More precisely, as illustrated in Fig. 4a, our results show that before applying the external magnetic field, the magnetization vectors are not saturated along the sample normal but tilted (see discussion in the Supplementary Note 2 and Supplementary Fig. 3). After applying the external field, the perpendicular component increases because of the large perpendicular magnetic anisotropy. InFig. 4c, we did not observe a preferential alignment of the magnetic moments (anti-)parallel to the field, as evidenced by the similar areas covered by the red and blue domains. Applying the field in the opposite direction (−1,000 Oe, not shown) yields identical results.

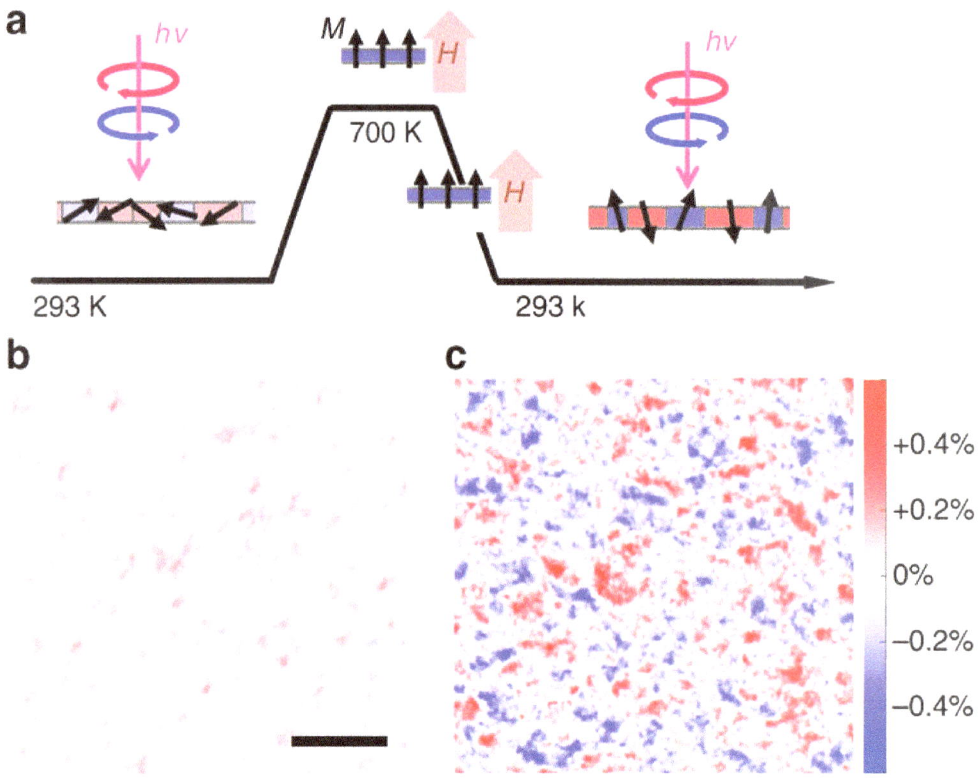

(**a**) Procedure of measurements. The geometry of the magnetic imaging is sensitive to the perpendicular component of the magnetization. The sample was first flash annealed at 1,000 K, and then measured at room temperature. Afterwards, the sample was heated up to 700 K to lower its coercivity. A magnetic field of about 1,000 Oe perpendicular to the sample surface was applied at 700 K, and maintained during cooling down to room temperature. The sample was finally measured in zero field at room temperature. (**b**) Magnetic image of an as-annealed sample before application of the external magnetic field. The scale bar corresponds to 1,000 nm. (**c**) Magnetic image of the same sample after the magnetic field was applied. The magnetic images are shown using the same colour scale. The magnetic contrast has increased by applying the magnetic field as the magnetization vectors of the different magnetic domains are tilted towards the sample normal.

Figure 4: Magnetic domain formation before and after application of an external magnetic field.

To explain these observations, we first discuss the initial state of our system, Fig. 4b. Owing to the competition between the perpendicular magnetic anisotropy and the in-plane shape anisotropy, which tends to reduce the demagnetization energy derived from the film-like shape of the magnetic layer, the initial state shows various magnetization distributions: some are in-plane, other out-of-plane or tilted. The application of a perpendicular magnetic field above the Curie temperature results in a single domain with magnetization perpendicular to the surface under the field. After switching off the field at room temperature, the magnetostatic energy of the total system is reduced by flipping some of the moments and thereby, breaks the single domain state. Such a behaviour often happens, especially in very thin films28.

DISCUSSION

Our results indicate that ferromagnetism in $SrTiO_3$ is closely related to oxygen vacancies. In this material, oxygen vacancy formation results in the creation of itinerant electrons, as well as localized electrons and magnetic moments[29,30]. At interfaces or surfaces, the interaction between itinerant and localized electrons determines the magnetic state (refs [31, 32, 33]). The interplay between these two types of electrons can lead to rich magnetic phase diagrams not only in $SrTiO_3$-based 2DEGs[34,35], but also in diluted magnetic semiconductors[36,37] Using the MCD laser-PEEM technique, neither the concentration of localized nor the one of itinerant electrons can be directly accessed. Hence, the microscopic mechanism for the creation of the magnetic moments cannot be deduced from our data. However, the measured high Curie temperature (T=900 K) can exclude RKKY interaction as the mechanism responsible for the onset of magnetism in the 2DEG at the surface of $SrTiO_3$, because the indirect exchange coupling between impurity moments in semiconductors via the RKKY interaction is very weak and forms ferromagnetic ordering at only very low temperatures (T<20 K)[37]

Our finding of ferromagnetic nano-domains at the surface of a non-magnetic oxide presents a playground for the engineering of innovative nano-devices with properties suitable for applications in oxide electronics and spintronics: high ferromagnetic transition temperature, conductive surface, highly insulating bulk matrix entailing optical transparency, negligible leakage currents, higher breakdown voltages, lower electronic noise and high-temperature and high-power operation.

METHODS

Sample preparation and characterization

We used non-doped $SrTiO_3$ (001) substrates that were TiO_2 terminated by a treatment in a buffered HF solution and subsequent annealing[15,16]. The purity of the crystals was 99.9866 %. The main components of impurities were Ba (129 p.p.m.) and Ca (5.4 p.p.m.). Components of transition metal elements were under detection limit: for example, Cr (<2 p.p.m.), Cu (<10 p.p.m.), Fe (<1 p.p.m.), Mn (<0.2 p.p.m.) and Ni (<5 p.p.m.). An atomically smooth surface with clear unit-cell-height steps was confirmed by atomic force microscopy. The $SrTiO_3$ crystals were then flash-annealed by rising their temperature up to 1,000 K during 10 min, and then keeping them at that temperature in ultra-high vacuum for 30–60 s to produce a metallic state with oxygen vacancies at the surface. The samples were heated from the backside using electron bombardment to avoid any adsorption of contaminants on the surfaces. The side to be measured was isolated from the backside to be heated by a mask made of non-magnetic molybdenum. After annealing, the bulk crystals were still transparent and showed an atomically flat surface with step-and-terrace structure (Fig. 2a).

Laser-PEEM MCD imaging

Our PEEM instrument has electron mirror optics and an energy analyser to reduce spherical and chromatic aberrations in the electron optics[21]. The best lateral resolution of the microscope is 2.6 nm. Threshold photoemission MCD, which we applied in this study, usually detects spin polarizations near the Fermi level around the Γ point of the sample surface. Therefore, this technique is especially suitable for samples with single crystalline surfaces, like $SrTiO_3$. The wavelength of the laser we used is 266 nm, which corresponds to a photon energy of 4.66 eV. A photo-elastic modulator was used to switch between left- and right-handed circular polarizations. If we use intense pulsed light sources, which are conventionally employed for magnetic imaging, the lateral resolution of PEEM significantly degrades due to the Coulomb repulsion between the ejected photoelectrons. To avoid such a space charge effect, we used a continuous wave laser as an excitation source. The intensity of the laser is 2×10^{18} photons per second. In spite of using such an intense laser, we did not see any deterioration of the resolution. The combination of very intense continuous wave laser and polarization-switching method enables us to precisely measure even a small MCD asymmetry with high lateral resolution. The accuracy of the MCD measurement is ~0.1 %, which was estimated from the noise level of Fig. 2d showing no MCD signal above the Curie temperature. MCD signals correspond to the magnetization direction projected to the incident direction of the light source. Therefore, if the illumination angle is perpendicular to the sample surface, this geometry shows magnetic components perpendicular to the surface. As the path of photoelectrons in our microscope is bent by a magnetic field after they pass through a hole of objective lens in front of the sample surface, we are able to perform the perpendicular incidence measurements without causing interference with components of the microscope. For the details of experimental geometry, see the Suuplementary Note 3 withSupplementary Fig. 4 and (ref. 21).

We did not observe any effect of the laser irradiation on the sample surfaces. If the number of oxygen vacancies increased with the laser irradiation, we would be able to observe it as an increase in photoelectron intensity because of the increase in the number of Ti 3*d*electrons (carriers) in the conduction band, as shown in Fig. 1a. However, we never observed any change in photoelectron intensity induced by laser irradiation. Note, furthermore, that the energy of our laser is very small compared with the threshold energy (~38 eV) necessary to efficiently create oxygen vacancies upon irradiation[38], whereas its photon flux is much smaller than the one of pulsed lasers.

REFERENCES

1. Ohtomo, A. & Hwang, H. Y. A high-mobility electron gas at the $LaAlO_3/SrTiO_3$ heterointerface. *Nature* 427, 423–426 (2004).
2. Thiel, S., Hammerl, G., Schmehl, A., Schneider, C. W. & Mannhart, J. Tunable quasi-two-dimensional electron gases in oxide heterostructures. *Science* 313, 1942–1945 (2006).
3. Reyren, N. *et al.* Superconducting interfaces between insulating oxides. *Science* 317, 1196–1199 (2006).
4. Ueno, K. *et al.* Electric-field-induced superconductivity in an insulator. *Nat. Mater* 7, 855–858 (2008).
5. Brinkman, A. *et al.* Magnetic effects at the interface between non-magnetic oxides. *Nat. Mater.* 6, 493–496 (2007).

6. Li, L., Richter, C., Mannhart, J. & Ashoori, R. C. Coexistence of magnetic order and two-dimensional superconductivity at $LaAlO_3/SrTiO_3$ interfaces. *Nat. Phys.* 7, 762 (2011).
7. Bert, J. A. *et al.* Direct imaging of the coexistence of ferromagnetism and superconductivity at the $LaAlO_3/SrTiO_3$interface. *Nat. Phys.* 7, 767 (2011).
8. Dikin, D. A. *et al.* Coexistence of superconductivity and ferromagnetism in two dimensions. *Phys. Rev. Lett.* 107, 056802 (2011).
9. Hwang, H. Y. *et al.* Emergent phenomena at oxide interfaces. *Nat. Mater.* 11, 103–113 (2012).
10. Santander-Syro, A. F. *et al.* Two-dimensional electron gas with universal subbands at the surface of $SrTiO_3$. *Nature* 469, 189 (2011).
11. Meevasana, W. *et al.* Creation and control of a two-dimensional electron liquid at the bare $SrTiO_3$ surface. *Nat. Mater.* 10, 114–118 (2011).
12. Santander-Syro, A. F. *et al.* Giant spin splitting of the two-dimensional electron gas at the surface of $SrTiO_3$. *Nat. Mater.* 13, 1085–1090 (2014).
13. Ariando *et al.* Electronic phase separation at the $LaAlO_3/SrTiO_3$interface. *Nat. Commun.* 2, 188 (2010).
14. Kalisky, B. *et al.* Critical thickness for ferromagnetism in $LaAlO_3/SrTiO_3$ heterostructures. *Nat. Commun.* 3, 922 (2012).
15. Plumb, N. C. *et al.* Mixed dimensionality of confined conducting electrons in the surface region of SrTiO3. *Phys. Rev. Lett.* 113, 086801 (2014).
16. Rödel, T. C. *et al.* Orientational tuning of the Fermi sea of confined electrons at the $SrTiO_3$ (110) and (111) surfaces. Phys. *Rev. Applied* 1, 051002 (2014).
17. Liu, Z. Q. *et al.* Reversible room-temperature ferromagnetism in Nb-doped $SrTiO_3$ single crystals. *Phys. Rev. B* 87, 220405(R) (2013).
18. Lopez-Bezanilla, A., Ganesh, P. & Littelwood, P. B. Research update: Plentiful magnetic moments in oxygen deficient SrTiO3.*APL Mater.* 3, 100701 (2015).
19. Lopez-Bezanilla, A., Ganesh, P. & Littelwood, P. B. Magnetism and metal-insulator transition in oxygen deficient $SrTiO_3$. *Phys. Rev. B*92, 115112 (2015).
20. Potzger *et al.* Defect-induced ferromagnetism in crystalline $SrTiO_3$. *J. Magn. Magn. Mater.* 323, 1551–1562 (2011).
21. Taniuchi, T., Kotani, Y. & Shin, S. Ultrahigh spatial resolution chemical and magnetic imaging by laser-based photoemission electron microscopy. *Rev. Sci. Instrum* 86, 023701 (2015).
22. Sthr, J. Exploring the microscopic origin of magnetic anisotropies with X-ray magnetic circular dichroism (XMCD) spectroscopy. *J. Magn. Magn. Mater.* 200, 470–497 (1999).
23. Lee, J.-S. *et al.* Titanium d_{xy} ferromagnetism at the $LaAlO_3/SrTiO_3$interface. *Nat. Mater.* 12, 703–706 (2013).
24. Salluzzo, M. *et al.* Origin of interface magnetism in $BiMnO_3/SrTiO_3$ and $LaAlO_3/SrTiO_3$ heterostructures. *Phys. Rev. Lett.* 111, 087204 (2013).
25. Nakagawa, T. & Yokoyama, T. Magnetic circular dichroism near the Fermi level. *Phys. Rev. Lett.* 96, 237402 (2006).
26. Nakagawa, T. & Yokoyama, T. Measurements of threshold photoemission magnetic dichroism using ultraviolet lasers and a photoelastic modulator. *Rev. Sci. Instrum* 78, 023907 (2007).
27. Pal, P., Kumar, P., Aswin, V., Dogra, A. & Joshi, A. G. Chemical potential shift and gap-state formation in $SrTiO_{3-\delta}$ revealed by photoemission spectroscopy. *J. Appl. Phys.* 116, 053704 (2014).
28. Johnson, M. T., Bloemen, P. J., der Broeder, F. J. A. & de Varies, J. J.Magnetic anisotropy in metallic multilayers. *Rep. Prog. Phys.* 59, 1409–1458 (1996).
29. Aiura, Y. *et al.* Photoemission study of the metallic state of lightly electron-doped $SrTiO_3$. *Surf. Sci.* 515, 61–74 (2002).
30. Jeschke, H. O., Shen, J. & Valentí, R. Localized versus itinerant states created by multiple oxygen vacancies in $SrTiO_3$. *N. J. Phys.*17, 023034 (2015).
31. Pavlenko, N., Kopp, T., Tsymbal, E. Y., Mannhart, J. & Sawatzky, G. A. Oxygen vacancies at titanate interfaces: two-dimensional magnetism and orbital reconstruction. *Phys. Rev. B.* 86, 064431 (2012).
32. Pavlenko, N., Kopp, T. & Mannhart, J. Emerging magnetism and electronic phase separation at titanate interfaces. *Phys. Rev. B* 88, 201104(R) (2013).
33. Lechermann, F., Boehnke, L., Grieger, D. & Piefke, C. Electron correlation and magnetism at the $LaAlO_3/SrTiO_3$ interface: A DFT+DMFT investigation. *Phys. Rev. B* 90, 085125 (2014).
34. Ruhman, J., Joshua, A., Ilani, S. & Altman, E. Competition between Kondo screening and magnetism at the $LaAlO_3/SrTiO_3$ interface.*Phys. Rev. B.* 90, 125123 (2014).
35. Behrmann, M. & Lechermann, F. Interface exchange processes in LaAlO3/SrTiO3 induced by oxygen vacancies. *Phys. Rev. B* 92, 125148 (2015).
36. Coey, J. M. D., Venkatesan, M. & Fitgerald, C. B. Donor impurity band exchange in dilute ferromagnetic oxides. *Nat. Materials* 4, 173–179 (2005).
37. Zhao, Y. J., Shishidou, T. & Freeman, A. J. Ruderman-Kittel-Kasuya-Yosida-like ferromagnetism in Mn_xGe_{1-x}. *Phys. Rev. Lett.*90, 047204 (2003).
38. Walker, S. M. *et al.* Carrier-density control of the $SrTiO_3$ (001) surface 2DEG studied by ARPES. *Adv. Mater.* 27, 3894–3899 (2015).

CHAPTER 5 MACROSPIN DYNAMICS IN ANTIFERROMAGNETS

D. Bossini, S. Dal Conte, Y. Hashimoto, A. Secchi, R.V. Pisarev, Th. Rasing, G. Cerullo & A.V. Kimel

INTRODUCTION

Experimental studies allowing to investigate correlated matter in general, and magnetism in particular, at the sub-nanometre length- and femtosecond timescales of the exchange interaction have recently developed into an exciting research area. The experiments involving ultrashort timescales provided intriguing results, like the femtosecond laser-induced transient ferromagnetic state of a ferrimagnet alloy[1] and even light-induced superconductivity[2]. However, in these cases the wavelengths of the photo-induced excitations lie orders of magnitude above the lengthscale of the exchange interaction. An alternative strategy consists in the investigation of the magnetic order induced by introducing impurities with atomic resolution in space[3,4,5], but without temporal resolution. A hitherto unexplored approach to the problem, which combines the femtosecond timescale with the nanometre lengthscale, consists in studying the ultrafast dynamics of the macroscale magnetic order parameter triggered by spin excitations with wavelength and period matching the length- and timescales of the exchange interaction.

These spin excitations correspond to magnons (or spin waves) with wavevector near the edges of the first Brillouin zone. In antiferromagnetic materials these magnons can be elegantly excited in the time domain, via a second-order impulsive stimulated Raman scattering process involving pairs of magnons with wavevectors almost equal in magnitude and opposite in sign. It was demonstrated that in the absence of an external magnetic field, such spin excitations can be generated only by a modification of the exchange interaction[6] and cannot be triggered by a change of the spin–orbit coupling. A light-induced modification of the exchange interaction generates pairs of magnons (as shown in the Methods section) and properly describes the Raman spectra[6,7]. The microscopic mechanism of perturbation of the exchange interaction by a laser pulse has been recently described in terms of an effective energy shift of the charge-transfer transition[8,9]. Although pairs of spin waves could in principle be generated throughout the whole Brillouin zone, the magnon density of states is largest in the high-frequency region near the zone edges, where the frequencies are defined mostly by the exchange interaction[6,7](Supplementary Note 1 and Supplementary Fig. 1). Note that the wavelength of these magnons is comparable to the distance of nearest-neighbour spins, whose interactions determine the macroscopic magnetic order. Therefore, studying the photo-induced dynamics of such magnons discloses the response of the magnetic system to excitations on the length- and timescales of the exchange interaction. A bound state of two high-energy, high-wavevector and counter-propagating spin waves, usually denoted as two-magnon (2M) mode, can be induced by a femtosecond light pulse. The frequency and wavevector of this magnetic excitation are the sums of the frequencies and wavevectors of the two magnons involved in the bound state[6,7,10,11]. Although an impulsive excitation of the 2M mode was reported, the subsequent dynamics of the magnetic order parameter has not been discussed[12]. A previous attempt to measure the light-induced evolution of the 2M mode directly in the frequency domain, via the femtosecond-stimulated Raman scattering technique, could reveal a modification of the Raman shift of the 2M peak only during the overlap of the laser pulses[13]. This signal is dominated by nonlinear purely optical effects, originating from the simultaneous interaction of three ultrashort intense laser pulses with the sample. Therefore, disentangling the optical and the magnetic components of the signal was possible only via a sophisticated data analysis and modelling. In femtosecond-stimulated Raman scattering no signal at delays subsequent the instrument response time was observed, which did not allow to disclose the dynamical response of the magnetic order to the 2M excitation.

The present work reports the femtosecond dynamics of the macroscopic magnetic order parameter, by means of an impulsive all-optical injection of the intrinsically highest-frequency spin excitations with wavevectors near the edges of the Brillouin zone. The wavelengths of the corresponding

Citation: D. Bossini, S. Dal Conte, Y. Hashimoto, A. Secchi, R. V. Pisarev, Th. Rasing, G. Cerullo & A. V. Kimel, Macrospin dynamics in antiferromagnets, doi:10.1038/ncomms10645.

magnons are of the order of 1 nm, and the frequencies in the 20 THz range. We demonstrate the manipulation of the order parameter via the coherent control of the short-range magnons: by changing the conditions of the photo-excitation we arbitrarily reverse the phase of the spin waves and enhance or quench their amplitude. Our time-resolved approach discloses even a dynamical coupling of the high-wavevector magnons with a lattice vibrational mode.

RESULTS

Experimental Approach

An excellent system for the all-optical excitation and detection of the dynamics of high-frequency and shortest-wavelength magnons is the cubic Heisenberg antiferromagnetic $KNiF_3$, which is ordered below the Néel point T_N=246 K. In this material the Raman cross-section of the 2M mode is so high that it dominates the whole spectrum[6,14]. A time-domain excitation of the 2M mode in $KNiF_3$ ($\nu_{2M}\approx$22 THz, period≈45 fs, wavevector≈10^7 cm^{-1} and wavelength≈1 nm) can be achieved by the impulsive stimulated Raman scattering mechanism, provided that the duration of the stimulus is shorter than the period of the magnetic mode[15,16]. A successful impulsive excitation of such high-frequency magnons therefore demands laser pulses with a duration significantly shorter than 40 fs. To meet these requirements we used linearly polarized sub-20-fs laser pulses, with a central photon energy of 2.2 eV. For the probe we employed equally short pulses centred around 1.3 eV and with a polarization perpendicular to that of the pump.

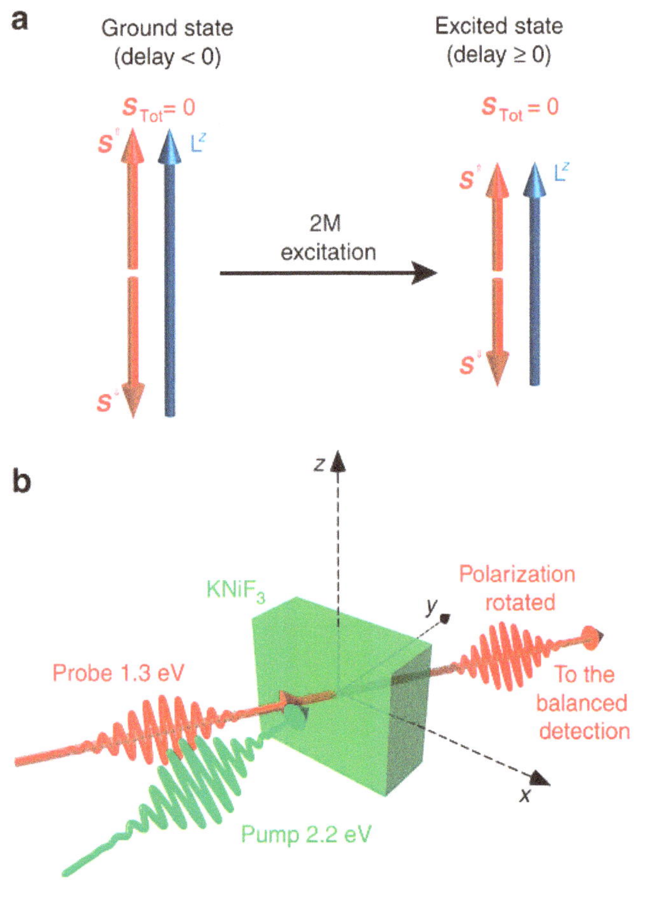

(**a**) The 2M excitation is equivalent to a spin-flip event per sublattice. Thus, the magnetization of each sublattice (*S* and *S*, represented by the two red arrows with opposite orientation) and, therefore, the antiferromagnetic vector (L^z, blue arrow) is decreased in the excited state. The sum of the spins of the two sublattices, thus the total magnetization, vanishes both in the ground and in the excited state. (**b**) Schematic representation of the experimental geometry. The pump (green pulse) photon energy was tuned to 2.2 eV in the transparency window of the material. This choice avoided contributions of laser-heated electrons and phonons to the spin dynamics[19]. The central photon energy of the probe beam (red pulse) was 1.3 eV. The arrows indicates the direction of propagation of the two laser beams.

Figure 1: 2M mode and experimental configuration.

Light scattering by a magnon pair can be visualized as two spin-flip events, one on each sublattice, such that the total spin remains unchanged[6,7,14] (Fig. 1a). Consequently, the transient magneto-optical Faraday and Kerr effects, which measure light-induced variations of the total spin, fail to track the dynamics of such a magnetic excitation. On the other hand, the 2M process is expected to be revealed by those second-order magneto-optical effects, which depend quadratically on the transversal spin components (that is, on the spin deviations), so that a pair of spin-flip events can be detected even if the total spin is unaffected (see discussion in Method section and equations (6) and (7)). Such a magneto-optical effect depends on the spin operators via the same spin-correlation function appearing in the Heisenberg term of the Hamiltonian17 (equation (9)). In particular, we define the antiferromagnetic linear dichroism (ALD) in terms of the symmetric components of the dielectric tensor $\epsilon_s^{\lambda\nu}$ for which $\epsilon_s^{\lambda\nu}=\epsilon_s^{\nu\lambda}$, where ν and λ are indices. The following definition holds:[17,18]

$$\epsilon_s^{\lambda\nu} = \sum_{ij}\sum_{\gamma\delta} \rho^{\lambda\nu\gamma\delta}\langle \hat{S}_i^{\gamma\Uparrow}\hat{S}_j^{\delta\Downarrow}\rangle \tag{1}$$

where $\rho^{\lambda\nu\gamma\delta}$ is a magneto-optical polar fourth rank tensor and $\lambda\nu\gamma\delta$ are spatial coordinate indices, while $\hat{S}_i^{\Uparrow}$ and $\hat{S}_j^{\Downarrow}$ are the spin operators located on two nearest-neighbour sites (i, j), belonging to oppositely oriented magnetic sublattices (and). Note that the ALD induces a rotation of the probe polarization in the experimental configuration shown in Fig. 1b (see Methods section). It is crucial to notice that the two spins in the correlation function belong to two different ionic sites on opposite sublattices. This fact differentiates the ALD from the more conventional magnetic linear birefringence and dichroism (see equation (9) and discussion in the Methods). The macroscopic magnetic order of an ideal Heisenberg antiferromagnet is conveniently described in terms of the antiferromagnetic vector **L**, which is the order parameter[18] and is defined as

$$\mathbf{L} = \sum_i \langle \hat{S}_i^{\Uparrow}\rangle - \sum_j \langle \hat{S}_j^{\Downarrow}\rangle = \mathbf{S}^{\Uparrow} - \mathbf{S}^{\Downarrow} \tag{2}$$

where *S* and *S* are the total spins of the two sublattices (Fig. 1a). Our considerations about the dynamics of the spin system triggered by the impulsive excitation of the 2M mode are based on the approximation of non-interacting magnons. In this framework it is straightforward to demonstrate that the z-projection of **L** has the same time dependence as the spin-correlation function, at the leading terms in the magnon operators (equations (11) and (12)). Here z is the direction parallel to the spins in equilibrium. The dynamics of L^z, at the leading order in the excitation intensity, is given by

$$\Delta L^z(t) \propto A\sin(\omega_{2M}t) \tag{3}$$

where ω_{2M} is the frequency of the 2M mode and A is the amplitude. Equation (3) describes a purely longitudinal, non-precessional, dynamics of the antiferromagnetic vector (see Supplementary Note 2for the complete derivation). Therefore, by detecting the femtosecond dynamics of the ALD, we access the time evolution of the macroscopic order parameter via the spin-correlation function, which ultimately defines also the dynamics of the exchange energy.

Femtosecond dynamics of the magnetic system

Figure 2a shows the typical result of a time-resolved measurement of the laser-induced spin dynamics. The transient rotation of the probe polarization shows oscillations in time with a period of ≈45 fs (that is, a frequency of ≈22 THz) that are damped on a 500-fs timescale. The oscillatory dynamics is superimposed on an incoherent increase of the background, as it is clear from the difference between the time trace and the zero line at longer delays (>500 fs). The spectrum of the oscillation measured in the time-domain experiment closely matches the 2M mode as measured by spontaneous Raman. To definitely assess the nature of the 22 THz mode, we performed temperature-dependent measurements (Fig. 3a) and we compared the temperature dependence of the time-domain signal with that of the spontaneous Raman spectra of the 2M bound state (Fig. 3b,c). Figure 3 shows that the frequency and the lifetime of the pump-induced oscillations decrease as the Néel point is approached, in qualitative and quantitative agreement with spontaneous Raman data[6,14]. Thus, Fig. 2a unambiguosly reveals the femtosecond spin dynamics triggered by the impulsive excitation of the 2M mode in $KNiF_3$, which is not accessible with any other experimental approach.

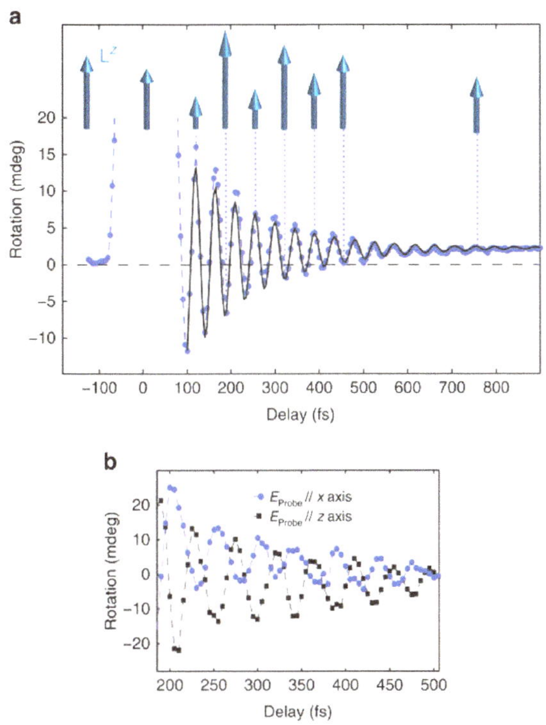

(**a**) The transient rotation of the probe polarization was measured with the electric fields of the pump and the probe beams linearly polarized along the z and x axes, respectively. The pump fluence was set to ≈8.6 mJ cm^{-2}. The corresponding dynamics of L^z (blue arrows) is schematically represented. When the pump pulses impinge on the sample (0 delay) L^zdecreases, due to the generation of magnons. At positive delays, oscillations at the frequency of the 2M mode are visible (equation (3)). The black line is a fit to the data (equation (13)). (**b**) The phase of the oscillation is shifted by π when the orientation of the electric field of the probe ($\boldsymbol{E}_{Probe}$) is rotated by 90°. The pump beam was linearly polarized along the z axis, the fluence was set to ≈12 mJ cm^{-2}. The measurements reported in both panels were performed at 80 K.

Figure 2: Laser-induced dynamics of the antiferromagnetic vector.

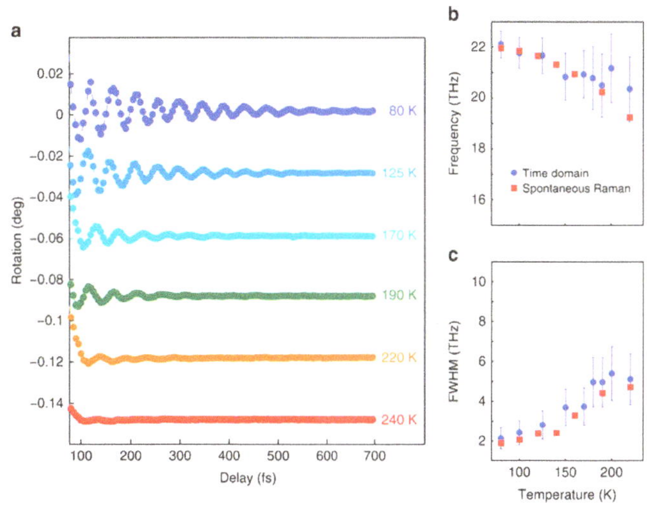

(**a**) Rotation of the probe polarization as a function of the temperature. The pump and the probe beams are linearly polarized along the z and y axes, respectively. The fluence is ≈8.6 mJ cm^{-2}. (**b**,**c**) The temperature dependence of the frequency and linewidth of the 2M Raman peak (red squares) compared with the trends obtained by applying the Fourier transform to the time-domain data (blue circles). The error bars in **b** and **c** for the time-domain data are defined as the half width at half maximum of the 2M frequency distribution. The error bars for the Raman data are defined by the instrumental sensitivity (≈2 cm^{-1}) and are not visible in this scale.

Figure 3: Temperature dependence of the ultrafast spin dynamics.

A fit to the data in Fig. 2a (see Methods section, equation (13)) gave τ_d=(167±4) fs (damping of the coherent oscillations) and τ_r=(255±16) fs (rise time of the incoherent background response). While τ_d represents the decoherence of the 2M band, we interpret τ_r as the characteristic demagnetization time of the two sublattices, solely driven by magnetic interactions[19]. As a laser pulse excites a continuum of magnons with different frequencies, the damping time τ_d of the oscillations observed in our experiment is actually due to the decoherence of the inhomogeneous ensemble of the coherently excited magnons. This is usually described[20] by means of the characteristic time T_2^*. The demagnetization of the sublattices is a result of the heating of spins, which is caused by the decoherence of single-magnon modes[19] in the ensemble, on a timescale generally indicated with T_2 ($T_2 > T_2^*$)[20]. Measuring the probe polarization dependence of the signal (Fig. 2b) allowed us to demonstrate that the detected rotation of the polarization in our experiment arises from the symmetric components of the $\epsilon^{\lambda,\nu}$ tensor and, in particular, from the ALD. In fact the change of sign and the different amplitude of the two data sets shown in Fig. 2b is predicted by the microscopic theory of the ALD (Supplementary Notes 2 and 3).

Coherent Control

The ability to arbitrarily manipulate the properties of spin waves is a necessary requirement for the development of magnon-based devices[21].Figure 4a demonstrates the control of the phase of the nanomagnons. The upmost time traces were measured exciting the sample with orthogonally polarized pump beams. A clear π shift of the phase of the oscillations is observed, if the polarization of the excitation beam is rotated by 90°. This phase shift is due to the non-equivalence of the excitations of the magnetic system induced by light linearly polarized along and orthogonally to the direction of the spins. This interpretation is substantiated by a symmetry analysis of the two-magnon excitation (Supplementary Note 4), based on the point group of $KNiF_3$. The coherent nature of the photo-induced spin dynamics enables the manipulation of the amplitude of the spin waves, by means of a sequence of laser pulses. We demonstrate this effect in Fig. 4aemploying two time-delayed excitation pulses. Tuning the delay we can either amplify or strongly quench the oscillations compared with the case when the spin dynamics is triggered by a single-pump pulse.Figure 4b reveals the periodic dependence of the amplitude of the first magnetic oscillation on the delay between the pump pulses. Such a coherent control of the femto-nanomagnons allows a direct and complete manipulation of the antiferromagnetic order parameter on the femtosecond timescale.

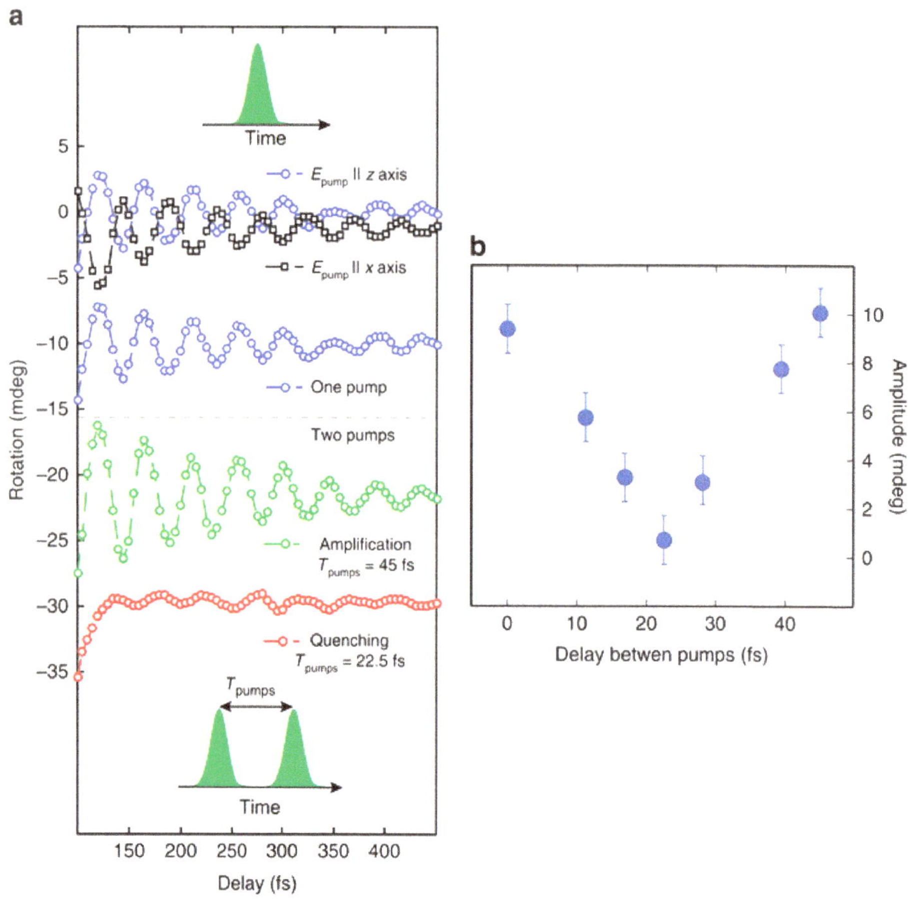

(**a**) The blue and black dotted traces were obtained using a single-pump beam linearly polarized along the z axis (blue circles) and the x axis (black squares). The phase of the oscillations is shifted by π. The single-pump measurement is repeated for the sake of comparison with the results of the double-pump experiments. The green-circle data set was observed by delaying two pump pulses by T_{pumps}=45 fs, that is, the period of the 2M mode. The coherent amplification of the signal is observed. The coherent oscillations were quenched, when the delay between the two pump pulses was set to T_{pumps}=22.5 fs. The fluence of both pump beams was set to $\approx$1.5 mJ cm^{-2}. The temperature was 80 K. The probe beam was polarized along the x axis. (**b**) Amplitude of the first oscillation as a function of the delay between the two pumps (T_{pumps}), the periodic trend is clear. The error bars are defined as twice the sensitivity of our set-up.

Figure 4: Coherent control of the spin dynamics.

Time-frequency Analysis

The real-time measurement of the ultrafast spin dynamics triggered by short-range magnons allows us to disclose another phenomenon not observable with equilibirum techniques. Figure 5a shows the spectrum of the time trace in Fig. 2a obtained via a Fourier transform (blue curve). On the same graph we plot the spectrum of the 2M mode measured via spontaneous Raman scattering at the same temperature (red curve). The zoom in the inset of Fig. 5a shows two sidebands at about 7.5 THz from the central 2M frequency, which appear to originate from a modulation of the 2M mode in the time domain. Figure 5b shows a two-dimensional spectrogram obtained by performing a time-frequency analysis[22,23] of the data in Fig. 2a (Supplementary Note 5). The colour map in Fig. 5b represents the time-dependent spectrum of the 2M mode.

The frequencies of the peak of the spectrum at different time delays are traced by a blue dotted line. This curve displays a periodic oscillation of ν_{2M}, which is consistent with the observation of the sidebands in the inset of Fig. 5a. We define the relative frequency shift as

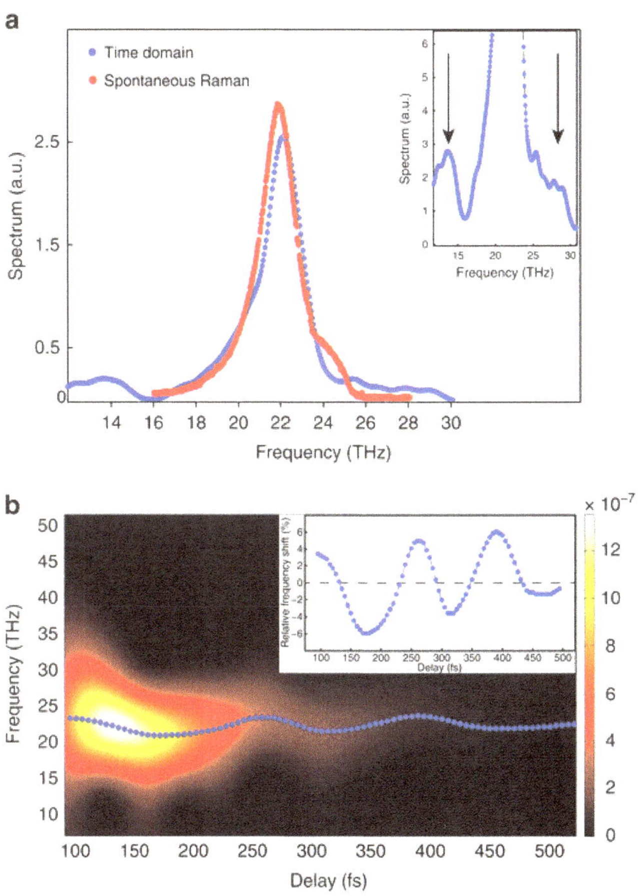

(**a**) The Fourier transform of the time trace measured at 80 K (blue curve) is compared with the spontaneous Raman spectrum obtained at the same temperature (red curve). In the inset a zoom of the Fourier transform reveals two sidebands at ≈±7.5 THz away from the peak frequency. (**b**) The squared modulus of the Wigner distribution of the signal is represented by the colour plot. At each time step a blue dotted line highlights the maximum of the spectrum, in which black represents zero intensity. In the inset the relative frequency shift (equation (4)) is plotted as a function of the delay showing ≈7.5 THz oscillations.

Figure 5: Spectrum of the ultrafast response of the antiferromagnetic vector.

$$\frac{\Delta\nu_{2M}}{\nu_{2M}} = \frac{\nu_{2M} - \langle\nu_{2M}\rangle}{\langle\nu_{2M}\rangle} \tag{4}$$

where $\langle\nu_{2M}\rangle$ is the average frequency in the temporal interval where the oscillations have a significant amplitude (0–500 fs). We plot the peaks of the spectra obtained by the time-frequency analysis as a function of the delay in the inset of Fig. 5b. The frequency of the modulation of ν_{2M} is ≈7.5 THz, which corresponds to the frequency of the infrared-active phonon24·25 (≈7.7 THz), assigned to the stretching vibration of the Ni–F–Ni bond. Unlike other phonon modes in this material, the frequency of the stretching mode is temperature-independent[24,25], which is consistent with the data (Supplementary Fig. 2). Considering only the term of the polarization of the first order in the electric field of light, it would be concluded that this stretching mode is not Raman-active[26]. However, the light–matter interaction at the next order (hyper-Raman scattering[26,27]) allows to excite lattice vibrations with the symmetry of the stretching mode (F_{1u}) in cubic crystals27. It is well established that a frequency modulation as the one shown in Fig. 5b can be modelled in terms of coupled oscillators[23]. Therefore, we assign the modulation of ν_{2M} to the interaction on the femtosecond timescale between the stretching mode and the 2M mode, which are simultaneously and coherently excited by the laser pulse. The 2M–phonon interaction was previously suggested[11], however our time-resolved experiment provides the first evidence of this effect.

DISCUSSION

We would like to underline that the dynamics here reported is totally different from the spin response previously observed after the photo-excitation of the low-energy magnons with wavevectors at the centre of the Brillouin zone in this material[19]. In fact, our previous investigation revealed spin oscillations with the period of 11 ps (that is, a 90-GHz frequency), which are still visible 200 ps after the photo-excitation. This spin dynamics was generated by a single spin-flip event (that is, one-magnon mode), thus the magnetic linear birefringence, which depends quadratically on the spin of a single atom but only linearly on the spin deviation (equation (9)), was employed to probe the response of the magnetic system. The interaction relevant for the low-energy magnons is the spin–orbit coupling, which defines the frequency of such collective excitations6. The use of 100-fs laser pulses prevented the access to high-frequency magnetic modes19. The present work describes a completely different regime of spin dynamics, whose characteristic timescale is determined by the exchange energy6. Moreover, the measurements of the time evolution of the spin-correlation function (via ALD) provide an access to the dynamics of the exchange energy. Within the approximations of our model, the spin-correlation dynamics corresponds to the longitudinal dynamics of the order parameter (equations (11) and (12) and Supplementary Note 2).

In summary, our results reveal the dynamics of the macroscopic order parameter triggered by impulsively excited nanometre-wavelength and femtosecond-period magnons. Moreover, our experiment pushes the coherent control of magnons, previously demonstrated at the centre of the Brillouin zone[28,29], to the edges of the Brillouin zone, laying the foundations for a magnon-based nanotechnology operating in the 20-THz regime. Our investigation concerned an ideal Heisenberg antiferromagnet, however the approach here employed to study the femtosecond dynamics of the macroscopic magnetic order parameter caused by short-range spin excitations can be employed in a broad group of multisublattice systems[6,7]. In particular, it will become possible to monitor the evolution of the exchange energy during a photo-induced phase transition and to probe the femtosecond dynamics of the sub-nanometre range spin correlations in strongly correlated materials. This might even elucidate the dynamical interplay between short-range spin excitations and high-temperature superconductivity in cuprates[30,31,32].

METHODS

Sample

Our sample was a 340-μm-thick (100) plane-parallel plate of $KNiF_3$, which has a perovskite crystal structure (point group m3m). Two equivalent Ni^{2+} sublattices are antiferromagnetically coupled below the Néel temperature T_N=246 K (ref. 19). This material is known to be a cubic Heisenberg antiferromagnet because of its very weak anisotropy. The positive sign of the cubic magnetic anisotropy constant determines the alignment of spins along the [001], [010] or [100] axes18. The measurements on the $KNiF_3$ sample are carried out at a minimum temperature of 77 K in a liquid nitrogen cryostat. The temperature of the sample is monitored by a thermocouple placed on the sample holder.

Light Source

For the pump–probe experiments we used a regeneratively amplified mode-locked Ti:Sapphire laser, providing 150-fs, 500-μJ pulses at 780 nm and 1 kHz repetition rate. The laser drives two non-collinear optical parametric amplifiers (NOPAs) operating in two different spectral ranges33. Both NOPAs are pumped by the second harmonic of the laser (that is, 390 nm) and seeded by the white-light continuum produced by focusing the 780-nm beam into a sapphire plate. The amplified pulse from the first NOPA, which initiates the dynamics (pump), has a spectrum spanning the 500–700 nm range and is compressed to nearly transform-limited duration (that is, 8 fs) by a pair of custom-made chirped mirrors. The amplified pulse, generated by the second NOPA (probe), covers the frequency range between 820 and 1050 nm, and is compressed to nearly transform-limited duration (that is, 13 fs) by a couple of fused silica prisms. The temporal resolution of the set-up has been characterized by the cross-correlation frequency-resolved optical gating technique and was below 20 fs (ref. 32). The pump and probe beams were focused on the sample by a spherical mirror down to approximately 100 and 70 μm spot sizes, respectively. The high temporal resolution is preserved by using a very thin (200 μm) fused silica window as optical access to the cryostat. The two pump pulses employed to show the coherent control of the 2M mode were generated in a Michelson interferometer scheme. One of these beams was reflected by mirrors mounted on a stepper-motor delay line, able to introduce a minimum time delay of 0.3 fs. The two orthogonal polarizations were obtained with a polarizer sheet (thickness ≈0.3 mm), to avoid the use of thicker waveplates and the consequent broadening of the pulse duration.

Detection Scheme

We measured the pump-induced rotation of the probe polarization employing a balanced-detection scheme. The transmitted probe is split by a Wollaston prism into two orthogonal linearly polarized beams, and focused on a couple of balanced photodiodes. The Wollaston prism is rotated to equalize the probe intensities on the two photodiodes. The pump-induced imbalance of the signal registered by the two photodiodes is measured by a lock-in amplifier, which is locked to the modulation frequency of the pump beam (that is, 500 Hz). Our apparatus was able to detect rotations of the polarization on the order of 1 mdeg.

Excitation Mechanism

As explained in the main text, a light-induced modification of the exchange interaction generates the 2M mode6·7. This concept can be easily verified by taking into account the transient value of the exchange in the energy of the magnetic medium

$$\hat{\mathscr{H}} = J_0 \sum_{i,j} \hat{S}_i^{\Uparrow} \cdot \hat{S}_j^{\Downarrow} + \sum_{i,j} \Delta J_{ij}(t) \hat{S}_i^{\Uparrow} \cdot \hat{S}_j^{\Downarrow} \tag{5}$$

where $\Delta J_{ij}(t)$ is the laser-induced modification of the exchange interaction, while J_0 is the value of the exchange in the ground state. The ij indices refer to the anisotropic coupling of light with the magnetic system, determined by the polarization of the laser beam, which breaks the isotropic symmetry of J_0 (Supplementary Note 2). The second term of the equation can be rewritten by expanding the dot product in the following way

$$\sum_{i,j} \Delta J_{ij}(t) \left(\hat{S}_i^{x\Uparrow} \hat{S}_j^{x\Downarrow} + \hat{S}_i^{y\Uparrow} \hat{S}_j^{y\Downarrow} + \hat{S}_i^{z\Uparrow} \hat{S}_j^{z\Downarrow} \right) \tag{6}$$

Defining z as the direction along which the spins are oriented at equilibrium, this equation shows that the modification of the exchange generates terms in the energy quadratic in the transversal spin components (that is, x and y). Such contributions describe the process of generation of two magnons[6] reported in Fig. 1a. This becomes evident by introducing the conventional ladder operators $\hat{S}^{\pm} = \hat{S}^x \pm i\hat{S}^y$, in terms of which we can rewrite equation (6) as

$$\sum_{i,j} \Delta J_{ij}(t) \left(\frac{1}{2} \left(\hat{S}_i^{+\Uparrow} \hat{S}_j^{-\Downarrow} + \hat{S}_i^{-\Uparrow} \hat{S}_j^{+\Downarrow} \right) + \hat{S}_i^{z\Uparrow} \hat{S}_j^{z\Downarrow} \right) \tag{7}$$

The physical interpretation of the first two terms of the last equation consists in two spin-flip processes, one on each sublattice. Once a positive direction of the quantization axis is chosen (direction of L^z in Fig. 1a), flipping a spin on the $\Uparrow(\Downarrow)$ sublattice results in a decrease (increase) of the total spin, taken into account by $\hat{S}_i^{-\Uparrow}(\hat{S}_j^{+\Downarrow})$. In conclusion, a transient modification of the exchange interaction, which breaks the isotropic symmetry of J_0, triggers two magnons described by the quadratic dependence of the Hamiltonian on the transversal spin components.

Antiferromagnetic Linear Dichroism

The Hamiltonian representing the interaction between light and the magnetic system can be written as6

$$\hat{\mathscr{H}} = \sum_{\lambda,\nu} \epsilon^{\lambda\nu}(\hat{S}) E^{\lambda} E^{\nu} \tag{8}$$

where $\epsilon^{\lambda\nu}$ is the dielectric tensor that depends on the spin ($\hat{S}$), while λ, vare indexes and E is the electric field of light. The tensor $\epsilon^{\lambda\nu}$ can be expanded in powers of the spin operators. Following a well-established approach[6,17], we may express the spin-dependent components of $\epsilon^{\lambda v}$ as

$$\epsilon^{\lambda v} = \sum_i \sum_\gamma K^{\lambda v \gamma} \langle \hat{S}_i^\gamma \rangle + \sum_i \sum_{\gamma\delta} G^{\lambda v \gamma\delta} \langle \hat{S}_i^\gamma \rangle \langle \hat{S}_i^\delta \rangle + \sum_{i,j} \sum_{\gamma\delta} \rho^{\lambda v \gamma\delta} \left\langle \hat{S}_i^{\gamma\Uparrow} \hat{S}_j^{\delta\Downarrow} \right\rangle \tag{9}$$

where higher-order terms are omitted. Each term in this expansion represents a magneto-optical effect and a Raman mode, given the intimate connection between the tensors defining the Raman scattering on magnons and the magneto-optical coefficients[6,34]. The first and second terms involve spin operators at a single ionic site i, they describe the linear magneto-optical effects: the complex Faraday effect and the magnetic linear birefringence(and dichroism), respectively. In terms of Raman scattering the first two terms of equation (9) are connected with the lowest energy magnetic excitations, correspondent to a single spin-flip process, the so called one-magnon modes. These spin waves have wavevectors close to the centre of the Brillouin zone and their frequencies are mostly defined by the spin–orbit coupling. Note that although the second term is quadratic in the spin, it describes a single-magnon excitation and a linear magneto-optical effect. In fact this term is linear in the spin deviation, that is, in the spin transversal component, which implies a linear dependence on the spin ladder operators and therefore the generation of a single magnon6 (see discussion in the previous section). The magnetic linear birefringence was employed to probe the spin dynamics in reference[19], which accordingly reports the excitation and the time evolution of the low-energy one-magnon mode in $KNiF_3$.

In the present study we measure the ALD, which is accounted for by the last term in equation (9): the quadratic dependence on the spin deviations (equations (6) and (7)) of this term allows to unravel the dynamics of the 2M mode, triggered by two spin-flip processes. Moreover, the two spin operators in this term belong to different sublattices, being placed on different ionic sites i, j. This is the pivotal difference between this magneto-optical effect and the others described so far: the ALD probes the correlations between spins on different sites. Since in $KNiF_3$ only the nearest-neighbour interaction is significant, the dynamics of the spin-correlation function in the last term of equation (9) defines also the dynamics of the exchange energy. The transient ALD, which consists of a different absorption for light beams linearly polarized along and orthogonally to the direction of the antiferromagnetic vector, results in the detected rotation of the probe polarization.

Dynamics of the order parameter

The dynamics of the spin-correlation function can be related to the dynamics of the order parameter in a straightforward way. The z-component of the local spins on sites i, j, belonging to the and sublattices respectively, can be written as

$$\hat{S}_i^{z\Uparrow} = S - \hat{b}_i^\dagger \hat{b}_i; \qquad \hat{S}_j^{z\Downarrow} = -S + \hat{a}_j^\dagger \hat{a}_j \tag{10}$$

where the operators $\hat{b}_i^\dagger \hat{b}_i$ and $\hat{a}_j^\dagger \hat{a}_j$ (introduced rigorously in Supplementary Note 2) represent the number of spin deviations from the maximum values of local spins for both sublattices ±S. We can then write the expectation value of the spin-correlation function on a time-dependent state $|\psi(t)\rangle$ as

$$\langle \psi(t) | \hat{S}_i^{z\Uparrow} \hat{S}_j^{z\Downarrow} | \psi(t) \rangle \approx -S^2 + S \langle \psi(t) | \left(\hat{b}_i^\dagger \hat{b}_i + \hat{a}_j^\dagger \hat{a}_j \right) | \psi(t) \rangle, \tag{11}$$

where only the leading terms in the creation and destruction operators have been considered. Recalling the definition of the antiferromagnetic vector in equation (2), the time-dependent z-component of the order parameter is given by

$$\begin{aligned} L^z(t) &= \sum_i \langle \psi(t) | \hat{S}_i^{z\Uparrow} | \psi(t) \rangle - \sum_j \langle \psi(t) | \hat{S}_j^{z\Downarrow} | \psi(t) \rangle \\ &= SN - \frac{2}{N} \sum_{i,j} \langle \psi(t) | \left(\hat{b}_i^\dagger \hat{b}_i + \hat{a}_j^\dagger \hat{a}_j \right) | \psi(t) \rangle, \end{aligned} \tag{12}$$

where N is the total number of spins. It stems from equations (11) and (12) that the time-dependent parts of the spin-correlation function and of the antiferromagnetic vector are proportional.

Fitting procedure

The transient rotation of the probe polarization was fitted employing the following function:

$$f(t) = C\sin[(\tilde{\omega}(t)t + \phi)]e^{(-t/\tau_d)} + H(t)D\left(1 - e^{(-t/\tau_r)}\right) \tag{13}$$

where C and D are amplitude coefficients, φ is the phase of the oscillations, H(t) is the Heaviside function, τ_d is the damping time of the oscillations and τ_r is the characteristic rise time of the incoherent contribution to the signal. Considering the outcome of the Wigner analysis in Fig. 5b, we employed a time-dependent frequency $\tilde{\omega}$ in the sinusoidal function, namely

$$\tilde{\omega}(t) = 2\pi v_{2M}[1 + G\sin(2\pi v_{mod} t + \psi)] \tag{14}$$

where v_{2M} is the 2M frequency, G is an amplitude coefficient, v_{mod} is the modulation frequency and ψ is the phase. From the Wigner analysis we set the following parameters: v_{2M}=22.12 THz; G=0.002; v_{mod}=7.5 THz;ψ=45°; and D=2.3·10^{-3} deg. The fit parameters allowing to reproduction at best the data (Fig. 2) are as follows: C=(2.5±0.1)·10^{-2} deg; φ=(220±1)°;τ_d=(167±4) fs; and τ_r=(255±16) fs.

REFERENCES

1. Radu, I. et al. Transient ferromagnetic-like state mediating ultrafast reversal of antiferromagnetically coupled spins. Nature472, 205–208 (2011).
2. Fausti, D. et al. Light-induced superconductivity in a stripe-ordered cuprate. Science 331, 189–191 (2011).
3. Loth, S., Baumann, S., Lutz, C. P., Eigler, D. M. & Heinrich, A. J.Bistability in atomic-scale antiferromagnets. Science 335, 196–199 (2012).
4. Khajetoorians, A. A. et al. Atom-by-atom engineering and magnetometry of tailored nanomagnets. Nat. Phys. 8, 497–503 (2012).
5. Khajetoorians, A. A. et al. Current-driven spin dynamics of artificially constructed quantum magnets. Science 339, 55–59 (2013).
6. Cottam, M. G. & Lockwood, D. J. Light Scattering in Magnetic Solids 1st ed. Wiley-Interscience (1986).
7. Fleury, P. & Loudon, R. Scattering of light by one-

and two-magnon excitations. Phys. Rev. 166, 514–530 (1968).

8. Secchi, A., Brener, S., Lichtenstein, A. & Katsnelson, M. Non-equilibrium magnetic interactions in strongly correlated systems.Ann. Phys. 333, 221–271 (2013).
9. Mentink, J. H. & Eckstein, M. Ultrafast quenching of the exchange interaction in a mott insulator. Phys. Rev. Lett. 113, 057201 (2014).
10. Balucani, U. & Tognetti, V. Theory of two-magnon raman scattering in the ordered region for cubic antiferromagnets. Phys. Rev. B 8, 4247–4257 (1973).
11. Cottam, M. G. Theory oftwo-magnon Raman scattering in antiferromagnets at finite temperatures. J. Phys. C Solid State Phys. 5, 1461–1474 (1972).
12. Zhao, J., Bragas, A., Lockwood, D. & Merlin, R. Magnon squeezing in an antiferromagnet: reducing the spin noise below the standard quantum limit. Phys. Rev. Lett. 93, 107203 (2004).
13. Batignani, G. et al. Probing ultrafast photo-induced dynamics of the exchange energy in a Heisenberg antiferromagnet. Nat. Photon. 9, 506–510 (2015).
14. Chinn, S., Zeiger, H. & O'Connor, J. Two-magnon Raman scattering and exchange interactions in antiferromagnetic $KNiF_3$and K_2NiF_4 and ferrimagnetic $RbNiF_3$. Phys. Rev. B 3, 1709–1735 (1971).
15. Kimel, A. V. et al. Ultrafast non-thermal control of magnetization by instantaneous photomagnetic pulses. Nature 435, 655–657 (2005).
16. Yan, Y.-X., Gamble, E. B. & Nelson, K. A. Impulsive stimulated scattering: general importance in femtosecond laser pulse interactions with matter, and spectroscopic applications. J. Chem. Phys. 83, 5391 (1985).
17. Ferre, J. & Gehring, G. A. Linear optical birefringence ofmagnetic crystals. Rep. Prog. Phys. 47, 513–611 (1984).
18. Landau, L. D. & Lifshitz, E. M. Electrodynamics of Continuous Media 2nd edn Vol. 8, Elsevier (1984).
19. Bossini, D., Kalashnikova, A. M., Pisarev, R. V., Rasing, T. & Kimel, A. V. Controlling coherent and incoherent spin dynamics by steering the photoinduced energy flow. Phys. Rev. B 89, 060405 (2014).
20. Yakovlev, D. R. & Bayer, M. in Spin Physics in Semiconductors, 1st edn, Ch. 6 135–178Springer (2008).
21. Chumak, A. V., Vasyuchka, V. I., Serga, A. A. & Hillebrands, B.Magnon spintronics. Nat. Phys. 11, 453–461 (2015).
22. Bartelt, H., Brenner, K.-H. & Lohmann, A. The wigner distribution function and its optical production. Opt. Commun. 32-38, 32 (1980).
23. Gambetta, A. et al. Real-time observation ofnonlinear coherent phonon dynamics in single-walled carbon nanotubes. Nat. Phys. 2, 515–520 (2006).
24. Sintani, K., Tomono, Y., Tsuchida, A. & Siratori, K. Influence of magnetic ordering on the lattice vibration of $KNiF_3$. J. Phys. Soc. Jpn 25, 99–108 (1968).
25. Tomono, Y., Takaoka, T., Yajima, M., Tanokura, Y. & Jinda, N.Temperature dependence of lattice vibration frequencies of KNiF3 in connection with antiferromagnetism. J. Phys. Soc. Jpn 59, 579–583 (1990).
26. Long, D. A. The Raman Effect: A Unified Treatment of the Theory of Raman Scattering by Molecules 1st edn Wiley (2002).
27. Cyvin, S. J., Rauch, J. E. & Decius, J. C. Theory of hyper-Raman effects (nonlinear inelastic light scattering): selection rules and depolarization ratios for the second-order polarizability. J. Chem. Phys. 43, 4083–4095 (1965).
28. Kanda, N. et al. The vectorial control of magnetization by light.Nat. Commun. 2, 362 (2011).
29. Kampfrath, T. et al. Coherent terahertz control of antiferromagnetic spin waves. Nat. Photon. 5, 31–34 (2010).
30. Le Tacon, M. et al. Intense paramagnon excitations in a large family of high-temperature superconductors. Nat. Phys. 7, 725–730 (2011).
31. Dal Conte, S. et al. Disentangling the electronic and phononic glue in a high-Tc superconductor. Science 335, 1600–1603 (2012).
32. Dal Conte, S. et al. Snapshots of the retarded interaction of charge carriers with ultrafast fluctuations in cuprates. Nat. Phys.11, 421–426 (2015).
33. Brida, D. et al. Few-optical-cycle pulses tunable from the visible to the mid-infrared by optical parametric amplifiers. J. Opt. 12, 013001 (2010).
34. Wettling, W., Cottam, M. G. & Sandercock, J. R. The relation between one-magnon light scattering and the complex magneto-optic effects in YIG. J. Phys. C Solid State Phys. 8, 211–228 (1975).

CHAPTER 6 SYNTHESIS OF MAGNETIC NANOFIBERS

Mohammed-Amin Alubaid, Krishnan Venkatakrishnan and Bo Tan

INTRODUCTION

Nanomaterials field is of current interest because it studies materials with morphological features on the nanoscale. Nanosized materials show distinctive properties compared with bulk materials [1, 2, 3]. In particular, magnetic nanostructures have recently attracted much attention because of their intriguing properties that are not displayed by their bulk or particle counterparts. These nanostructures are potentially useful as active components for ultrahigh-density data storage, as well as in the fabrication of sensors and spintronic devices [4].

The growth of nanofibers using ultrafast laser offers advantages of high resolution, high throughput, uniformity, localized heating, simplicity, and reproducibility [5, 6, 7, 8]. The time scale of materials heating and cooling of traditional thermal processes is significantly higher than that with femtosecond laser irradiation [9]. The rapid absorption of energy leads to efficient material removal before significant heat diffusion to the substrate occurs. Femtosecond laser radiation has already been used to fabricate nano-sized spikes of semiconductor [10], metallic [11, 12], and dielectric surfaces [13] in vacuum.

Magnetic neodymium-iron-boron (NdFeB) nanofibers and nanoparticles have become one of the hotspots in the research field of magnetic materials to meet the demand for miniaturization of electronic components in recent years, and have been successfully prepared by various routes like the sol-gel auto-combustion method [14], co-precipitation [15], hydrothermal method [16], reverse micelles [17], microemulsion method [18], alternate sputtering [19], pulsed-laser deposition [20], and so on. However, until now there have been no reports on the synthesis and magnetic properties of NdFeB ferrite nanofibers in literatures.

In the present study a magnetic weblike fibrous nanostructure is formed due to the agglomeration of the bulk quantity of nanoparticles created during laser ablation at mega hertz pulse frequency. A distinct characteristic of the fibrous nanostructures is that particles are fused and the agglomeration shows certain degree of organization, unlike the random stacking of particles observed at femtosecond laser ablation at pulse frequency in kilohertz and hertz regime. The effect of pulse repetition rate on the nanofibers size and hence the magnetization was also investigated. The nanostructures were characterized by scanning electron microscopy (SEM), transmission electron microscopy (TEM), energy-dispersive X-ray (EDX), X-ray diffraction (XRD), and magnetic force microscopy (MFM). The mechanism of formation is explained by the well-established theory of vapor condensation induced by ultrafast laser ablation. Also, the fibrous nanostructures have relatively uniform diameters (30-90 nm) and did not observe a wide range of variation in size distribution. This agrees with the characteristics of nanoparticle formation through homogenous nucleation, which tends to generate monosized nanoparticles.

Experimental details

The laser source is a diode-pumped Yb-doped fiber oscillator/ amplifier system (Clark MXR Inc.) capable of producing an average power of 15.5 W with pulse repetition frequency between 200 kHz and 25 MHz. A neodymium-iron-boron magnetic specimen of 1" ± 0.008» length by 1» ± 0.008»; width by 0.1» ± 0.005» thickness was cut into four pieces of same size. The tetragonal $Nd_2Fe_{14}B$ crystal structure has exceptionally high uniaxial magnetocrystalline anisotropy. This gives the compound the potential to have high coercivity. To generate magnetic nanofibers, the first piece of the magnetic specimen was irradiated with laser using 1040 nm wavelength with 15 W power and a pulse repetition rate of 4 MHz. The experiment was repeated to generate nanofibers on the specimen using the same power and wavelength with frequencies of 8, 13, and 26 MHz. The irradiated sample was characterized using SEM, TEM, EDX, and XRD analysis.

Citation: Mohammed-Amin Alubaidy, Krishnan Venkatakrishnan and Bo Tan, Synthesis of magnetic nanofibers, DOI: 10.1186/1556-276X-6-375.

RESULTS AND DISCUSSION

The energy of the femtosecond laser is delivered into the material in a short time scale that absorption occurs at nearly solid-state. The energy is first deposited in the electronic subsystem within a layer of thickness of tens of nanometer. Enough energy is absorbed to produce macroscopic ablation when the density of the free electrons exceeds a certain threshold [21]. The ionized material is removed away from the surface in the form of expanding high pressure plasma. The plasma remains confined close to the specimen surface at atmospheric pressure. Condensation of vapor in the plume leads to the generation of nanoparticles. Some of these nanoparticles aggregate and then get deposited on the surface of the specimen [8]. Vapor condensation starts with nucleation, proceeds with growth of supercritical nucleus and come to a halt due to quenching. For nanoparticles to aggregate and form fibrous structure, a continuous supply of vapor is required to the expanding plume to maintain the nucleus density. Hence nanoparticles generated from the successive laser pulse are fused to the particles created from the previous laser pulse that are still above the melting temperature and grow as nanofibrous like structure as shown in Figure 1. Dipole-dipole interactions then trigger anisotropic chain growth under the influence of serendipitous Brownian collisions, attractive van der Waals, as well as the residual electrostatic repulsions that maintain colloidal stability [22]. The energy barrier to surface reorganization is overcome over the very high temperature, resulting in the rapid onset of self-assembly of the nanoparticle chains (or nanofibers).

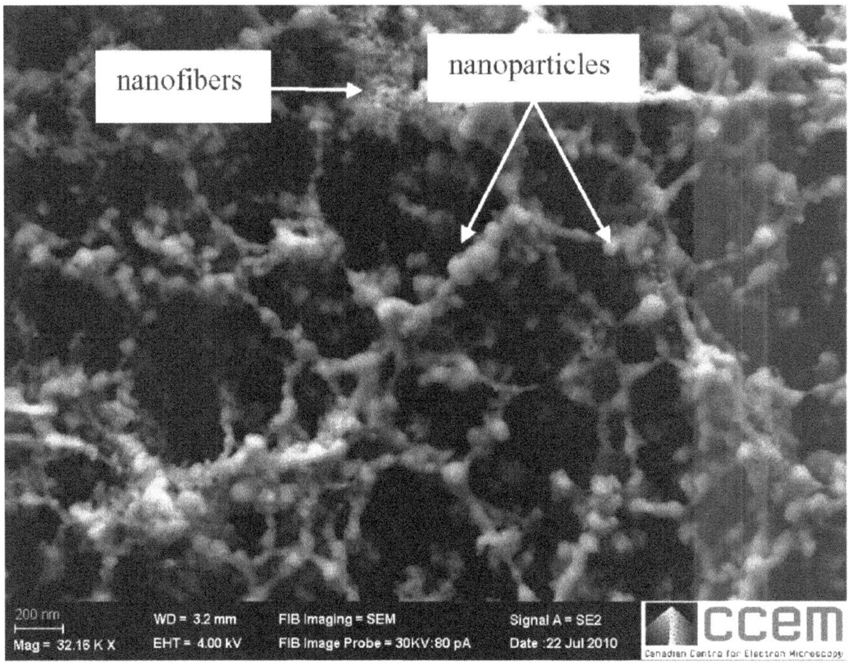

Figure 1: SEM image of magnetic nanofibrous structure and nanoparticles on NdFeB substrate irradiated with femtosecond laser at 26 MHz repetition rate and 15 W average power.

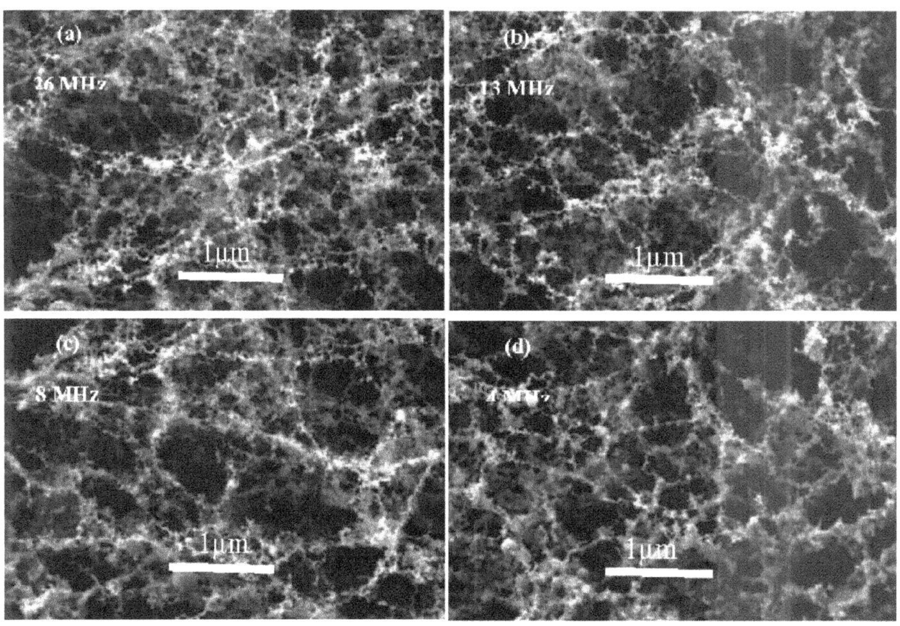

Figure 2: SEM images of the generated nanofibers (a) 26 MHz, (b) 13 MHz, (c) 8 MHz, and (d) 4 MHz.

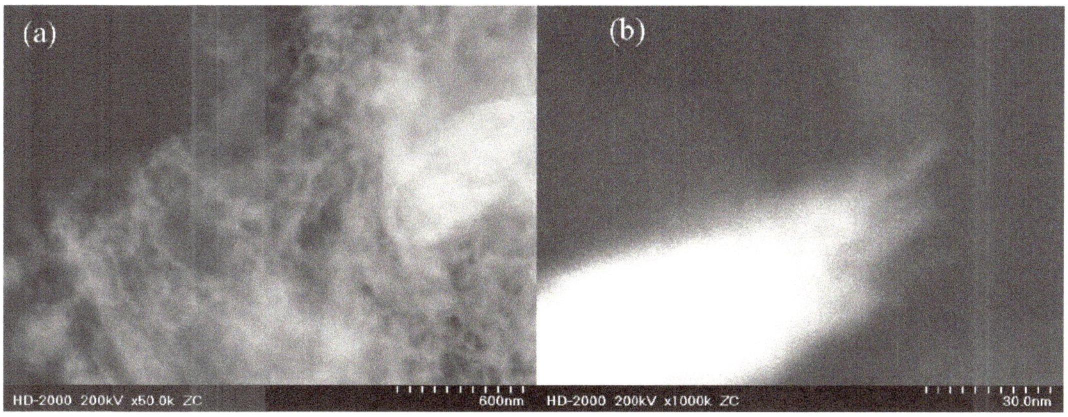

Figure 3: TEM images of magnetic nanofibers generated by femtosecond laser at 26 MHz pulse repetition rate and 15 W power.

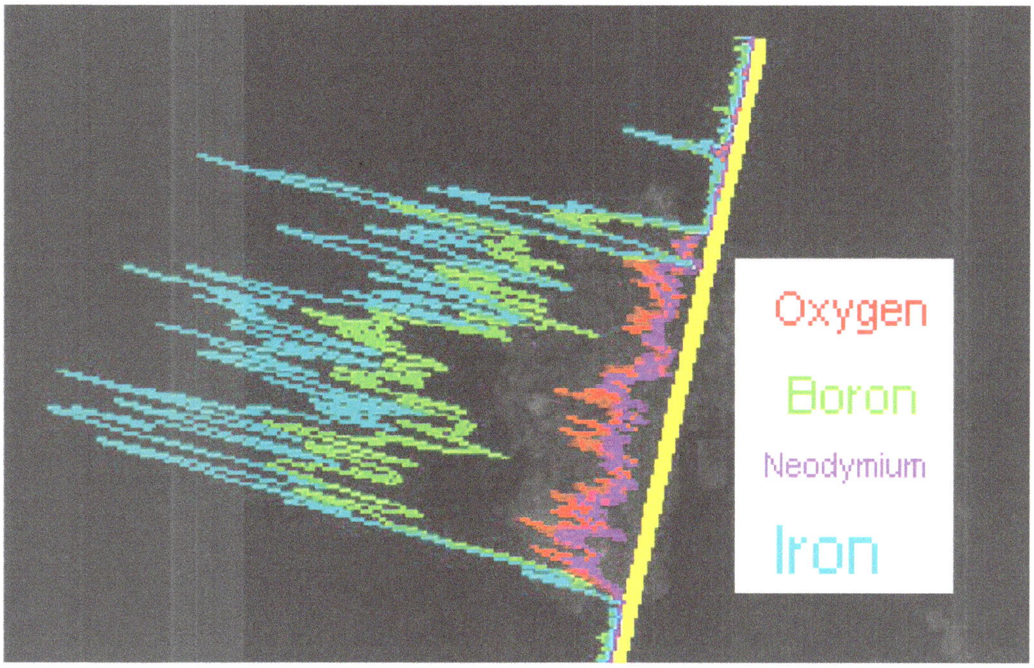

Figure 4: EDX analysis of magnetic nanofibers structures.

The laser pulse repetition rate plays a critical role in the formation of nanofibrous like structure. Figure 2 shows SEM images of the magnetic weblike nanofibers generated at 4, 8, 13, and 26 MHz pulse repetition rate. The average diameters of the generated nanofibers were around 70, 60, 45, and 30 nm, respectively. Figure 3a shows the TEM image of magnetic nanofibers at 26 MHz pulse repetition rate and Figure 3b shows a single magnetic nanofiber generated by femtosecond laser. It depicts that magnetic nanofibers possess weblike structures with diameter not more than 30 nm. Further EDX analysis of the irradiated surface shows existence of oxygen as shown in Figure 4which indicates, besides the percentage of oxygen to neodymium-iron-boron, the existence of oxidized magnetic nanofibers [23].

During ablation, the ionized material is removed away from the surface in the form of expanding high pressure plasma. The temperature of the plasma is above the melting temperature and hence Curie temperature of the magnet. Thus the irradiated spot and the surrounding area where temperatures above 400°C will be demagnetized while the rest of the sample remains magnet. The plasma remains confined close to the specimen surface at atmospheric pressure. Condensation of vapor in the plume leads to the generation of nanoparticles which move in the direction where the paramagnetization potential energy is minimized [24]. The nanoparticles travelled perpendicular to the direction of the magnetic field and then aggregate and get deposited on the surface of the specimen [25]. The generated nanofibers remagnetized when its temperature reduced below the Curie temperature of the sample to form magnetic nanofibers. The total magnetization of a nanofiber is given by the vectorial sum of all single magnetic moments of the atoms [24]. As for the atomic magnetic moments in generated nanofibers, the average magnetization will be zero in the absence of magnetic field since all magnetic moments are randomly directed in space. When a magnetic field is applied by the substrate, the magnetic moments will orient in the direction of the field and give rise to a net magnetization of the nanofibers. Magnetic field microscopy, from NT-MDT, (MFM) image of the weblike nanofibers structures generated at 26 MHz is shown in Figure 5. The NdFeB nanofibers exhibit magnetic properties (darker parts) as shown in the MFM image of Figure 5 which are distinguishable from the background (brighter parts).

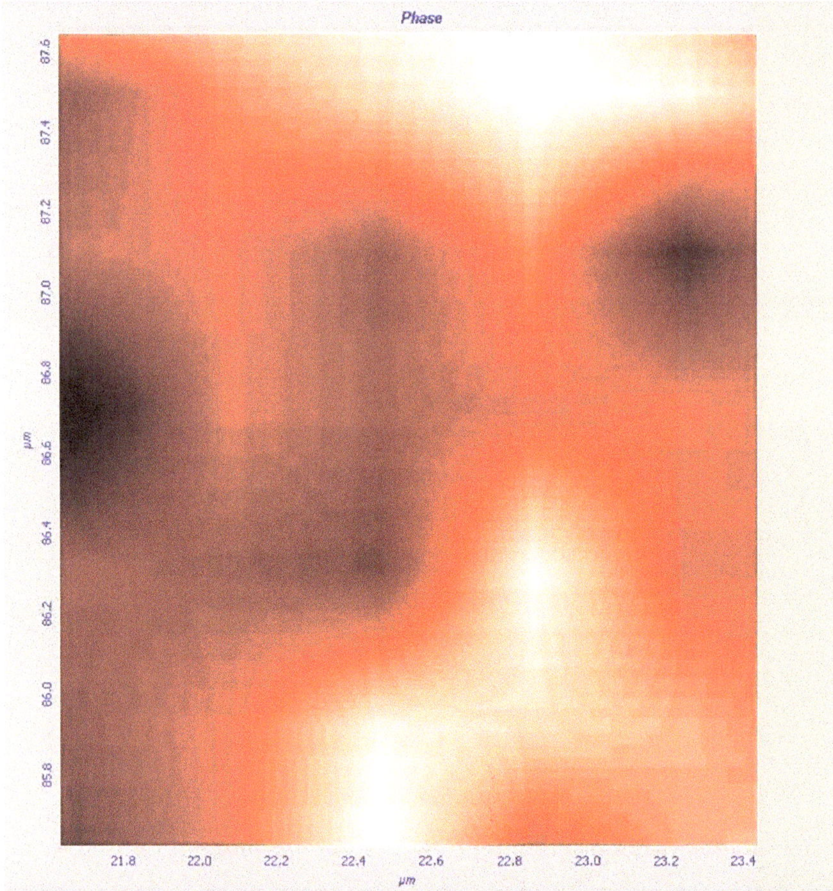

Figure 5: MFM image of NdFeB nanofibrous structures formed upon irradiation of laser at 26 MHz pulse repetition rate.

The laser pulse repetition rate plays a critical role in the formation of magnetic nanofibrous structure [26]. In order for nanoparticles to aggregate and form fibrous structure, a continuous supply of vapor is required to maintain the nucleus density of the expanding plume. Nanoparticles generated from the successive laser pulse are fused to the particles created from the previous laser pulse that are still above the melting temperature and grow as nanofibrous like structure as shown in Figure 1. As the pulse repetition rate of the femtosecond laser increases, the time between successive pulses decreases which gives less time for clusters to agglomerate and generate nanofibers with smaller diameter. It is evident from the SEM images shown in Figure 2a-d that smaller size nanofibers was generated with the increase of the laser pulse repetition rate.

Characterization was performed using XRD as a function of femtosecond laser pulse repetition rate. Figure 6 shows XRD pattern of NdFeB magnetic nanofibers generated by femtosecond laser at 26 MHz and a power of 15 W. The average nanofibers size is about 28.5 nm estimated from the XRD peaks using the Scherrer formula [25]. This value is consistent with nanofiber size obtained by TEM analysis as shown in Figure 3. In comparison, the size of nanofibers prepared using the conventional methods is around 40 nm which is slightly bigger than our method and do not have the weblike structure [27]. Figure 7 shows the XRD patterns for magnetic nanofibers generated at 4, 8, 13, and 26 MHz, respectively. For the non-irradiated area in Figure 7, no diffraction peaks indexed by the Nd2Fe14B phase were observed. However, the peaks from Nd2Fe14B phase can be observed clearly in the samples irradiated with femtosecond laser. For the area irradiated with laser at 26 MHz, the peak from α-Fe was mainly found. Therefore, it is considered that the α-Fe peak is attributed to the surface oxidation and it is existed on the surface of the sample.

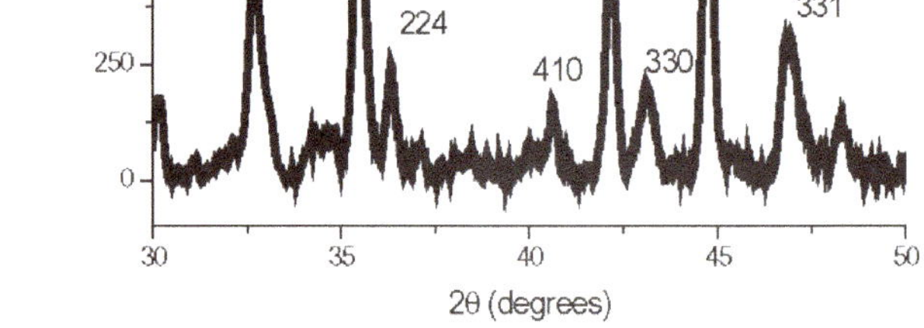

Figure 6: XRD pattern of NdFeB magnetic nanofibers generated at 26 MHz and 15 W.

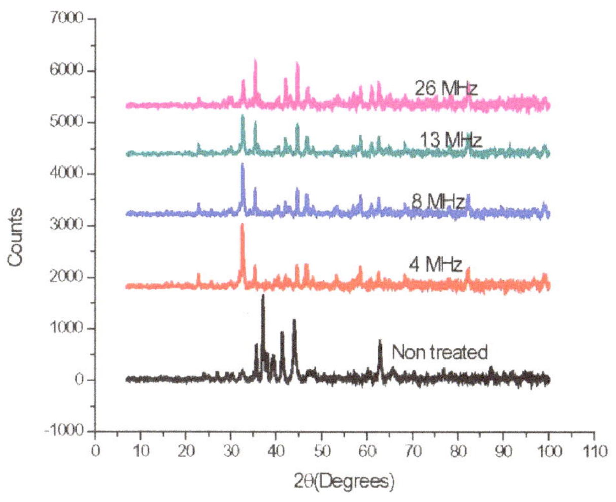

Figure 7: XRD patterns for NdFeB magnetic nanofibers generated at 4, 8, 13, and 26 MHz.

Figure 8 shows the experimental and theoretical relationship between laser pulse repetition rate and magnetic nanofibers size. The nanostructures were generated as a result of nanoparticle agglomeration. As the laser pulse repetition rate increases, the pulse to pulse duration will be shorter and hence less time for agglomeration process is available which results in smaller size fibrous nanostructure [23]. The average nanofiber size can be estimated from the Sherrer equation [28]:

$$r = \frac{0.9\lambda}{B\cos\theta} \qquad (1)$$

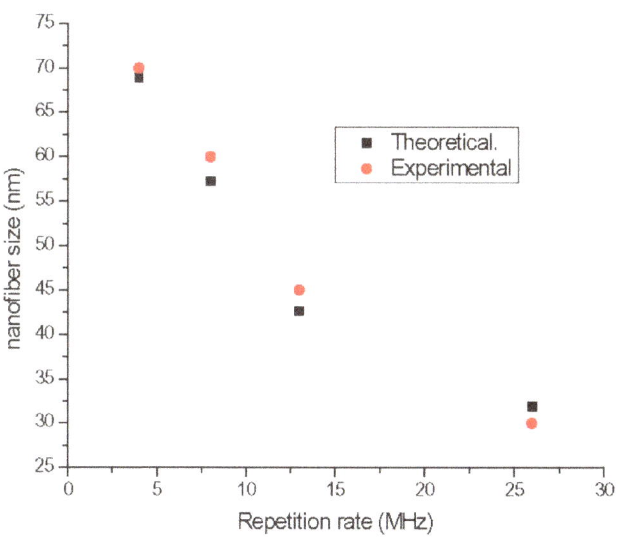

Figure 8: Theoretical and experimental magnetic nanofibers size as a function of femtosecond laser pulse repetition rate.

where r is the nanofiber size, λ is the X-ray wavelength, B is the full width at half maximum of the peak (FWHM), and θ is the diffraction angle. From the diffraction peaks in Figure 7, the average nanofiber size was estimated using the above equation and plotted in Figure 8. Those calculations are close to our experimental results as shown in the figure.

The metastable Nd-rich phase is a grain-boundary phase which has an FCC structure. This grain boundary phase exhibits a characteristic contrast which is similar to a metastable high-pressure phase observed previously as FCC γNd [29]. The structure of the phase is, however, closely related to that of NdO and it was frequently reported that oxygen content is fundamental in the formation of this phase [30]. However, oxygen-containing FCC phases as shown in Figure 4 were observed only at high temperatures. Therefore, oxygen presence is not critical for the formation of the FCC phase, although at higher temperature this phase may absorb oxygen more easily than other phases because of the high Nd content. Moreover, oxygen can probably stabilize this metastable phase and at higher temperature it can transform into the stable NdO oxide. It was noticed, however, more than three phases can coexist at a given temperature (e.g., at melting point) only if the fourth element was introduced into the ternary system, i.e., oxygen in Nd-Fe-B system [31]. The FCC phase is presumably a metastable phase with a structure close to the short-range order in the Nd-rich amorphous phase [32]. It probably forms from the undercooled substrate with lower melting point than $Nd_2Fe_{14}B$ or from the amorphous phase produced at grain boundaries during the laser ablation process.

Figure 9 shows the typical variations of magnetic strength M as a function of laser repetition rate for the NdFeB nanofibers grown at room temperature. The thickness of the generated fibers layer in all of the four pieces were the same [23], however, the morphology of the nanostructures would be changed because of the change in nanofibers size caused by the change in repetition rate. The data were for the samples measured with a Guassmeter along the in-plane direction. The figure indicates that at higher repetition rates, the M of the nanofibrous structure get lower due to the presence of an abundant amorphous phase which also shows lower coercivity. The relatively large coercivities of nanofibrous structures were due primarily to their specific morphology. Theory has predicted that a system containing magnetic dipoles that are arranged into a linear chain will exhibit an increase in coercivity [33]. Our results seemed to be consistent with this prediction as long as dipole-dipole interactions between grains played the dominant role in the magnetization process. The NdFeB grains contained in each nanofiber were actually aligned along its long axis, and the dipole-dipole interactions between grains tended to line up all magnetic dipoles along the same axis.

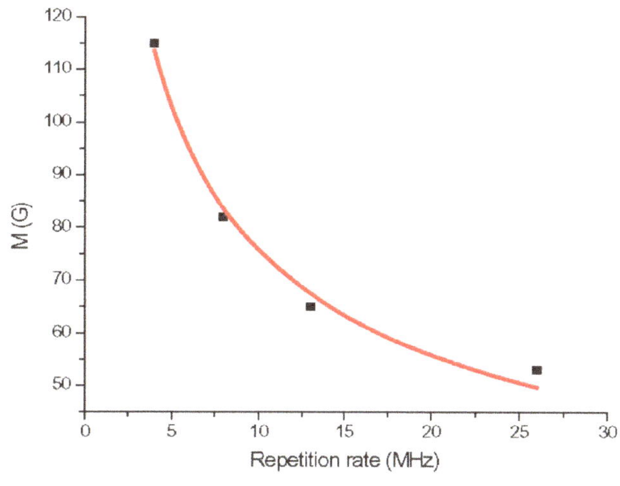

Figure 9: Magnetic strength M as a function of laser pulse repetition rate.

CONCLUSIONS

We introduced synthesis of NdFeB magnetic fibrous nanostructure and nanoparticle on bulk substrate using femtosecond laser radiation under ambient conditions. The phase structures and microstructures have been investigated using XRD, SEM and EDX analysis. The magnetic nanofibers were grown in the order of few nanometers and organized themselves in weblike structures. Magnetic nanoparticles with diameter in the order of few nanometers were attached to the nanofibrous structure. Increasing the repetition rate of the femtosecond laser results in increasing the number of pulses and hence decreases size of the generated magnetic nanofibers. Increasing repetition rate of the femtosecond laser results in generating smaller size magnetic nanofibers. The magnetic strength of the generated nanofibers can be controlled by changing the repetition rate of the femtosecond laser. These magnetic nanofibers may be utilized in many applications, such as magnetic devices, carriers, tissue engineering materials, and drug delivery.

REFERENCES

1. Lieber C: Nanoscale science and technology: building a big future from small things. *MRS Bull* 2003, 28: 486–491. 10.1557/mrs2003.144
2. Xia Y, Yang P: Chemistry and physics of nanowires. *Adv Mater* 2003, 15: 351. 10.1002/adma.200390086
3. Sander M, Prieto A, Gronsky R, Sands T, Stacy A: Fabrication of high-density, high aspect ratio, large-area bismuth telluride nanowire arrays by electrodeposition into porous anodic alumina templates. *Adv Mater* 2002, 14: 665. 10.1002/1521-4095(20020503)14:9<665::AID-ADMA665>3.0.CO;2-B
4. Thurn-Albrecht T, Schotter J, G Kastle A, EmLey N, Shibauchi T, Krusin-Elbaum L, Guarini K, Black C, Tuominen M, Russell T: Ultrahigh-density nanowire arrays grown in self-assembled diblock copolymer templates. *Science* 2000, 290: 2126–2129. 10.1126/science.290.5499.2126
5. Georgiev D, Baird R, Avrutsky I, Auner G, Newaz G: Controllable excimer-laser fabrication of conical nano-tips on silicon thin films. *Appl Phys Lett* 2004, 84: 4881. 10.1063/1.1762978
6. Korte F, Koch J, Chichkov B: Formation of microbumps and nanojets on gold targets by femtosecond laser pulses. *Appl Phys A Mater Sci Process* 2004, 79: 879–881.
7. Tan B, Venkatakrishnan K: Synthesis of fibrous nanoparticle aggregates by femtosecond laser ablation in air. *Opt Exp* 2009, 17: 1064–1069. 10.1364/OE.17.001064
8. Cui B, Wu L, Chou S: Fabrication of high aspect ratio metal nanotips by nanosecond pulse laser melting. *Nanotechnology*2008, 19: 345303. 10.1088/0957-4484/19/34/345303
9. Ovsianikov A, Jacques V, Chichkov B, Oubaha M, MacCraith B, Sakellari I, Giakoumaki A, Gray D, Vamvakaki M, Farsari M, Fotakis C: Ultra-Low shrinkage hybrid photosensitive material for two-photon polymerization microfabrication. *ACS Nano*2008, 2: 2257–2262. 10.1021/nn800451w
10. Mallick M: *Fiber-reinforced composites: materials, manufacturing, and design*. New York: Dekker; 1993.
11. Preuss S, Demchuk A, Stuke M: Sub-picosecond UV laser ablation of metals. *Appl Phys A* 1995, 61: 33–37. 10.1007/BF01538207
12. Bonse J, Rudolph P, Krueger J, Baudach S, Kautek W: Femtosecond pulse laser processing of TiN on silicon. *Appl Surf Sci*2000, 154–155: 659–663. 10.1016/S0169-4332(99)00481-X
13. Jee Y, Becker MF, Walser RM: Laser-induced damage on single-crystal metal surfaces. *J Opt Soc Am B* 1988, 5: 648.
14. Azadmanjiri J: Preparation of Mn-Zn ferrite nanoparticles from chemical sol-gel combustion method and the magnetic properties after sintering. *J Non-Cryst Solids* 2007, 353: 4170–4173. 10.1016/j.jnoncrysol.2007.06.046
15. Mathur P, Thakur A, Singh M: Effect of nanoparticles on the magnetic properties of Mn-Zn soft ferrite. *J Magn Magn Mater*2008, 320: 1364–1369. 10.1016/j.jmmm.2007.11.008
16. Nalbandian L, Delimitis A, Zaspalis V, Deliyanni E, Bakoyannakis D, Peleka E: Hydrothermally prepared nanocrystalline Mn-Zn ferrites: Synthesis and characterization. *Microporous Mesoporous Mater* 2008, 114: 465–473. 10.1016/j.micromeso.2008.01.034
17. Shultz M, Allsbrook M, Carpenter E: Control of the cation occupancies of MnZn ferrite synthesized via reverse micelles. *J Appl Phys* 2007, 101: 09M518. 10.1063/1.2713694
18. Makovec D, Košak A, Drofenik M: The preparation of MnZn-ferrite nanoparticles in water-CTAB-hexanol microemulsions.*Nanotechnology* 2004, 15: S160. 10.1088/0957-4484/15/4/009
19. Wang L, Bai J, Li Z, Cao J, Wei F, Yang Z: The influence of substrate on the magnetic properties of MnZn ferrite thin film fabricated by alternate sputtering. *Phys Status Solidi A* 2008, 205: 2453–2457. 10.1002/pssa.200723228
20. Etoh H, Sato J, Murakami Y, Takahashi A, Nakatani R: Magnetic properties of Mn-Zn ferrite thin films fabricated by pulsed laser deposition. *J Phys: Conf Ser* 2009, 165: 012031.
21. Venkatakrishnan K, Tan B, Stanley P, Lim L, Ngoi B: Femtosecond pulsed laser direct writing system. *Opt Eng* 2002, 41: 1441–1445. 10.1117/1.1476324
22. Li M, Johnson S, Guo H, Dujardin E, Mann S: A generalized mechanism for ligand-induced dipolar assembly of plasmonic gold nanoparticle chain networks. *Adv Funct Mater*
23. Mahmood A, Sivakumar M, Venkatakrishnan K, Tan B: Effect of laser parameters and assist gas on spectral response of silicon fibrous nanostructure. *Appl Phys Lett* 2009, 95: 34107. 10.1063/1.3168499
24. Harilal S, O'Shay B, Tillack M, Bindhu C, Najmabadi F: Fast photography of a laser generated plasma expanding across a transverse magnetic field. *IEEE Trans Plasma Sci* 2005, 33: 474–475.
25. Luk'yanchuk B, Marine W: On the delay time in

photoluminescence of Si-nanoclusters, produced by laser ablation. *Appl Surf Sci* 2000, 154: 314–319. 10.1016/S0169-4332(99)00472-9

26. Pereira A, Cros A, Delaporte P, Georgiou S, Manousaki A, Marine W, Sentis M: Surface nanostructuring of metals by laser irradiation: effects of pulse duration, wavelength and gas atmosphere. *Appl Phys A Mater Sci Process* 2004, 79: 1433–1437.

27. Li D, Wang Y, Xia Y: Electrospinning of polymeric and ceramic nanofibers as uniaxially aligned arrays. *Nano Lett* 2003, 3: 1167–1171. 10.1021/nl0344256

28. Cullity B: *Elements of X-Ray Diffraction*. Reading, MA: Addison-Wesley; 1956.

29. Li D, Herricks T, Xia Y: Magnetic nanofibers of nickel ferrite prepared by electrospinning. *Appl Phys Lett* 2003, 83: 4586. 10.1063/1.1630844

30. Piermarini G, Weir C: Allotropy in some rare-earth metals at high pressures. *Science* 1964, 144: 69–71. 10.1126/science.144.3614.69

31. Love B: The metallurgy of Yttrium and the rare earth metals. *Wadd Tech Rep* 1961, 37: 61.

32. Weizhong T, Shouzeng Z, Run W: On the neodymium-rich phases in Nd-Fe-B magnets. *J Less-Common Met* 1988, 141: 217–223. 10.1016/0022-5088(88)90407-9

33. Jacobs I, Bean C: An approach to elongated fine-particle magnets. *Phys Rev* 1955, 100: 1060–1067. 10.1103/PhysRev.100.1060

CHAPTER 7 FEMTOSECOND LASER PULSES

Oleg R. Pavlyniuk and Vitaly V. Datsyuk

Keywords: Electromagnetic forces, Laser-induced periodic surface structure, Surface plasmon polariton

BACKGROUND

Femtosecond laser-pulse interaction with matter can lead to formation of laser-induced periodic surface structures [1, 2, 3]. Various applications of the laser-induced periodic surface structures (LIPSSs) have been proposed. They can be used for building microfluidic channels, controlling over laser marking and changing the color of materials [4,5], overcoming the diffraction limit in laser nanomachining [6], modifying local electrical properties, improving the efficiency of solar cells [7], grating production, and optical data storage [8].

Though LIPSS was first observed in 1965 [9], the origin of the ripple structures is still debated [10]. Two types of basic formation mechanisms explained the observed structures. The first was a resonant mechanism based on (i) periodic electromagnetic-energy deposition due to roughness of the surface [11] or (ii) excitation of surface plasmon polaritons (SPPs) [1, 2, 6, 12, 13, 14]. The second non-resonant mechanism was related with thermal consequences of laser irradiation of the target, for example, capillary waves formed in the melted layer [15].

The classical resonant model [11] of the LIPSS formation assumed scattering of the incident laser wave by surface roughness and interference of the laser beam with refracted light. This interference causes inhomogeneous electromagnetic energy absorption just beneath the surface. According to this model, the period of the energy deposition can be smaller than the laser wavelength. However, initial roughness at the surface and several laser pulses are required to form LIPSS in this case [6]. In [16], the finite-difference time-domain method was applied to study the inhomogeneous energy absorption of a linearly polarized laser beam below the rough surface. The numerical results confirmed the resonant scenario. A resonant mechanism for ripple formation involved excitation of the surface electromagnetic wave by the incident laser irradiation [1, 2, 6, 11, 12, 13, 14, 15, 16, 17, 18]. According to [1, 6, 12, 14, 15, 17, 18], interference of the laser light with SPP creates periodic field distribution at the interface, which in some way leads to growth of a structure with the same period.

The SPPs can be excited at a metal surface. Therefore, LIPSS-covered metals were explained by SPP excitation [13, 14]. In addition, LIPSSs were observed at surfaces of semiconductors [1, 2, 12, 15, 17] and dielectrics [2, 6, 12]. Therefore, the possibilities and conditions of SPP excitation in semiconductor and dielectric materials were studied in [1, 2, 6, 12, 15].

Among the non-resonant mechanisms, it can be mentioned a theoretical model comprising heat transfer and hydrodynamic components [15]. The heat transfer component described particle dynamics, carrier excitation, and heat conduction phenomena. The hydrodynamic component described the molten material dynamics and the process of capillary wave solidification.

It should be noted that the timescale of the femtosecond laser pulses and SPPs ($\simeq$fs) is significantly less than the expected timescale of material movement ($\simeq$ps) [18] in the processes described above. Moreover, some authors [1,17, 18] have demonstrated the possibility of the LIPSS creation just with a single femtosecond pulse. All these facts allow us to assume that mechanical action of the electromagnetic field can play an important role in the ripple formation. Therefore, the goal of our study is to determine the electromagnetic forces exerted on the material surface allowing for interference of the incident electromagnetic field with the excited SPP. We define a distribution of the electrically induced normal pressure and tangential stress at metal surface and estimate their amplitudes. In addition, we evaluate a density of the volume forces inside the surface layer. Then, we check correspondence between our calculations and known characteristics of the ripple structures.

METHODS

We applied the classical electromagnetic theory both to determine the properties of the SPPs and to calculate the force

Citation: Oleg R. Pavlyniuk and Vitaly V. Datsyuk, Laser Pulses, doi: 10.1186/s11671-015-1224-5.

acting on a metal surface. For simplicity, we consider the interaction of a femtosecond laser pulse with a nonmagnetic metal ($\mu_2=1$ at z>0) in air ($\varepsilon_1=1$, $\mu_1=1$ at z<0).

Surface Plasmon Polaritons

In most cases, the ripples were observed to be perpendicular to the electric field vector [1, 2, 3, 6, 11, 12, 13, 14]. Therefore, their origin was attributed to SPPs excited by a TM polarized mode [1, 3, 6, 7, 12, 14, 15, 17, 18].

The properties of the SPPs are described by the classical electromagnetic theory (see, for example, [19, 20, 21]). Namely, the fields that decrease exponentially with increasing distance from the metal-dielectric interface are found from the Maxwell's equations (see Appendix 1 and Fig. 1). The dispersion relation for the SPP propagation constant k_p follows from the Maxwell's boundary conditions giving

$$k_{\mathrm{p}} = k_0 \left(\frac{\epsilon_1 \epsilon_2}{\epsilon_1 + \epsilon_2} \right)^{1/2}, \tag{1}$$

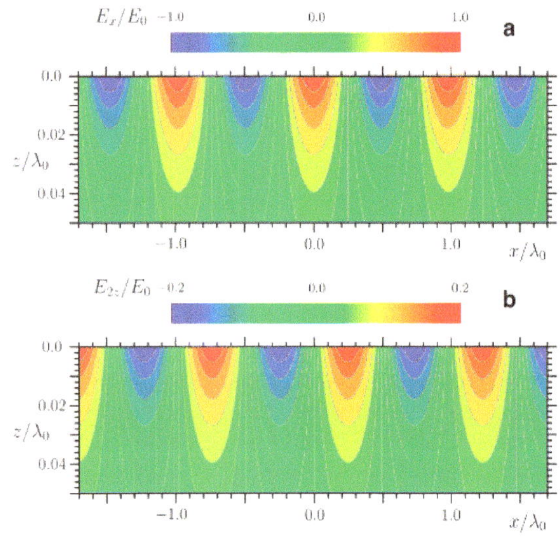

Figure 1: Electric field distribution of the SPP on copper surface. Calculated distribution of **a** x-component and **b** z-component of the electric field of the SPP excited at the air-copper interface at a frequency corresponding to a free-space wavelength of $\lambda_0=800$ nm

where $k_0=\omega/c$ is the vacuum wave number, c the speed of light in vacuum, and ε_1 and ε_2 are the permittivities of the media at frequency ω.

According to Eq. 1 and conditions of SPP existence (one of dielectric constants is negative with an absolute value exceeding that of another) [19, 20], k_p is always larger than k_0. Therefore, the SPP can be excited if there are nanoparticles at the surface and a localized plasmon excited around it [20]. The required nanoparticles can be created by electromagnetic forces acting on nanobubbles under the surface of metal [22].

Maxwell Stress

A force acting on a dielectric body was first defined in Maxwell's treatise on electricity and magnetism [23]. The time-averaged force can be expressed through the integral over a closed surface surrounding the body

$$\mathbf{F} = \oint \langle \mathbf{T} \rangle \cdot \mathbf{n}\, \mathrm{d}A, \tag{2}$$

where **n** is the external unit vector normal to the surface, **T** the Maxwell stress tensor, and brackets denotes averaging over time. Among several forms of the Maxwell stress tensor, we chose one introduced by Lorentz

$$\mathbf{T} = \epsilon_0 \mathbf{E} \otimes \mathbf{E} + \mu_0 \mathbf{H} \otimes \mathbf{H} - \frac{1}{2}\left(\epsilon_0 E^2 + \mu_0 H^2\right) \mathbf{I}, \tag{3}$$

where ε_0,μ_0 are the permittivity and permeability of vacuum, ⊗ denotes the outer product, and **I** is the unit matrix. The Maxwell-Lorentz tensor takes into account electrostriction and agrees with experiments better than other stress tensors [24].

Since the normal component of the electric field is discontinuous at the boundary between the media, the scalar product of **T** with **n** has a jump. The difference between ⟨**T**⟩ ·**n** in two media gives the surface force FFFF which is measured in pascals. In order to relate the z-component of FFFF with the normal pressure P on metal surface, we use the following definition:

$$\mathcal{F} = P\hat{\mathbf{z}} + \mathbf{S} = \left[\langle \mathbf{T}^{(2)} \rangle - \langle \mathbf{T}^{(1)} \rangle\right] \cdot \hat{\mathbf{z}}, \tag{4}$$

where $\hat{\mathbf{z}}$ is a unit vector in the direction of the z-axis; the tangential component of FFFF is called the tangential stress **S**. Formulas for calculating P and **S** are given in Appendix 2.

Application of the Gauss' theorem transforms the right hand side of Eq. 2 into integral over the volume of the body

$$\mathbf{F} = \int \nabla \cdot \langle \mathbf{T} \rangle\, \mathrm{d}V. \tag{5}$$

The quantity under the integral sign is measured in newton per cubic meters and can be considered as the volume density of the electromagnetic force

$$\mathbf{f} = \nabla \cdot \langle \mathbf{T} \rangle. \tag{6}$$

RESULTS AND DISCUSSION

The normal pressure and tangential stress can be determined through the electric field **E** just beneath the surface (see Appendix 2). This field is a superposition of the transmitted electric field and the electric field of the surface plasmon wave $\mathbf{E}_2$. The normal pressure is expressed through the square of the normal component of **E** averaged over time. Here, we consider the incidence of a plane electromagnetic wave with the magnetic field parallel to the interface z=0, i.e., the case of p-polarization. In this case, the normal component of the electric field is nonzero and $E_n=E_z$. We neglect the reflected wave and took into account that the frequency ω of the incident wave is close to the plasmon frequency ω_s (usually, the central wavelength of the femtosecond laser is at 800 nm). Under these assumptions, we obtain

$$\langle E_n^2 \rangle = \frac{1}{2}\left(\left|\frac{\epsilon_1}{\epsilon_2}\right|^2 |E_{1z}|^2 + |E_{2z}|^2 \right) + \left|\frac{\epsilon_1}{\epsilon_2}\right| |E_{1z}|\,|E_{2z}| \cos[(k_{\mathrm{p}} - k_{1x})x], \tag{7}$$

where E_{iz} and k_{ix} are the z-component of the electric field and the x projection of the wavevector **k** of the incident wave, respectively; E_{2z} is the z-component of the SPP field defined by Eq. 19. Unlike the normal pressure, the tangential stress depends on both the tangential and normal components of the electric field. Therefore, we calculate

$$\begin{aligned}\langle E_t E_n\rangle = & -\frac{1}{2}\left|\frac{\epsilon_1}{\epsilon_2}\right| |E_{ix}|\,|E_{iz}| \\ & +\frac{1}{2}\left|\frac{\epsilon_1}{\epsilon_2}\right| |E_x|\,|E_{iz}|\sin[(k_p-k_{ix})x] \\ & +\frac{1}{2}|E_{ix}|\,|E_{2z}|\cos[(k_p-k_{ix})x]\,.\end{aligned} \quad (8)$$

If θ is the incident angle, then $k_{ix}=k_0 \sin\theta$ and

$$k_p - k_{ix} = \frac{2\pi}{\lambda_0}\left(\frac{\lambda_0}{\lambda_s} - \sin\theta\right), \quad (9)$$

where $\lambda_0=2\pi/k_0$ is the wavelength of the incident wave and

$$\lambda_s = \lambda_0\left(\frac{\epsilon_1\epsilon_2}{\epsilon_1+\epsilon_2}\right)^{-1/2} \quad (10)$$

is the wavelength of the SPP.

Equations 7–9 show that the period of the normal pressure and tangential stress is equal to

$$\Lambda = \frac{\lambda_0}{\frac{\lambda_0}{\lambda_s} \pm \sin\theta}, \quad (11)$$

where + corresponds to backward-propagating SPPs. The found periodicity is the same as the periodicity of the LIPSSs measured and estimated by a number of authors [6, 11, 12, 14, 15, 17, 18].

By using the formulas for P and S of Appendix 2 and Eqs. 7 and 8, we evaluated the magnitude of the surface force. All calculations were performed using the permittivity copper equal to $\varepsilon_2=-25.07+2.54$ i at a wavelength of 800 nm. According to [1, 6, 15, 17], the amplitude of the electric field of the SPP can be of the order of the electric field of the incident wave. We took

$$E_x = E_i \quad (12)$$

and expressed E_i through the laser parameters

$$\frac{\Phi}{\tau} = \frac{\epsilon_0 c}{2}\langle|\mathbf{E}_i|^2\rangle, \quad (13)$$

where Φ is the fluence of the laser pulses and τ is the pulse duration. Lasers with τ=140–150 fs, λ_0=800 nm and Φ in a range of 0.2−1.5 J/cm^2 are commonly used to generate LIPSSs. In numeric estimates, we used τ=150 fs and Φ=1 J/cm^2 giving

$$P_0 = \frac{\Phi}{\tau c} \simeq 0.2 \text{ GPa}. \quad (14)$$

Textbooks in optics identify P_0 as the pressure of light on a perfectly absorbing medium with zero reflection coefficient [20]. If the reflection coefficient is equal to unity, the pressure would be equal to 2 P_0. If ε_1=1, light always pushes the absorbing medium exerting positive pressure, according to the common understanding of the radiation pressure. In this letter, we applied the Maxwell-Lorentz stress tensor to define the pressure P acting on a surface layer of a solid. In the case of normal incidence of light without plasmon excitation, one gets P=0 since E_n=0. The normal component of the electric field is not zero if SPPs are generated. Using Eq. 12, we got P$\simeq$−12 P_0 and a periodic S_t varying with a period $\Lambda=\lambda_s\simeq$770 nm and amplitude 5 P_0 at the surface of copper.

A picture of the light-matter interaction to be more complete if one calculates the volume force **f**. Thus, we found a distribution of **f** in copper for the case of normal incidence of a plane electromagnetic wave following by SPP excitation. The result of the calculation is presented in Fig. 2 which represents **f** in units of $f_0 = \frac{1}{2}\epsilon_0|\mathbf{E}_i|^2$ ×109 m^{-1}. The integral $\int_0^\infty f dz$ gives as the pressure on a tin semi-infinite cylinder with the axis coincident with the z axis. The value of this pressure was found to be 50 P_0. The volume force is distributed in a surface layer with a depth of about 25 nm. Our calculation thus show that the magnitudes of the surface and volume pressures can be as large as a few gigapascals. Such pressures should result in inelastic deformation of copper according to [25]. Figure 2 implies that a depth of the grooves created by a single femtosecond pulse could be about 25 nm.

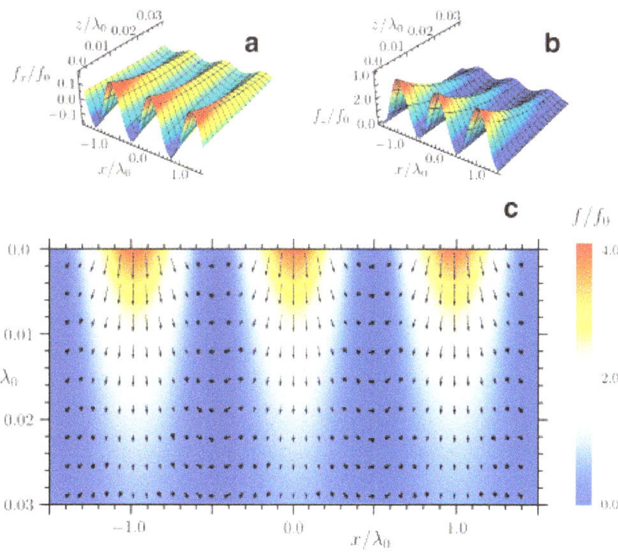

Figure 2: Distribution of the normalized volume electromagnetic forces inside copper. Calculated distribution of **a** x-component, **b** z-component, and **c** field of the volume force f inside copper in the case of normal incidence of a laser beam with the free-space wavelength λ_0=800 nm

We also established the period of the volume forces in the x direction. It is equal to 770 nm that is less than λ_0=800 nm. The former number coincides with the period λ_s of the tangential stress found from Eq. 11 for the normally incident wave when θ=0.

CONCLUSIONS

In this letter, we have proposed an electrostrictive mechanism of the LIPSS formation. We have calculated the electrically induced normal pressure and tangential stress exerted on the air-copper interface under interference of the incident laser field with the excited SPP. In addition, we determined the electrostrictive volume forces to a metal surface layer. The interfacial stress and the pressure on the surface layer have been found to reach a few gigapascals which is sufficient for copper surface modification. These calculations demonstrate that electromagnetic forces could be very important in the process of the LIPSS formation. The predicted period of the surface force distribution is in good agreement with the experimentally observed LIPSS periods and those given by other theoretical models. In the framework of the proposed theory, the depth of the grooves created with a single femtosecond laser shot has been also estimated.

APPENDICES

Appendix 1. Electric Field of Surface Plasmon Polariton

SPP is an electromagnetic excitation that may exist at the interface of two media with dielectric constants of opposite signs (for instance, dielectric and metal) [19, 20]. The electric and magnetic fields satisfy the Maxwell's equations and ordinary Maxwell's boundary conditions. The amplitudes of the fields decrease exponentially with increasing distance from the interface in both media. The electric field **E** has two components: E_x along the propagation direction of the wave and E_z perpendicular to the surface; the magnetic field **H** has only y-component.

The distribution of each component A of the electric and magnetic fields in a surface wave, propagating along the x-axis, has the following form:

$$A = A_0 \exp(\pm k_{1,2} z) \exp[\, i(k_p x - \omega_s t)], \tag{15}$$

where A_0 is the amplitude, k_p is the propagation constant of the SPP, $k_1>0$ and $k_2>0$ are the damping factors of the surface electromagnetic wave in mediums 1 and 2, respectively, and t is the time. The upper "+" sign refers to medium 1 where z<0 and the lower "–" sign refers to medium 2 where z>0. The coefficients k_1 and k_2 are defined as

$$k_1 = k_p \left(-\frac{\epsilon_1}{\epsilon_2}\right)^{1/2}, \; k_2 = k_p \left(-\frac{\epsilon_2}{\epsilon_1}\right)^{1/2}. \tag{16}$$

For a given amplitude H of the magnetic field, it is possible to express the amplitudes of the electric fields as follows:

$$E_x = \frac{i}{[-\epsilon_1 - \epsilon_2]^{1/2}} Z_0 H, \tag{17}$$

$$E_{1z} = -\left[\frac{\epsilon_2}{\epsilon_1 (\epsilon_1 + \epsilon_2)}\right]^{1/2} Z_0 H, \tag{18}$$

$$E_{2z} = \left[\frac{\epsilon_1}{\epsilon_2 (\epsilon_1 + \epsilon_2)}\right]^{1/2} Z_0 H, \tag{19}$$

where $Z_0 = (\mu_0/\epsilon_0)^{1/2}$.

Appendix 2. Normal Pressure and Tangential Stress

The components P and **S** of the surface force FFFF defined by Eq. 4 are readily found using the boundary conditions for the tangential components of **E** and **H** and normal components of **D** and **B**. Using the Maxwell-Lorentz stress tensor of Eq. 3, we obtain

$$P = \frac{\epsilon_0}{2}\left(1 - \left|\frac{\epsilon_2}{\epsilon_1}\right|^2\right)\langle E_n^2\rangle + \frac{\mu_0}{2}\left(1 - \left|\frac{\mu_2}{\mu_1}\right|^2\right)\langle H_n^2\rangle, \tag{20}$$

$$S_t = \epsilon_0 \left\langle \left(1 - \frac{\epsilon_2}{\epsilon_1}\right) E_t E_n \right\rangle + \mu_0 \left\langle \left(1 - \frac{\mu_2}{\mu_1}\right) H_t H_n \right\rangle. \tag{21}$$

In these equations, the the values of P and **S** are expressed through the electric field in medium 2. In the case of $\mu_1=\mu_2$, the surface force is independent of the magnetic field.

DECLARATIONS

Acknowledgements

The authors thank Saulius Juodkazis and M. G. Zubrilin for a bulk of experimental data on LIPSSs and discussions of the mechanisms of their formations.

REFERENCES

1. Bonse J, Rosenfeld A, Krüger J (2009) On the role of surface plasmon polaritons in the formation of laser-induced periodic surface structures upon irradiation of silicon by femtosecond laser pulses. J Appl Phys 106: 104910.
2. Bonse J, Krüger J, Höhm S, Rosenfeld A (2012) Femtosecond laser-induced periodic surface structure. J Laser Appl 24(4): 042006.
3. Bonse J, Höhm S, Rosenfeld A, Krüger J (2012) Sub-100-nm laser-induced periodic surface structures upon irradiation of titanium by Ti:sapphire femtosecond laser pulses in air. Appl Phys A 110: 547–551.
4. Vorobyev AY, Guo C (2008) Colorizing metals with femtosecond laser pulses. Appl Phys Lett 92: 041914.
5. Vorobyev AY, Guo C (2008) Femtosecond laser blackening of platinum. J Appl Phys 104: 053516.
6. Derrien TJ-Y, Itina TE, Torres R, Saints M (2013) Possible surface plasmon polariton excitation under femtosecond laser irradiation of silicon. J Appl Phys 114: 083104.
7. Mahmood AS, Venkatakrishnan K, Tan B (2013) 3-D aluminium nanostructure with microhole array synthesized by femtosecond laser radiation for enhanced light extinction. Nanoscale Res Lett 8: 477.
8. Mansuripur M, Zakharian AR, Lesuffleur A, Oh SH, Jones RJ, Lindquist NC, Im H, Kobyakov A, Moloney JV (2009) Plasmonic nano-structures for optical data storage. Opt Express 17(16): 14001–14014.
9. Birnbaum M (1965) Semiconductor surface damage produced by Ruby laser. J Appl Phys 36: 3688–3689.
10. Reif J, Varlamova O, Uhlig S, Varlamov S, Bestehorn M (2014) On the physics of self-organized nanostructure formation upon femtosecond laser ablation. Appl Phys A

117: 179–184.

11. Sipe JE, Young JF, Preston JS, van Driel HM (1983) Laser-induced periodic surface structure. I, Theory. Phys Rev B 27(2): 1141–1154.
12. Huang M, Zhao F, Cheng Y, Xu N, Xu Z (2009) Origin of laser-induced near-subwavelength ripples: interference between surface plasmon and incident laser. ACS Nano 3(12): 4062–4070.
13. Sakabe S, Hashida M, Tokita S, Namba S, Okamuro K (2009) Mechanism for self-formation of periodic grating structure on a metal surface by a femtosecond laser pulse. Phys Rev B 79: 033409.
14. Garrelie F, Colombier JP, Pigeon F, Tonchev S, Faure N, Bounhalli M, Reynaud S, Parriaux O (2011) Evidence of surface plasmon resonance in ultrafast laser-induced ripples. Opt Express 19(10): 9035–9043.
15. Tsibidis GD, Barberoglou M, Loukakos PA, Stratakis E, Fotakis C (2012) Dynamics of ripple formation on silicon surfaces by ultrashort laser pulses in subablation conditions. Phys Rev B 86: 115316.
16. Skolski JZP, Römer GRBE, Obona JV, Huis in't Veld AJ, Hosson JTMD (2012) Laser-induced periodic surface structures: fingerprints of light localization. Phys Rev B 85: 075320.
17. Tsibidis GD, Stratakis E, Aifantis KE (2012) Thermoplastic deformation of silicon surfaces induced by ultrashort pulsed lasers in submelting conditions. J Appl Phys 111: 053502.
18. Kafka KRP, Austin DR, Li H, Yi A, Cheng J, Chowdhury EA (2015) Time-resolved measurement of single pulse femtosecond laser-induced periodic surface structure formation induced by a pre-fabricated surface groove. Opt Express 23(15): 19432–19441.
19. Bonch-Bruevich AM, Libenson MN, Makin VS, Trubaev VV, Vavilov SI (1992) Surface electromagnetic waves in optics. Opt Engin 31(4): 718–730.
20. Novotny J, Hecht B (2006) Principles of nano-optics. Cambridge University Press, New York.
21. Zakharian AR, Moloney JV, Mansuripur M (2007) Surface plasmon polaritons on metallic surfaces. Opt Express 15(1): 183–197.
22. Buividas R, Rosa L, Šliupas R, Kudrius T, Šlekys G, Datsyuk V, Juodkazis S (2011) Mechanism of fine ripple formation on surface of (semi)transparent materials via a half-wavelength cavity feedback. Nanotechnol 22: 055304.
23. Maxwell JC (1873) A treatise on electricity and magnetism. Vol. II. Clarendon Press, Oxford.
24. Datsyuk VV, Pavlyniuk OR (2015) Maxwell stress on a small dielectric sphere in a dielectric. Phys Rev A 91(2): 023826.
25. Hebesberger T, Stüve HP, Vorhauer A, Wetscher F, Pippan R (2005) Structure of Cu deformed by high pressure torsion. Acta Mater 53: 393–402.

CHAPTER 8

LASER-DRIVEN MAGNETIC FIELDS

J J Santos, M Bailly-Grandvaux, L Giuffrida, P Forestier-Colleoni, S Fujioka, Z Zhang, P Korneev, R Bouillaud, S Dorard, D Batani, M Chevrot, J E Cross, R Crowston, J-L Dubois, J Gazave, G Gregori, E d'Humières, S Hulin, K Ishihara, S Kojima, E Loyez, J-R Marquès, A Morace, P Nicolaï, O Peyrusse, A Poyé, D Raffestin, J Ribolzi, M Roth, G Schaumann, F Serres, V T Tikhonchuk, P Vacar and N Woolsey

Keywords: Strong magnetic field, laser-driven coil targets, laser-plasma interaction, B-dot probing, faraday rotation, proton-deflectometry, plasma magnetization

INTRODUCTION

The properties of matter on all scales (atoms, molecules, condensed matter, plasmas) are severely modified when exposed to strong magnetic fields (*B*-fields) [1]. The possibility of imposing a strong, laser-driven *B*-field to a variety of samples opens interesting perspectives for laboratory studies of magnetized plasma-[2], atomic- [3] and nuclear-physics [4]. We foresee great progress on the understanding of systems of astrophysical scale [5–10], on the improvement of inertial fusion energy schemes [11–14] and for various applications of magnetically-guided particle beams [15, 16], among many other applications.

State-of-the-art magnets nowadays allow generation of *B*-fields in the 10–300 T range, depending on their static/pulsed or destructive/non-destructive character [17]. However, reaching, or exceeding the 1 kT level, as required for some high-energy-density or atomic physics applications, is much more challenging, unless resorting to large-scale *Z*-pinch machines [18, 19] or explosive experiments [20]. Likewise, the heavy technical and infrastructure constraints posed by high-performance pulsed magnets (exceeding 100 T) make them ill-suited to the compactness of laser experiments. This motivates the development of portable, all-optical magnetic generators that may be easily implemented in any high-energy and/or high-power laser facility. Relativistic laser interaction with dense targets and the issuing intense currents over the target surface or into the target bulk can generate super-strong *B*-fields [21, 22], on the range of 10kT for current short-pulse laser parameters, but these are rather transient as they evolve on the time-scale of ~10 ps. Quasi-static magnetic-field production coupled to laser facilities has been explored so far by the development of capacitor-bank pulsed discharges in solenoids (magnetic pulsers) [23, 24], but the specific physical limitations restrain the maximum generated fields to the range of~40 T (lowered to a more safe level~20 T in usual operation) [8, 25].

Instead, the use of powerful lasers interacting with so-called capacitor-coil targets—first proposed by Daido *et al* back in the 1980s [26, 27] and recently explored to higher levels by Fujioka *et al* [28]—give unprecedented quasi-static (time-scale of a few ns) high *B*-field amplitudes for such a compact system (~mm) and an energy laser pulse driver of 1 kJ: they reported a *B*-field of ≈1.5 kT at 0.65 mm away from the U-turn coil centre, measured by Faraday rotation of the polarization of a probe laser beam in a SiO_2 sample. But the reported value would yield, according to a magnetostatic simulation of the U-shaped capacitor-coil target [29], a non-realistic magnetic energy, greater than the invested laser energy. The problem probably lies in that the tabulated Verdet constants of the birefringent crystals are of questionable validity in the presence of strong and rapidly changing *B*-fields, or maybe the crystal properties begin to be affected by x-rays and fast particles due to the close laser–target interaction.

Besides reaching high *B*-field strengths, advantages of laser-driven coil-targets is that they can be of relatively low cost if mass produced and are adaptable to laser sources with

Citation: J J Santos, M Bailly-Grandvaux, L Giuffrida, P Forestier-Colleoni, S Fujioka, Z Zhang, P Korneev, R Bouillaud, S Dorard, D Batani, M Chevrot, J E Cross, R Crowston, J-L Dubois, J Gazave, G Gregori, E d'Humières, S Hulin, K Ishihara, S Kojima, E Loyez, J-R Marquès, A Morace, P Nicolaï, O Peyrusse, A Poyé, D Raffestin, J Ribolzi, M Roth, G Schaumann, F Serres, V T Tikhonchuk, P Vacar and NWoolsey, Laser-driven platform magnetic fields, doi:10.1088/1367-2630/17/8/083051.

higher repetition rate. Most importantly these coils have an open geometry providing easy access for several diagnostic views and to the magnetization of secondary samples eventually driven by secondary laser or particle beams. In this study, we present an accurate characterization of the *B*-field produced by laser-driven capacitor-coil targets, using (i) high-frequency pickup B-dot probes, (ii) Faraday rotation of laser probe beam polarization (we placed the birefringent crystals sufficiently away from the coil to measure *B*-fields weaker than those reported by Fujioka *et al* [28], but using much more sensitive crystals) and (iii) deflectometry of an energetic proton beam. This combination of three independent diagnostics confirms that high power laser facilities may be used for *B*-field production in the range of several hundred Tesla with a controlled and reproducible rise time, peak value and spatial distribution.

METHODS AND EXPERIMENTAL RESULTS

The experiments were conducted at the LULI pico 2000 laser facility with a $1.057\ \mu m$ wavelength ($1\omega_0$), 500 ± 30 J, 1 ns flat-top long-pulse laser beam (≈ 100 ps rise time), focused to intensities of $(1.0 \pm 0.1) \times 10^{17}\ \mathrm{W\,cm^{-2}}$. The targets were made of two parallel disks (3500 diameter, $50\ \mu m$ thickness, with a hole in the front one enabling focusing of the laser pulse into the rear disk's surface), connected by a coil-shaped wire (coil radius $a = 250\ \mu m$, squared rod section of $50\ \mu m \times 50\ \mu m$): see figure 1. The targets were made of Cu, Ni or Al (1st experiment, with a $1750\ \mu m$-diameter hole in the front disk), or exclusively of Ni (2nd experiment, with a $1000\ \mu m$-diameter hole). The target parameters were reproducible within a $\pm 1\ \mu m$ precision thanks to accurate laser cutting of an initial 2D metallic form, the only effectively variable parameter being the distance between the disks, $d_0 = 900 \pm 200\ \mu m$, as a consequence of the manual target folding.

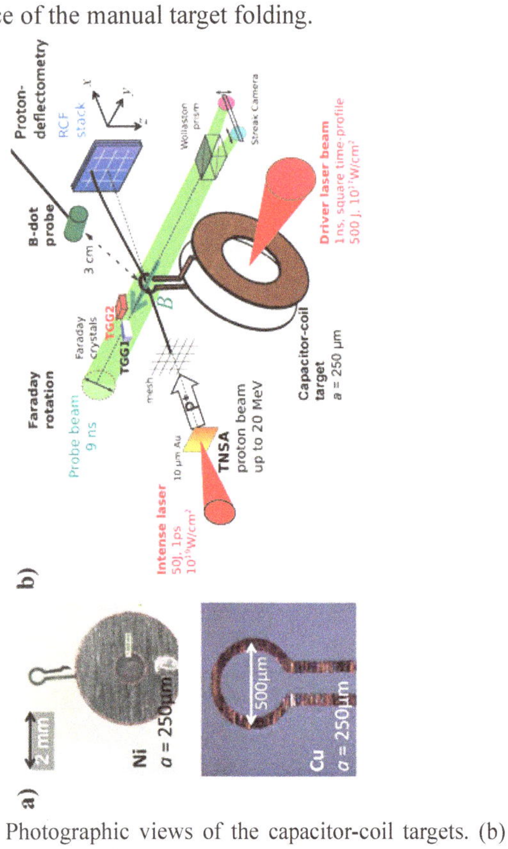

Figure 1: (a) Photographic views of the capacitor-coil targets. (b) Sketch of the experimental setup.

The laser pulse irradiates the rear disk passing through the hole of the front disk and creates the supra-thermal electrons that are escaping the potential barrier [30, 31]. A fraction of them are captured by the opposite disk. The target reacts like an RL-circuit to the potential difference between the disks and the subsequent discharge current through the coil-shaped wire. The laser-driven target charging and the discharge current process simultaneously (steady-state regime), during the laser-pulse irradiation, establishing a quasi-static current I looping in the space between the disks and the connecting wire. The coil, of radius a, concentrates the magnetic flux yielding a quasi-static, dipole-like *B*-field over a time-scale of a few ns. The amplitude of the *B*-field near the coil center scales like $B_0 \approx \mu_0 I/2a$, where μ_0 is the vacuum permeability. The target geometry, namely the distance between the disks, defines the short-circuiting time of the system, which is of the order of 1–2 ns: It happens when the plasma plume ejected from the irradiated disk reaches the opposite disk or when thermally expanding conducting wires start to overlap. This time is likely to be reduced by the front disk thermal expansion induced by x-ray irradiation from the rear disk plasma.

The experimental setup is sketched in figure 1(b).

B-dot probing

The B-dot probe axis was positioned parallel to the target's coil axis, either in the plane of the coil at a 30 ± 0.5 mm distance from the coil center and at a 33° angle above equator (1st experiment), or approximately along the coil axis at 70 ± 1 mm distance from the coil center, with a 10 ± 0.5 mm horizontal offset to clear the path for the probe beam used for Faraday rotation and shadowgraph (2nd experiment). The B-dot probe (and associated electronics) has a 2.5 GHz acquisition bandwidth, and the chosen sampling frequency yielded signals with time resolution of ≈ 50 and ≈ 10 ps respectively in the 1st and 2nd experiments.

Figure 2 details the analysis of a signal obtained in the 1st experiment with a Cu target: (a) a raw detected signal (attenuators excluded), proportional to the time-derivative of the *B*-field at the probe position, (b) the signal corresponding spectrum, calculated by the fast-Fourier-transform (FFT) of the signal (thick red curve; the thin black curve is the FFT of the signal for t<-50ps , that is the noise spectrum).

Figures 2(c) and (d) show the temporal evolution of the *B*-field, on the level of a few mT. This was obtained by integration of the signal in panel (a) for different configurations of bandpass filtering: the optimized result, represented by the thicker red curves, was obtained with the minimum frequency of $\nu_{min} = 1$ MHz to cut the dc component and $1/\nu$ noise, and a maximum frequency of $\nu_{max} = 1.5$ GHz to cut the high frequency parasitic EMP emission from higher frequency ground discharge currents [30, 31] (1st experiment; we used $\nu_{max} = 1$GHz for the 2nd experiment data, according to the reached spectral resolution). Thinner curves correspond to variations either on ν_{min} (panel (c)) or on ν_{max} (panel (d)) testing the result sensibility to these parameters: raising of ν_{min} yields an offset of the *B*-field at $t = 0$ (start of laser irradiation), yet the amplitude of the signal is consistent for ν_{min} up to 20 MHz. For fixed $\nu_{min} = 1$ MHz, the *B*-field rise-time and maximum yield do not significantly change for ν_{max} down to 1 GHz. Therefore, the large spectral peak at a few hundreds MHz in panel (b) corresponds to the main B-field signal due to the target discharge through the coil-shaped wire, the second large peak around 1 GHz determining the second B-field peak oscillation at $t \gtrsim 2$ ns (see other measurements in figure 4).

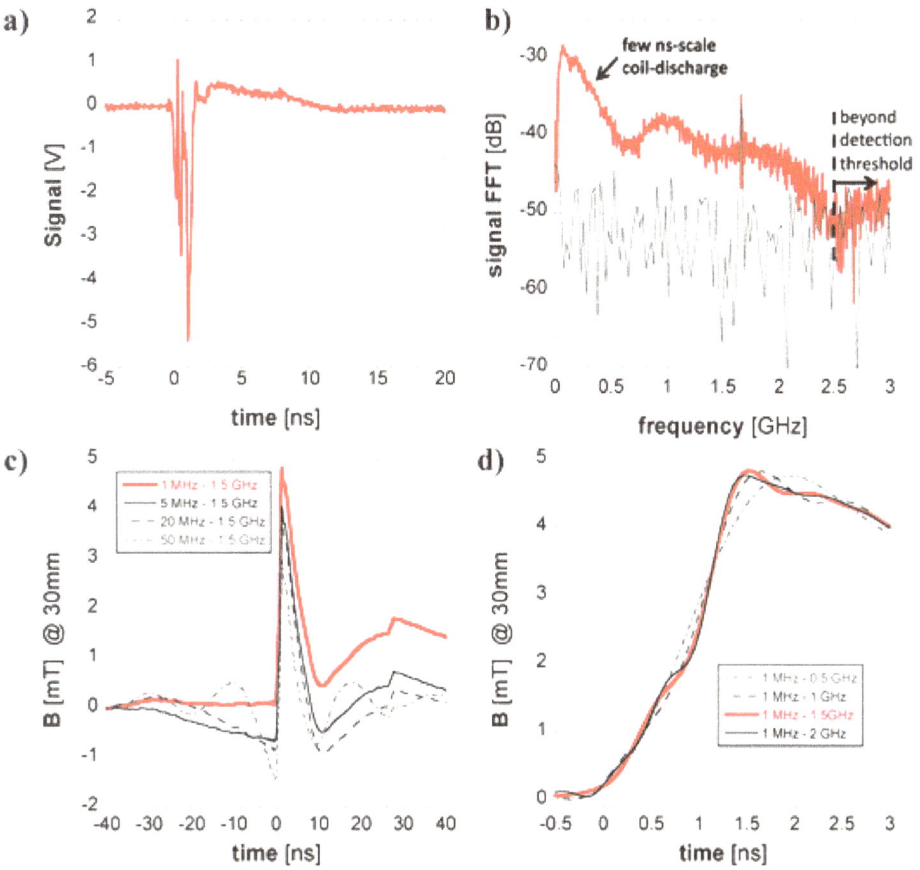

Figure 2: Sample B-dot signal and analysis obtained on the 1st experiment, for a Cu-$a = 250\ \mu m$ target, with the probe at $30\ mm$ from the coil centre. (a) Signal measured by the 2.5GHz bandwidth detection system. (b) FFT of the signal (thicker red curve). The thin black curve is the spectrum noise evaluated from the signal FFT for $t < -50$ ps. (c), (d) B-field signal after integration of the B-dot signal: thicker red curves for the default frequency bandpass from $\nu_{min} = 1$ MHz to $\nu_{max} = 1.5$ GHz, thinner black curves for test variations on (c) ν_{min} and (d) ν_{max}.

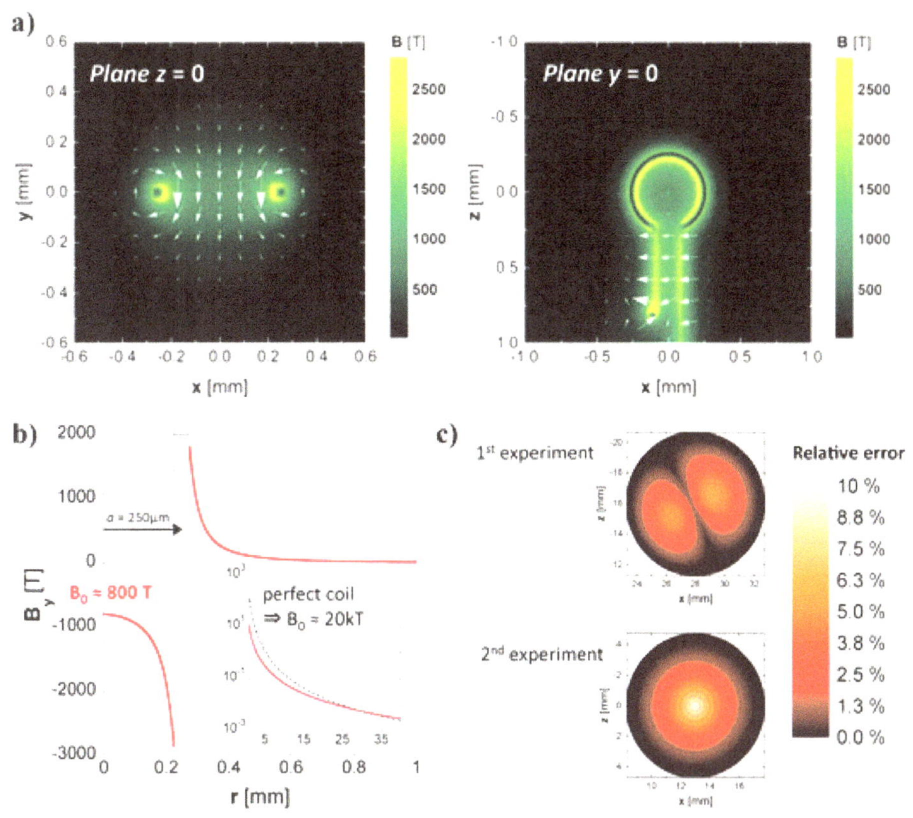

Figure 3: (a) 2D projections of the 3D Radia magnetostatic code extrapolation of the B-field produced by a $a = 250\ \mu m$ capacitor-coil target

for a circulating current I=340 kA: plane $z = 0$ on the left, plane $y = 0$ on the right (the coil centre is at the frame origin). (b) Corresponding axial component of the *B*-field against distance from the coil centre, along the line between coil centre and the B-dot probe position for the 1nd experiment, for a capacitor-coil target (solid red curve) and for a perfect coil of the same radius *a* (dashed black curve), for the same current. (c) Relative uncertainty of the *B*-field measurements by the B-dot probes at their positions, respectively in the 1st (top) and the 2nd (bottom) experiments.

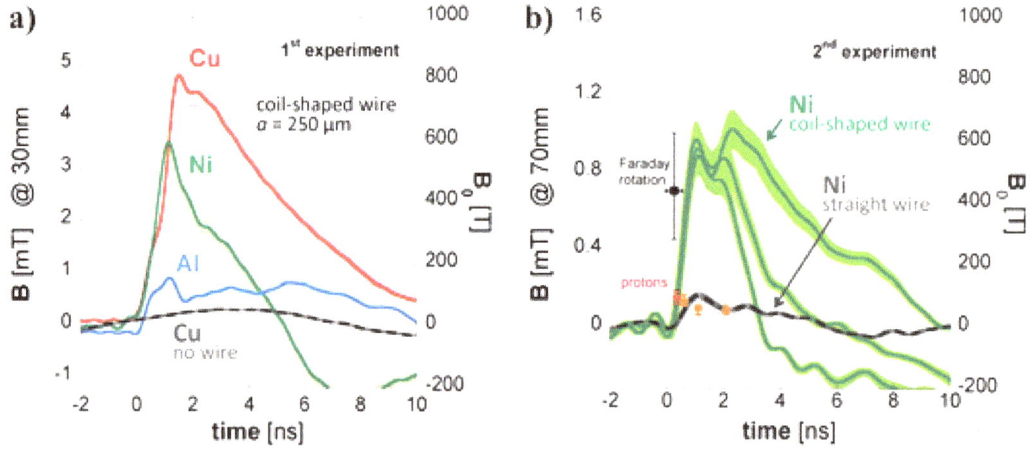

Figure 4: Capacitor-coil target results of *B*-field against time measured by the B-dot probes (solid curves), for the *B*-field at the probe positions (left-hand side ordinates) and the corresponding B_0 values at the coils center (right-hand side ordinates): (a) 1nd experiment—targets of different materials. (b) 2nd Experiment—Ni targets. The grey dashed curves correspond respectively to a shot on the rear disk holding the two Cu disks parallel at the distance $d_0 = 900\ \mu m$ but without any connecting wire, and to a Ni target with a straight wire between the disks (no coil): their values refer exclusively to the *B*-field at the probe position (left-hand side ordinates). The symbols refer exclusively to the *B*-field at the coil centre, B_0 (right-hand side ordinates), and are the measurements obtained for Ni capacitor-coil targets by Faraday rotation (square, at t=0.2 ns) and by proton-deflectometry (red circle, at t=0.35 ns). The smaller orange circles represent *B*-field estimates from proton-deflectometry images obtained at later times: the discrepancy with B-dot probe results is explained in the text by electrostatic effects due to electron trapping near the coil.

The 3D magnetostatic code Radia [29] was used to simulate the spatial distribution of the *B*-field taking into account the target and coil geometry, and their connection wires. We used the current *I* as a free parameter, varying it to obtain an equality between the measured and simulated *B*-field values at the B-dot spatial positions. The boundary condition consisted in imposing a closed circuit with the same current in the coil and between the two disks. Figure 3(a) represents two perpendicular planar sections of the *B*-field map, one horizontal containing the coil axis, $z = 0$ (left), the other vertical corresponding to the coil plane, $y= 0$ (right) (see figure 1(b) for the axis orientation: the *y*-axis is the coil axis and the origin is here at the coil centre). The field was calculated with an injected current of $I = 340$ kA and a coil radius $a = 250\ \mu m$, yielding the measured *B*-field peak value for the Cu target (given the $\approx 10\ \mu m\ ns^{-1}$ coil rod expansion velocity measured by time-resolved optical shadowgraphy, no rod expansion was considered in the magnetostatic calculations): the *B*-field norm is on color scale, the arrows represent the *B*-field vector projections on the two planes. The spatial distribution on the $z = 0$-plane clearly evidences, as expected, a dipole-like *B*-field. One can also see on the $y = 0$ plane that the *B*-field norm is quite homogenous over the space inside the coil. Poloidal fields around the coil rod and straight parts of the wire are quite strong, and they mostly determine the *B*-field distribution below the coil region.

For the same calculation, figure 3(b) shows the amplitude of the *B*-field component parallel to the coil-axis as a function of the perpendicular distance, along the line connecting the coil and the B-dot probe centres, respectively at $r = 0$ and r=30mm (solid red curve). The vertical dashed lines delimit the $50\ \mu m$-thick coil rod position. For a measured 5mT at r=30mm, the extrapolated value at the coil centre is $B_0 = 800$ T for the capacitor-coil target, with the uncertainty range of (795, 825 T) accounting for target-to-target eventual variations of $\pm 2\ \mu m$ on the separation between the two connection points of the wire circular part with the vertical rods. An over-estimated extrapolation *B*-field value of $B_0 = 20$ kT (for a I=8.5 MA current) is obtained when simulating a perfect circular coil of the same radius (dashed black curve in the insert): such unrealistic value illustrates the importance of an accurate modeling of the real target, in particular the discontinuity on the coil circular part. Yet, our target production and characterization are sufficiently precise to have only a very small contribution to the result uncertainty.

As for the B-dot probes distance with respect to the coil, figure 3(c) shows the calculated relative uncertainties of the *B*-field measurements at the probe positions: the values are plotted over the probes circular section and account for the 3D *B*-field gradients over the probes cylindrical volume. The estimated *B*-field errors, which remain under the 6% and the 10% respectively for the 1nd and 2nd experiments (two different probe orientations and distances relative to the coil, as described above), are mainly determined by the gradients along *x* and *z* in the first case (top), and along *y* for the second case (bottom), as expected from the setup geometry.

The typical final results for the produced *B*-field as a function of time are summarized in figure 4. The solid curves are the B-dot probe results obtained with capacitor-coil targets for the *B*-field at the probe positions (left-hand side ordinates) and the corresponding B_0 values at the coil'(s) centre (right-hand side ordinates, as extrapolated from the Radia magnetostatic calculations). In the 1nd experiment, panel (a), the peak values of the *B*-field depend on the target material, yielding $B_0 \approx 800$, ≈ 600 and ≈ 150 T (±6%), respectively for Cu (red), Ni (green) and Al targets (cyan). In the 2nd experiment, panel (b), the B-dot probe measurements are showing a shot-to-shot reproducible charging using Ni targets: the *B*-field production is synchronous with the beginning of the laser irradiation (at $t = 0$, within a ± 100 ps calibration uncertainty) and the rise time of ≈ 1 ns is consistent with the duration of the laser pulse (this is the system charge time). The reproducible measurements of a peak *B*-field of ≈ 1 mT are extrapolated to the coil centre peaking at $B_0 \approx 600$ T (±10%). This is in very good agreement with the result obtained with the Ni target in the 1nd experiment, where the probe was at a different position relative to the coil.

The targets used in the 2nd experiment had a reduced front disk hole (diameter of 1000 instead of $1750\ \mu m$), seeking for capturing more non-thermal electrons. However, a comparison of the results leads to the conclusion that the hole size is not a determining parameter for the peak *B*-field strength, yet the second *B*-field peak oscillation at $t \gtrsim 2$ ns is more pronounced. The typical duration of the pulsed *B*-field is of a few ns, with fluctuations mostly due to the varying distance d_0. All signals present slower amplitude variations at later times (non-represented), probably corresponding to electromagnetic coupling with objects around the target (secondary targets, diagnostics, chamber walls), converging to zero at ~150 ns.

The efficiency of the coils was tested by B-dot measurements in two *blank* shots: grey dashed curves in figures 4(a) and (b), corresponding respectively to two Cu disks without any connecting wire (held separately), and to a Ni target with a straight wire between the disks (no coil). The dashed curves' values correspond exclusively to the measurements at the probe positions (left-hand side ordinates). One concludes that (i) the irradiated disk charging by laser–target interaction and current of escaping electrons have a negligible contribution to the integrated signals (as well as any neutralizing current from the ground through the disk stalk) unless there is a physical connection between the disks, and (ii) the contributions from currents flowing through other target segments besides the coil (in particular the straight parts of the connecting wire) are sufficiently weak if compared to the results obtained with coil-shaped wires: the coil is the dominant source of the magnetic flux.

The B-dot results for $t \lesssim 0.4$ ns were confirmed in the 2nd experiment by measurements of the Faraday rotation of the polarization direction of a linearly-polarized probe laser beam, and of the laser-accelerated proton deflections: respectively full square and circular-symbols in figure 4(b). The *B*-field peak values could not be measured by these two diagnostics due to laser-plasma effects, as described in the next sections.

FARADAY ROTATION

The Faraday rotation measurements, using a 9 ns-duration probe laser at 533 nm wavelength incident along the coil-axis, were performed with two $500\ \mu m$-thick birefringent terbium gallium garnet (TGG) crystals with its centre placed at 3.5 mm from the coil plane: TGG1 (0.5 mm-wide) was centred on the coil axis and TGG2 (1 mm-wide) at a 1.9 mm perpendicular offset. The TGG Verdet constant, $11.35°\ T^{-1}\ mm^{-1}$, is 38 times higher than that of SiO_2 used in the earlier work by Fujioka *et al* [28], allowing to be sensitive to weaker *B*-field strengths. No Faraday rotation measurements were successful with the crystals located neither closer to the coil, because of signal blackout quasi-synchronous with the laser irradiation of the target—due to very rapid crystal ionization by the hard x-rays and fast particles issuing from the interaction region—nor further away from the coil where the local *B*-field was too weak. The Faraday Effect was measured by using a time-resolved polarimeter, constituted of a Wollaston prism to separate the two perpendicular components of the probe beam field, and a streak camera (see figure 1(b)). The time-resolution of ≈ 300 ps was limited by the camera time-resolution and by the laser jitter. The quantification of the rotation angle was previously calibrated, without *B*-field, by quantifying the polarization ratio of the two transmitted perpendicular-polarization signals, defined as $R = E_{\parallel}^2/(E_{\parallel}^2 + E_{\perp}^2)$, as a function of the orientation of the incident light polarization. The obtained linear fit of the experimental data was consistent with the Malus law by an average $\chi^2 \approx 0.002$ adjustment. For the shots with *B*-field, the incident polarization was setup at -45° and +45 relative to $E_{\parallel}$ and $E_{\perp}$ respectively, yielding a ratio $R = 0.50 \pm 0.01$ for the situation without *B*-field.

Figure 5 shows the analysis of a successful shot for the Faraday rotation effect measurement, as a function of time: transmitted $E_{\parallel}^2$ (solid black) and $E_{\perp}^2$ (dashed black), and corresponding *R* (solid red) are plotted for (a) TGG1 and (b) TGG2 (the plotted values correspond to averages over the respective crystals' width). The crystals' blackout times are indicated by the light grey dashed vertical lines. Panel (c) plots the corresponding local axial component of the *B*-field averaged over the crystals thickness and width, solid black for TGG1, dashed red for TGG2, obtained assuming that the TGG Verdet constant does not change with time due to the increase in crystal temperature. We see that the *B*-field inferred from the TGG2 measurements is too weak and remains at the signal noise level, of ± 0.03 T. As for the TGG1 measurements, the *B*-field clearly rises up to the blackout time. To compare with B-dot probe measurements we selected the time $t_1 \approx 0.2$ ns, indicated in both panels (a) and (c), where the averaged *B*-field is evaluated to $\approx 0.22 \pm 0.05$ T along the coil axis at a 3.50 ± 0.25 mm distance from its centre. The black square symbol in figure 4(b) corresponds to the B_0 strength (right-hand side ordinate axis) extrapolated from that measurement, using the same magnetostatic code [29] and a similar protocol as for the B-dot data. In spite of the diagnostic uncertainties—related to the crystal thickness, accessible width and equivalent height of the streak slit on the crystal plane, and the *B*-field gradients over such crystal volume—the obtained value for the extrapolated B_0 is fairly consistent with the B-dot probe measurements.

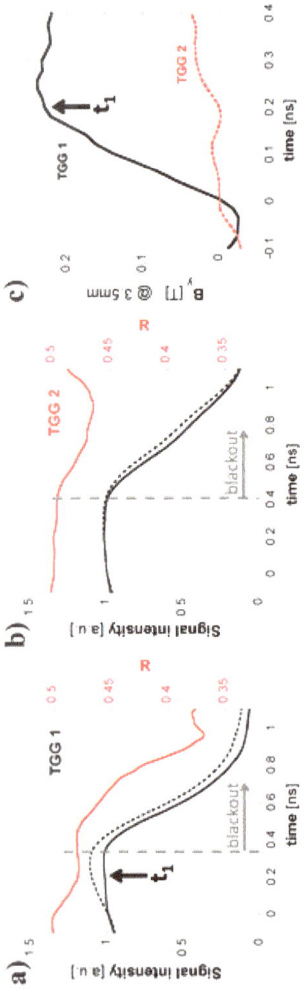

Figure 5. Faraday rotation effect on probe laser light over two TGG crystals, placed at 3.5 mm from the coil plane, TGG1 centred on the coil axis and TGG2 at a 1.9mm perpendicular offset, evaluated from a time-resolved polarimeter: (a) detected signals of perpendicular polarizations: $E_{\parallel}^2$ (solid black) and $E_{\perp}^2$ (dashed black), and the corresponding polarization ratio *R* (red), for TGG1. (b) Idem, for TGG2. (c) Axial component of the *B*-field averaged over the crystals $500\ \mu m$ -thickness: solid black for TGG1, dashed red for TGG2.

Proton-deflectometry

The proton-deflectometry technique allows to measure the *B*-field directly in the coil region. A proton beam was created with a short laser pulse (50J on target, 1ps FWHM at $1\omega_0$) focused onto $10\ \mu m$-thick Au foils at $\approx 10^{19}$ W cm^{-2} intensity. Proton beams of ~ 20 MeV maximum energy were generated by the target normal sheath acceleration (TNSA) mechanism at the foils' rear surface [32, 33], located at a distance of 5mm from the target coil. The proton beam propagation axis was perpendicular to the coil axis, and it was detected 45 mm away from the coil by a 15-layers radiochromic film (RCF) stack. The proton deflections due to the *B*-field were quantified with the help of a $42\ \mu m$-pitch mesh, positioned 3 mm before the coil (see figure 1(b)). RCF proton imprint images correspond to a magnification of 10 for the coil plane and of 25 for the mesh plane. A $195\ \mu m$-Al protection film before the RCF stack limited the detection to protons of energy $\epsilon_p \gtrsim 5$ MeV and electrons with energies $\epsilon_e \gtrsim 260$ keV. The proton imprint signal on each RCF layer corresponds to a narrow energy range of their spectrum, due to the Bragg peak energy absorption. Accounting for the time-of-flight (TOF) between the proton source and the coil, each shot, with a chosen delay $\Delta\tau$ between the main and proton-driving lasers, scanned the effects of the *B*-field on the proton-trajectories over the time range of $t = \Delta\tau + \mathrm{TOF}$, with TOF between 80 and 160 ps.

Figure 6(a) shows a typical image of the RCF layer corresponding to the imprint of protons of energy $\epsilon_p = 13 \pm 1$ MeV, obtained in a shot with $\Delta\tau = 0.25$ ns. The corresponding probing time is $t \approx 0.35$ ns. The mesh-shadow deformations are detected for distances till $\sim 500\ \mu m$ from the coil centre (the given spatial scale corresponds to the plane of the coil centre, perpendicular to the proton beam axis, where the unperturbed mesh shadow has a pitch of $105\ \mu m$). However, the most outstanding feature is the centred bulb region void of any proton imprint due to the very strong *B*-field. Figure 6(b) shows the result of a Monte-Carlo simulation of the trajectories of randomly injected protons within the energy range of the experimental signal in (a). Both the mesh-shadow deformations and bulb size compare very well with the experimental image when imposing a coil current I=40 kA, yielding the *B*-field in the coil centre $B_0 = 95$ T. Figure 6(c) shows the map of the experimental mesh vertical deformations (measured from the film corresponding to figure 4(a)), fairly indicating a dipole-like *B*-field spatial distribution, with a typical length scale of 1 mm. In figure 6(d) the horizontal lineout of the experimental vertical deformation map at $z = 0 \pm 105\ \mu m$ (circles) is compared to the corresponding synthetic deformations for different strengths of the *B*-field (curves), leading to the evaluation of $B_0 = 95 \pm 20$ T at the corresponding probing time, $t \approx 0.35$ ns. This value, solid red circle in figure 4(b) (right-hand side ordinates), agrees with the evaluation from the B-dot measurements.

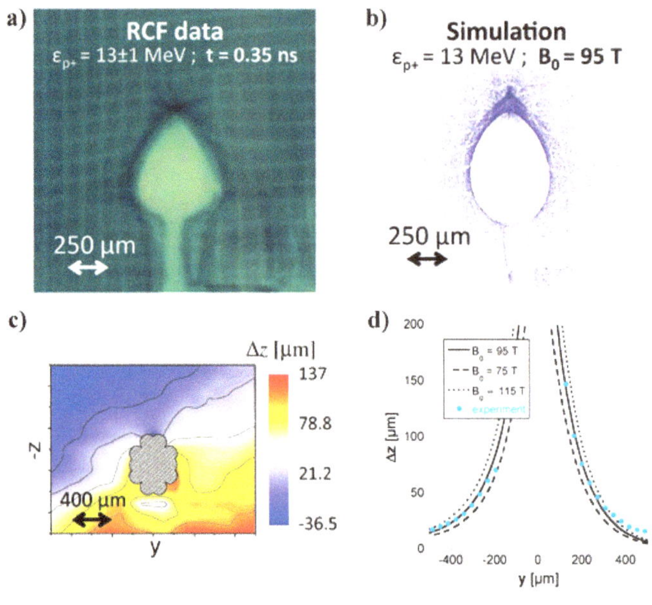

Figure 6: (a) Sample RCF image of proton-deflectometry measurements in an early probing time $t = 0.35$ ns after ns-laser light starts to irradiate the B-loop target's rear disk. The RCF position corresponds to the imprint of 13 ± 1 MeV protons. (b) 13 ± 1 MeV proton imprint on detector plane given by a Monte-Carlo simulation of their trajectories over a 3D *B*-field map with $B_0 = 95$ T(previously calculated by the magnetostatic code). (c) Map of experimental vertical deformations of the mesh shadow. The spatial scale on each image corresponds to the plane of the coil centre. (d) Corresponding horizontal lineouts of the experimental (circles) and synthetic (curves, for different *B*-field strengths) vertical deformations of the mesh imprint.

Figure 7(a) shows the RCF images obtained for $\epsilon_p = 13 \pm 1$ MeV protons with different delays $\Delta\tau$ between the laser pulses. Surprisingly, the bulb size and the overall mesh imprint deformations decrease with time: protons of the same energy undergo smaller deflections if injected at later times. The measurement of the void bulb size and mesh-imprint deformations for t>0.35 ns, following the same protocol as before, would hold the B_0 values and uncertainties represented by the small orange circles in figure 4(b): the B_0 decreasing behaviour as a function of time is in contradiction with the B-dot probe measurements, and does not agree with the laser-charging process up to t=1 ns. Figure 7(b) shows images from RCF layers corresponding to $\epsilon_p = 16.8 \pm 0.8$ MeV protons, for the same first two delays $\Delta\tau$: for each delay, the usual mesh imprint and void-bulb proton signatures are smaller, consistently with the higher proton energy. Yet, we detect a second particle species producing a large circular halo on both images, superposed to the proton signature. For each shot, such a halo is clearly visible with the same size and shape over the successive last six layers of the RCF stack, identifying it as the signature of relativistic electrons, accelerated at the Au-foil front surface by the short laser pulse, and relatively homogeneously deposing energy over the successive RCF layers. Accounting for their spectrum (inferred from particle-in-cell simulations of the short pulse laser interaction), the foil potential barrier and the Al-filtering of the RCF stack, the halo signal should correspond to electrons with energies between 3 and 5 MeV. Comparing images from different shots (see figure 7(b)), we observe that the halo size increases for increasing $\Delta\tau$.

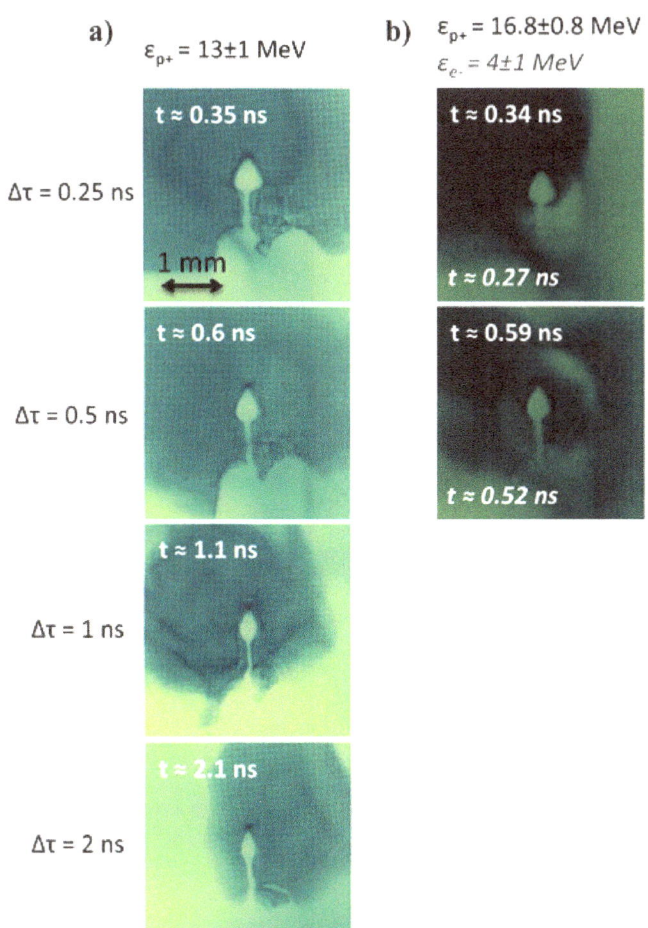

Figure 7: Sample RCF data from different shots with varying delay $\Delta\tau$: (a) for 13 ± 1 MeV protons, and (b) for 16.8 ± 0.8 MeV protons and 4 ± 1 MeV electrons. The corresponding probing times for protons are shown at the upper part of the images (straight t) and for electrons in the lower part (italic *t*). The spatial scale corresponds to the plane of the coil centre.

The signature of the relativistic electrons evolves in opposition to the proton signature. This can be explained by a monotonous increase, over the ns time scale of the main laser irradiation, of magnetized plasma electrons in regions of strong *B*-field around or at the vicinity of the coil. The main-laser interaction with the target rear disk creates a plasma 3 mm below the coil. As determined by x-ray spectrometry along with a comparison with calculated atomic spectra, its temperature is of the order of $T \approx 1.5 \pm 0.1$ keV, with a high-energy electron component of $T_e \approx 40 \pm 5$ keV. A fraction of these electrons can stream upwards to the coil-wire region in less than 30 ps, where they can be magnetized: their Larmor radius is $< 10\ \mu m$ ($B_0 \approx 100$ T is achieved at $t \approx 300$ ps (see figure 4)), much smaller than the typical size of the strong *B*-field region, $\sim 2a = 500\ \mu m$. The progressively increasing charge of magnetized electrons at the coil vicinity can produce an electrostatic effect sufficient important to influence the deflection trajectories of the probing particles, with increasing *focusing* or *defocusing* contributions respectively for the TNSA protons and for the relativistic electrons issuing from the short laser pulse interaction with the Au foil. A trapped electron charge as small as 5×10^{-7} C yields an electrostatic potential at a distance r~a from the coil centre already comparable to the20 MeV proton maximum energy. This charge corresponds to a density of magnetized electrons $n_e^{mag} \sim 5 \times 10^{16}\ cm^3$ in a sphere of radius *a*, which is about 5% of the density of the supra-thermal electrons expected to stream through the coil region. This would be enough to electrostatically balance the main magnetic force on the probing TNSA protons and relativistic electrons. The magnetization condition for the main plasma electrons, $\omega_{ce} \gtrsim \omega_{pe}$ (assuring that the potential stays localized at the coil vicinity), determines a maximum $n_e^{mag} \lesssim 10^{17}\ cm^{-3}$ (for $B_0 \sim 100$ T), consistent with the previous estimation.

Other plasma effects can influence proton-deflectometry measurements, but their effect is expected to be negligible compared to the electrostatic one: (i) the pressure of the magnetized electrons in the coil region $P_e = n_e^{mag} T_e \sim 6 \times 10^7$ Pa, which remains much smaller than the density of magnetic energy $P_B = B_0^2/2\mu_0 \sim 4 \times 10^9$ Pa, for $B_0 = 100$ T. (ii) Diamagnetic currents of the trapped electrons $j_1 \sim e n_e^{mag} v_e$, yield a magnetic field over the characteristic length *a* of $B_1 \sim \mu_0 j_1 a \sim 10$ T, eventually opposed but much smaller than the B_0 inferred by the other diagnostics.

CONCLUSIONS

In conclusion, our experimental results were obtained simultaneously by three independent diagnostics showing a reproducible quasi-static *B*-field generation by laser interaction with capacitor-coil targets, typically with a few ns duration and a 1 mm^3-volume, yielding peak strengths of several hundred Tesla, depending on the target material: the dipole-like magnetic field total energy is of the order of 8.3% ± 1.5%, 4.5% ± 0.8% and 0.35% ± 0.05% of the invested laser pulse energy, respectively for Cu, Ni and Al targets. The observed differences with the different target materials may be attributed to (i) the different resistive behaviour at low temperature, though the current rise time and consequent wire heating is identically rapid erasing resistivity differences, and, probably more important, to (ii) the plasma temperature and hydrodynamics yielding different short-circuiting times: this will be object of further investigations. The correct extrapolation of the B-field amplitude at the centre of the coils from distant measurements (at a few mm for Faraday rotation, and at a few cm for the pick-up coil probes) is performed by an accurate coding of the target shape and magnetostatic computation of the current intensity looping in the capacitor-coil targets. Results from Faraday rotation and proton-deflectometry are consistent with the B-dot probe measurements at the early stages of the target charging, up to $t \approx 0.35$ ns, and then are disturbed by radiation and plasma effects, respectively the blackout of the birefringent crystal and, as we concluded, a negative electrostatic potential. The latter is formed likely due to electron magnetization around the coil region, summing up to the main B-field effect and explaining the decreasing proton deflections fort >0.4 ns.

While the typical mm^3-volume and ns-duration of the produced B-field pulses are small compared to the parameters achieved in the large-scale experiments and the state-of-the-art magnets mentioned in the introduction, they are characterized by an unprecedentedly high conversion efficiency (approaching the range of 10%) of the driver energy into magnetic energy. Moreover, this all-optical technique lends itself to the magnetization and accurate probing of high-energy-density samples driven by secondary powerful laser or particle beams: given the ≈ 10 $\mu m\ ns^{-1}$ expansion velocity of the coil rod, the strong B-field region is accessible for several ns.

ACKNOWLEDGMENTS

We gratefully acknowledge the support of the LULI pico 2000 staff during the experimental run. This work was performed through funding from the French National Agency for Research (ANR) and the competitiveness cluster Alpha—Route des Lasers, project number TERRE ANR-2011-BS04-014. The authors also acknowledge support from the COST Action MP1208 'Developing the physics and the scientific community for Inertial Fusion' through four STSM visit grants. The work was also partially supported by MEXT/JSPS KAKENHI Grant Number 24684044 and JSPS Fellows Grant Number 14J06592. The research was carried out within the framework of the 'Investments for the future' program IdEx Bordeaux LAPHIA (ANR-10-IDEX-03-02) and of the EUROfusion Consortium, which received funding from the European Union's Horizon 2020 research and innovation program under grant agreement number 633053. The views and opinions expressed herein do not necessarily reflect those of the European Commission. JJS gratefully acknowledges fruitful discussions with L Gremillet and K Mima.

REFERENCES

1. Lai D 2001 Rev. Mod. Phys. 73 629–61
2. Liang E, Nishimura K, Li H and Gary S P 2003 Phys. Rev. Lett. 90 085001
3. Murdin B N et al 2013 Nat. Commun. 4 1469
4. Lerner E J et al 2011 J Fusion Energy 30 367
5. Engel D and Wunner G 2008 Phys. Rev. A 78 032515
6. Nordhaus J et al 2011 Proc. Natl Acad. Sci. USA 108 3140
7. Ciardi A et al 2013 Phys. Rev. Lett. 110 025002
8. Albertazzi B et al 2014 Science 346 325
9. Korneev Ph et al 2014 Phys. Plasmas 21 022117
10. Matsumoto Y, Amaro T, Kato T N and Hoshino M 2015 Science 347 975
11. Perkins L J et al 2013 Phys. Plasmas 20 072708
12. Strozzi D J et al 2012 Phys. Plasmas 19 072711
13. Wang W-M, Gibbon P, Sheng Z-M and Li Y-T 2015 Phys. Rev. Lett. 114 015001
14. Johzaki T et al 2015 Nucl. Fusion 55 053022
15. Roth M et al 2014 Proc. 5
16. Chen H et al 2014 Phys. Plasmas 21 040703
17. Debray F et al 2013 C. R. Phys. 14 2
18. Gomez M R et al 2014 Phys. Rev. Lett. 113 155003
19. Schmit P F et al 2014 Phys. Rev. Lett. 113 155004
20. Bykov A I et al 2001 Physica B 294–295 574
21. Pérez F, Kemp A J, Divol L, Chen C D and Patel P K 2013 Phys. Rev. Lett. 111 245001
22. Korneev P, d'Humières E and Tikhonchuk V 2015 Phys. Rev. E 91 043107
23. Pollock B B et al 2006 Rev. Sci. Instrum. 77 114703
24. Froula D H et al 2007 Phys. Rev. Lett. 98 135001
25. Albertazzi B et al 2013 Rev. Sci. Instrum. 84 043505
26. Daido H et al 1986 Phys. Rev. Lett. 56 846
27. Courtois C et al 2005 J. Appl. Phys. 98 054913
28. Fujioka S et al 2013 Sci. Rep. 3 1170
29. http://esrf.eu/Accelerators/Groups/InsertionDevices/Software/Radia
30. Dubois J-L et al 2014 Phys. Rev. E 89 013102
31. Poyé A et al 2015 Phys. Rev. E 91 043106
32. Snavely R A et al 2000 Phys. Rev. Lett. 85 2945
33. Wilks S C et al 2001 Phys. Plasmas 8 542

CHAPTER 9

EXPERIMENTAL DEMONSTRATION OF A SYNTHETIC LORENTZ FORCE BY USING RADIATION PRESSURE

N. Šantić, T. Dubček, D. Aumiler, H. Buljan & T. Ban

INTRODUCTION

Experiments on synthetic magnetic/gauge fields for neutral atoms[1,2,3,4,5,6,7,8,9,10,11] have enabled realizations of the Hall effect[4], famous Hamiltonians such as the Harper[7,8] and the Haldane Hamiltonian[10], intriguing topological effects[9,10], and the observation of synthetic Dirac monopoles[11]. There are a few recent reviews on this promising theoretical and experimental progress in synthetic magnetic/gauge fields[12,13,14,15]. The first implementation of synthetic magnetism was in rapidly rotating Bose-Einstein condensates (BECs), employing the analogy between the Lorentz force and the Coriolis force[1,2]. The methods based on laser-atom interaction employ the analogy between the Berry phase in atomic systems[16], and the Aharonov-Bohm phase for charged particles[13]. The first of them was realized in the NIST group with spatially dependent Raman optical coupling between internal hyperfine atomic states in bulk BECs[3]. Methods generating synthetic magnetic fields in optical lattices engineer the complex tunnelling matrix elements between lattice sites[5,6,7,8,9]. They include shaking of the optical lattice[6], laser assisted tunnelling in optical superlattices realizing staggered synthetic magnetic fields[5], in tilted lattices realizing homogeneous fields[7,8], and an all-optical scheme which enables flux rectification in optical superlattices[9]. Interestingly, radiation pressure has not yet been used among the methods for synthetic magnetism, that is, to create the analogue of the Lorentz force. Here we experimentally demonstrate the synthetic Lorentz force based on radiation pressure in cold atomic gases. We measure the dependence of the transverse radiation pressure force (analogous to the transverse Hall deflection) on the velocity of a cold atomic cloud by observing the centre-of-mass (CM) motion. The observed force is perpendicular to the velocity, and zero for the atomic cloud at rest. This concept based on radiation pressure, theoretically proposed in Ref. [17], is straightforward to implement in a large volume (e.g., volumes 1 mm^3–1 cm^3 are easily accessible)[18], is applicable for a broad range of velocities, and can be extended to different geometries. The main reason for the absence of radiation pressure from the previously used methods of synthetic magnetism is the associated heating due to spontaneous emission. However, this is not an obstacle for atomic gases cooled and trapped in a Magneto-Optical Trap (MOT), where our experiments are performed.

Citation: N. Šantić, T. Dubček, D. Aumiler, H. Buljan & T. Ban, Experimental Demonstration of a Synthetic Lorentz Force by Using Radiation Pressure, doi:10.1038/srep13485.

RESULTS

The Basic Idea

The idea behind our experiment is to drive two-step two-photon transitions between three atomic levels $|0\rangle \rightarrow |1\rangle \rightarrow |2\rangle$, with mutually perpendicular laser beams as illustrated in Fig. 1. Two counterpropagating laser beams aligned with the x-axis drive the $|0\rangle \rightarrow |1\rangle$ transition, whereas the $|1\rangle \rightarrow |2\rangle$ transition is driven by counterpropagating beams aligned on the y-axis. Due to the Doppler effect and the perpendicular configuration of the laser beams, both components of the radiation pressure force depend on both components of the atomic velocity: $F_x = F_x(v_x, v_y)$ and $F_y = F_y(v_x, v_y)$. This gives us the opportunity to design the detuning values of our lasers such that F_y is positive/negative for atoms with negative/positive velocity component v_x, and that the total force is zero for an atom at rest: $F(v = 0) = 0$. These are the characteristics of the synthetic Lorentz force that we experimentally demonstrate.

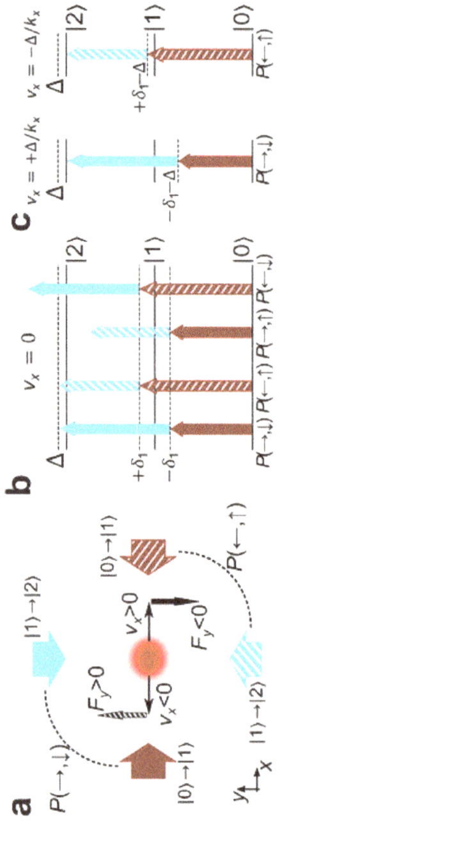

Figure 1: The scheme used to obtain the synthetic Lorentz force via radiation pressure.

It is based on two-step two-photon transitions involving three atomic levels. (**a**) Two counter-propagating laser beams on the x-axis (red arrows) drive the $|0\rangle \rightarrow |1\rangle$ transition, while the $|1\rangle \rightarrow |2\rangle$ transition is driven by two counterpropagating beams on the y-axis (blue arrows). The transverse radiation pressure force arising from the $|1\rangle \rightarrow |2\rangle$ transitionF_y, depends on the velocity v_x as indicated. (**b**) Four possible excitation pathways and a sketch of the detuning values: P(→, ↓) denotes absorption of a $|0\rangle \rightarrow |1\rangle$ photon going towards the positive x-direction, followed by absorption of a $|1\rangle \rightarrow |2\rangle$ photon travelling towards the negative y-direction, and so on. The total detuning value for P(→, ↓) and P(←, ↑) is Δ, whereas it is much larger in magnitude for P(→, ↑) and P(←, ↓); the two latter pathways are thus negligible in this configuration. (**c**) The Doppler shift provides F_y as a function of v_x as sketched in (**a**): pathway P(→, ↓) for an atom with positive velocity $v_x = +\Delta/k_x$ ($v_y = 0$) is on resonance, providing negative F_y. Likewise, P(←, ↑) is on resonance for an atom with negative velocity $v_x = -\Delta/k_x$, providing positive F_y. See text for details.

The design of the detuning values of the lasers is crucial in obtaining the desired result. The beams driving the first step of the transition $|0\rangle \rightarrow |1\rangle$ are detuned by the same magnitude, but with the opposite sign. The one towards the positive x-direction is red-detuned by $\delta_{\rightarrow} = -\delta_1 < 0$, while the other is blue-detuned by $\delta_{\leftarrow} = \delta_1 > 0$. Their intensities are equal. Thus, if just these two lasers were present, the net force on atoms (of any velocity) would be zero. However, the population of level $|1\rangle$ would depend on the velocity v_x, which implies that the rate of transitions $|1\rangle \rightarrow |2\rangle$ giving the transverse force will depend on v_x. The detuning values of the beams driving the second step of the transition, $|1\rangle \rightarrow |2\rangle$are denoted by $\delta_{\uparrow}$ and $\delta_{\downarrow}$, for the beam propagating in the positive and negative y direction, respectively. For now, let us set these values such that $\delta_{\rightarrow} + \delta_{\downarrow} = \delta_{\leftarrow} + \delta_{\uparrow} = \Delta > 0$, as indicated in Fig. 1(b).

The two-step two-photon transitions, where absorption of a $|1\rangle \rightarrow |2\rangle$photon follows absorption of a $|0\rangle \rightarrow |1\rangle$ photon with perpendicular momentum, yield the synthetic Lorentz force via momentum transfer from photons to atoms. Given the fact that we have two counterpropagating beams for each transition, we have four excitation pathways for the two-step two-photon transition, denoted by P(→, ↑),P(←, ↑), P(→, ↓), and P(←, ↓), see Fig. 1. The arrows correspond to the direction of the photon's momentum, for example, P(→, ↓) denotes the pathway where absorption of a photon travelling in the +x direction is followed by absorption of a photon in the −y direction and so on. Since the detuning magnitudes of the first step are identical for all pathways ($|\delta_{\rightarrow}| = \delta_{\leftarrow}$), the relevant quantity is the total detuning for the two-step two-photon transitions. It is important to note that the total detuning Δ for pathways P(→, ↓) and P(←, ↑), is much smaller in magnitude than the detuning values of P(→, ↑) and P(←, ↓). The last two are thus negligible in the setup of Fig. 1.

To understand the origin of the synthetic Lorentz force, we take into account the Doppler shift. For an atom moving along the x-axis with velocity v_x ($v_y = 0$), the P(←, ↑) pathway is detuned by $\Delta + k_x v_x$, i.e., it is on resonance when $v_x = -\Delta/k_x < 0$. Because photons from the second step of P(←, ↑) impart momentum towards the positive y-direction, there will be a positive force F_y for atoms with negative v_x. In the same fashion, the P(→, ↓) pathway will be on resonance when $v_x = +\Delta/k_x > 0$ yielding negative F_y for positive v_x.

The Experiment

In the experiment, we use cold ^{87}Rb atoms. For the $|0\rangle \rightarrow |1\rangle$ transition we use the D2 transition in ^{87}Rb: $|5S_{1/2}; F = 2\rangle \rightarrow |5P_{3/2}; F' = 3\rangle$ at 780 nm[19]. The easily accessible transition, $|5P_{3/2}; F' = 3\rangle \rightarrow |5D_{5/2}; F'' = 4\rangle$ at 776 nm20, is used for the second step $|1\rangle \rightarrow |2\rangle$. The linewidths of the states are $2\pi \times 6.1$ MHz for the $|5P_{3/2}; F' = 3\rangle$ state19, and $2\pi \times 0.66$ MHz for the $|5D_{5/2}; F'' = 4\rangle$ state[20].

The ^{87}Rb atoms are cooled and trapped in a standard glass vapour cell magneto-optical trap (MOT), arranged in a retro-reflected configuration. In this configuration, three orthogonal retro-reflected beams are used to create the total of six beams needed for the MOT. A pair of anti-Helmholtz coils provides a quadrupole magnetic field, which together with the laser beams creates a trapping potential (for example, see Ref. [18]). Fluorescence imaging of the cloud is performed with a camera aligned along the z-axis. In typical experimental conditions we obtain a cloud of 0.4 mm in diameter, which contains about 10^8 atoms of ^{87}Rb, at a temperature of 50 μK (for details of the experimental setup see Methods). The four beams implementing the synthetic Lorentz force, arranged as in Fig. 1, are of much smaller intensity than the MOT beams. Therefore, they are negligible when the MOT beams are ON. For the experimental detection of the synthetic Lorentz force, we turn the MOT beams OFF, as described in detail below.

We need to measure the transverse force F_y in dependence of the velocity v_x of the atomic cloud. Thus, we must prepare a cloud with a given centre of mass (CM) velocity. For this purpose we use an additional pair of current coils that produce a bias magnetic field along the symmetry axis of the anti-Helmholtz coils, x in our notation here. The bias field moves the centre of the trap (the point where $\mathbf{B}=0$), which displaces the cloud approximately 1 mm along the x-axis.

The measurement protocol is as follows. (i) We load the trap with the bias magnetic field on. (ii) At $t=-2$ ms we reverse the bias magnetic field by reversing the current, which suddenly shifts the centre of the trap. This introduces a force on the cloud due to the MOT beams. During the next 2 ms the cloud accelerates along the x-axis towards the new trap centre. (iii) Next we turn off the MOT cooling laser and all real magnetic fields. This moment corresponds to $t=0$ in our presentation. The system is now simplified because Zeeman splitting of hyperfine levels is absent and the radiation force left, arising solely from the lasers implementing the synthetic Lorentz force, is not spatially dependent (it is only velocity-dependent). (iv) The cloud starts to expand because the trapping is absent, but it also moves in the xy plane due to both the initial velocity, $v_x(t=0)$, and the radiation pressure force. (v) After some delay time t, the cooling laser is suddenly turned on and the cloud is imaged with the camera. From the trajectory traversed by the CM of the cloud (x(t), y(t)), we can find the force acting on the atoms. If we wish to image a cloud initially at rest we skip steps (i) and (ii). For a given delay time t, we repeat the measurement protocol 20 times in identical conditions, and subsequently average to obtain x(t) and y(t). The gravity is along the z-axis in our system; free fall of atoms due to gravity does not affect the motion in the xy plane. It should be stated that we perform these measurements first with $|1\rangle \rightarrow |2\rangle$ lasers OFF, and then with these lasers ON, keeping all other parameters identical. The difference in the path y(t), with lasers $|1\rangle \rightarrow |2\rangle$ ON and OFF, gives us the transverse motion due to the synthetic Lorentz force.

The Experimental Proof of the Synthetic Lorentz Force by Radiation Pressure

The results of the experiment are illustrated in Fig. 2. We show the trajectory of the cloud (x(t), y(t)), in the presence of the synthetic Lorentz force, for three initial velocities: $v_x=-0.3$ m/s (squares), $v_x=0$ m/s (diamonds), and $v_x=0.6$ m/s (circles); $v_y=0$ at $t=0$ in each run of the experiment. There is a difference in the magnitude of the initial v_x for the positive and negative velocity [circles and squares in Fig. 2(a)], which is a result of our MOT retro-reflected geometry, and the way we accelerate the cloud in step (ii) of the protocol. In order to prepare a cloud with positive v_x, the cloud is accelerated with the incoming MOT beam (coming from the laser side of the setup), whereas acceleration in the opposite direction is performed with the reflected beam which has smaller intensity. The reflected beam intensity is smaller due to the losses, which are a result of the passage of the incoming beam through the dense cloud (absorption), and partially due to reflection. Consequently, the negative initial velocity is smaller than the positive velocity.

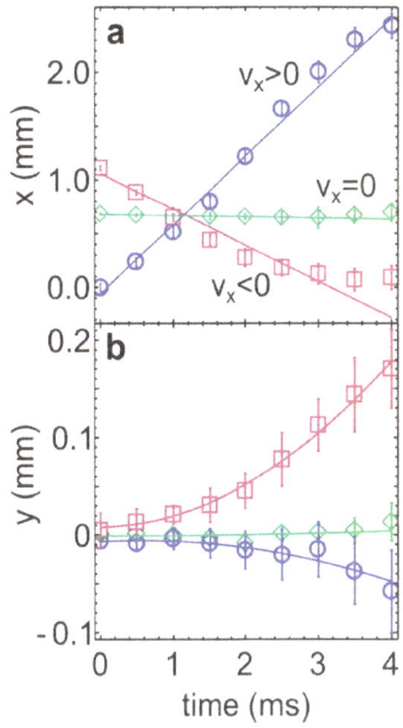

Figure 2: The trajectories of the CM of the atomic cloud in the presence of the synthetic Lorentz force.

(**a**) x(t), and (**b**) y(t) for three different initial velocities, $v_x=0.6$ m/s >0 (circles), $v_x=-0.3$ m/s <0 (squares), and $v_x=0$ m/s (diamonds); initial component of $v_y=0$ in all measurements. Accelerating motion along y is the signature of the transverse force F_y, which depends on v_x. The lines are fitted to the experimental data, linear fits for $v_x(t)$, and quadratic for $v_y(t)$.

For the lasers implementing the synthetic Lorentz force, we use the following detuning values: for the first step at 780 nm $\delta_{\leftarrow}=-\delta_{\rightarrow}=2\pi\times 6$ MHz, and for the second step at 776 nm $\delta_{\uparrow}=-2\pi\times 3.5$ MHz and $\delta_{\downarrow}=2\pi\times 7.1$ MHz. The two operational two-step pathways are P(←, ↑) and P(→, ↓), whereas the other two are far from resonance. The intensities of the beams used are $I_{780}=0.060$ mW/cm^2 and $I_{776}=2.9$ mW/cm^2, giving Rabi frequencies $\Omega_{780}=2\pi\times 1.2$ MHz and $\Omega_{776}=2\pi\times 0.94$ MHz.

By inspection of Fig. 2, we see that the cloud travels along the x-axis by inertia, whereas it accelerates along the y-axis due to the synthetic Lorentz force. The direction of the force depends on the sign of the velocity v_x (negative/positive v_x gives positive/negative F_y), and the force is zero for a cloud at rest. We observe an asymmetry in the forceF_y, for the positive and negative velocity. In order to justify the exact choice of the detuning values $\delta_{\uparrow}$ and $\delta_{\downarrow}$, and further investigate the observed asimmetry in the measured synthetic Lorentz force, we perform measurements in a slightly simplified configuration.

Measurements in an Auxiliary Configuration

We inspect the force along the y direction arising from the two-step two-photon resonances, by using a configuration with three laser beams illustrated in Fig. 3. We block the beam pointing towards the negative y-direction, and measure F_y arising from the remaining beam (the positive y-direction), which drives the $|1\rangle \rightarrow |2\rangle$ transition. The force is measured for an atomic cloud with velocity zero, as a function of the detuning $\delta_{\uparrow}$, see Fig. 3. Measurements are done for three different detuning values of the first-step 780 nm beams $\delta_{\leftarrow}=-\delta_{\rightarrow}=2\pi\times 4$, 6, 8 MHz. The intensities of the lasers driving the transitions are now $I_{780} = 0.043\,mW/cm^2$ and $I_{776} = 4.8\,mW/cm^2$, giving Rabi frequencies $\Omega_{780} = 2\pi \times 1.1$ MHz and $\Omega_{776} = 2\pi \times 1.2$ MHz. The two maxima in Fig. 3 are profiles of the two-step two-photon resonances: the peak in the vicinity of $\delta_\uparrow = -\delta_\rightarrow > 0$ corresponds to the excitation pathway P(→, ↑), and the peak close to $\delta_\uparrow = -\delta_\leftarrow < 0$ corresponds to P(←, ↑).

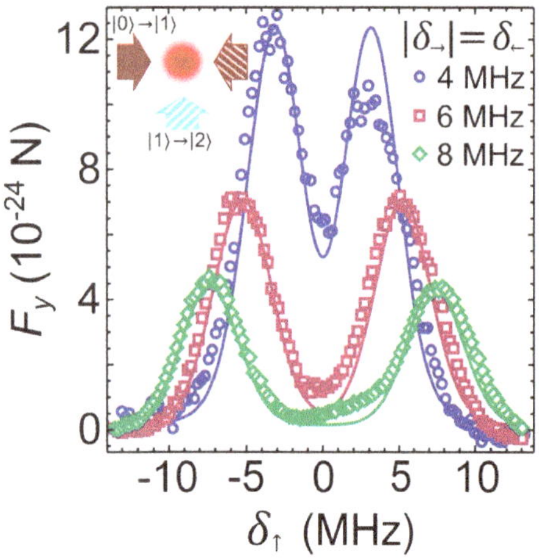

The calculated (solid lines) and measured force F_y as a function of $\delta_\uparrow$ for the excitation with just three beams as shown. Measurements are performed for the cloud with initial velocity zero. The plots show resonances for three values of the detuning $\delta_\leftarrow = -\delta_\rightarrow = 2\pi \times 4$ MHz (circles), 6 MHz (squares) and 8 MHz (diamonds).

Figure 3: Frequency scan of the two-step two-photon resonance in the auxiliary configuration.

Solid lines in Fig. 3 show the theoretically calculated profiles of F_y (see Methods for details of the theoretical calculation). The agreement between theory and experiment is evident. All parameter values used in the theoretical calculation are taken from the experiment, except that the Rabi frequencies are reduced by 20%. This is reasonable because in the experiment, the absorption of laser beams across a dense atomic cloud reduces their intensity[21].

It should be pointed out that the peaks in $F_y(\delta_\uparrow)$ are slightly displaced from the values $\delta_\uparrow = \pm|\delta_\rightarrow|$, towards $\delta_\uparrow = 0$. For example, for $|\delta_\rightarrow|$ at 6 MHz, the maxima are at ±5.3 MHz. Moreover, the FWHM of the peaks is larger than the linewidth of the state $|2\rangle$. These two observations are a consequence of the laser linewidth, which is larger than the linewidth of the state $|2\rangle = |5D_{5/2}; F'' = 4\rangle$ (see Methods). The finite linewidth of $|0\rangle \rightarrow |1\rangle$lasers at 780 nm distorts the peaks as follows: the side of the peak closer to $\delta_\uparrow = 0$ is lifted up in comparison the opposite side of the peak. This follows from the fact that the two-step two-photon resonance is stronger when $|\delta_\rightarrow|$ is closer to zero, which is evident from Fig. 3. The exact positions of the peaks at ±5.3 MHz, explain the chosen detuning values for the second step at 776 nm, which were used to obtain Fig. 2:$\delta_\uparrow = -2\pi \times 3.5$ MHz and $\delta_\downarrow = 2\pi \times 7.1$ MHz. They are chosen such that the operational pathways are effectively equally detuned from the two-step two-photon resonance: P(→, ↓) is detuned by $2\pi \times (-5.3 + 7.1)$ MHZ $= 2\pi \times 1.8$ MHZ, and P(←, ↑) is detuned by $2\pi \times (+5.3 - 3.5)$ MHZ $= 2\pi \times 1.8$ MHZ.

DISCUSSION

Suppose that we repeat measurements corresponding to Fig. 3, but for an atomic cloud with mean velocity v_x different from zero. The results of such measurements would be identical as for $v_x = 0$, but the positions of the peaks would correspond to the Doppler shifted detuning values$\delta_\rightarrow - kv_x$ and $\delta_\leftarrow + kv_x$. Thus, because detuning can be mapped to velocity space, Fig. 3 can be reinterpreted as measurements for a fixed value of$\delta_\leftarrow = -\delta_\rightarrow$, and for three different velocities $v_x < 0$ (4 MHz), $v_x = 0$ (6 MHz), and $v_x > 0$ (8 MHz). This is sketched in Fig. 4, where we see that the two peaks separate (approach) each other for $v_x > 0$ ($v_x < 0$, respectively).

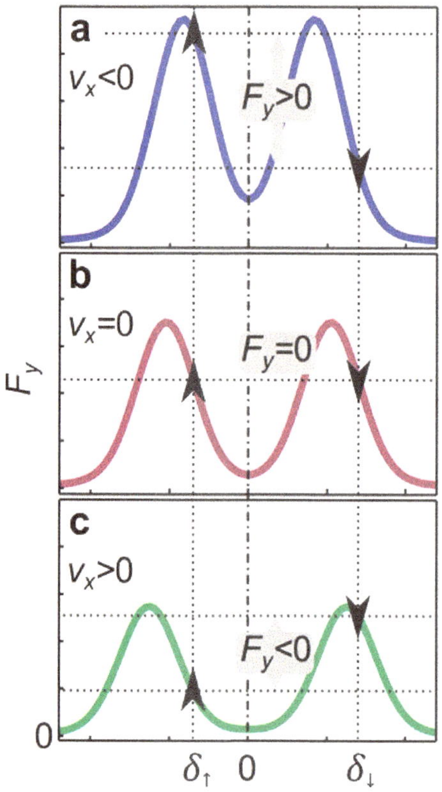

Figure 4: Interpretation of the synthetic Lorentz force from Figs 1and 2, via two-step two-photon resonances presented in Fig. 3

Sketch of the resonant peaks that would be obtained in the setup shown in Fig. 3 for three atomic velocities: (**a**) $v_x < 0$, (**b**) $v_x = 0$, and (**c**) $v_x > 0$. Vertical dotted lines illustrate the values of the detuning used for the lasers aligned on the y-axis in Fig. 1. Arrows denote the direction of the force exerted by those lasers, and illustrate the way F_y observed in Fig. 2arises as a function of velocity v_x. See text for details.

We use Fig. 4 for a detailed explanation of the synthetic Lorentz force measured in Fig. 2. In measurements shown in Fig. 2, we have $\delta_\uparrow < 0$, which means that the positive force $F_y > 0$ in Fig. 2 results from the left resonance peak in Fig. 4. Likewise, because $\delta_\downarrow > 0$ was used for Fig. 2, the negative force $F_y < 0$ results from the right resonance peak in Fig. 4. The transverse force F_y measured in Fig. 2, can be approximately identified with $[F_y(\delta_\uparrow) - F_y(\delta_\downarrow)]_{v_x}$, as illustrated in Fig. 4.

The choice of $\delta_\uparrow$ and $\delta_\downarrow$ in Fig. 2 is such that the strength of the forces arising from the two peaks balance each other, giving $F_y = 0$ for $v_x = 0$ [Fig. 4(b)]. Moreover, $\delta_\uparrow$ and $\delta_\downarrow$ are on the slopes of the two peaks in Fig. 4(b) for an atom with $v_x = 0$, where the maximal magnitude of $\Delta F_y/\Delta v_x$ is expected (detuning translates into velocity space via Doppler effect). For an atom with $v_x < 0$ [see Fig. 4(a)], the two peaks approach each other, yielding greater force from the left peak, which results in $F_y > 0$ for $v_x < 0$, and the opposite for $v_x > 0$ which yields $F_y < 0$ as shown in Fig. 4(c).

It should be pointed out that our scheme is inherently asymmetric. The intensity of the resonance peaks shown in Fig. 4 decreases when they separate (for $v_x > 0$), in contrast to when they approach each other (for$v_x < 0$). Thus, the net force along y is larger for negative v_x, than for the velocity of the same magnitude but with a positive sign. By using a scheme with four atomic levels[17], one could remedy the asymmetry in the force, present in the three-level scheme. In addition, the initial density of the atomic cloud is not identical prior to acceleration in the positive/negative direction due to the slight asymmetry of the displaced trapping potential. This also affects the transverse force due to absorption and multiple scattering[21]. The inherent asymmetry in the force $F_y(v_x)$ and in the initial density of the cloud, is reflected in the asymmetry of the observed motion in Fig. 2 for $v_x < 0$ and $v_x > 0$. Importantly, all of these details only quantitatively affect the results, but not qualitatively.

CONCLUSION

The presented experiment, which demonstrates the synthetic Lorentz force by using radiation pressure, is performed in a classical cold atomic gas, prepared in a MOT, where the heating due to spontaneous emission does not prevent the observations. The concept of a force is natural in our classical laser cooled system[18]. We would like to emphasize that our method is entirely different from the Berry phase approach, where the connection with the Lorentz force can be made in a semiclassical approximation[22].

In the outlook, we foresee many intriguing novel experiments based on the presented method. First, one could develop experiments using more sophisticated schemes, involving more atomic levels (see Ref. [17]), to create a uniform synthetic magnetic field. Next, we plan to adjust our system for the observation of the predicted rotation of the cloud during expansion[17]. One of the goals of this research is to build up a toroidal trap for cold atoms with a toroidal synthetic magnetic field, which holds potential to emulate the plasma in a tokamak. The proposed concept could be used for velocity selection of atomic beams, or for developing a novel kind of mass spectrometer for neutral atoms. We believe that our concept or an analogous scheme could be applicable in other systems, such as suspended nanoparticles.

METHODS

Experimental setup

The ^{87}Rb MOT is set up in the standard $\sigma^+ \sigma^-$ retro-reflected configuration, with beam diameters of 2 cm. The trap is vapour loaded in a glass cell which facilitates fast switching of the magnetic field. Cooling and repumper lasers are external cavity diode lasers (ECDL) delivering total powers of 80 mW and 20 mW, respectively. The cooling laser is typically $2\pi \times 24$ MHz (4Γ) red-detuned from the $|5S_{1/2}; F = 2\rangle \rightarrow |5P_{3/2}; F' = 3\rangle$ hyperfine transition. The repumper laser is in resonance with the $|5S_{1/2}; F = 1\rangle \rightarrow |5P_{1/2}; F' = 2\rangle$ hyperfine transition, thus keeping most of population in the $|5S_{1/2}; F=2\rangle$ ground level. The quadrupole magnetic field gradient is 13 G/cm. The number of atoms in the trap is deduced by measuring the cloud fluorescence with a calibrated photodiode.

For the implementation of the synthetic Lorentz force two additional ECDL lasers were introduced in the experiment: one at 780 nm driving $|5S_{1/2}; F = 2\rangle \rightarrow |5P_{3/2}; F' = 3\rangle$ transition, and other at 776 nm driving $|5P_{3/2}; F' = 3\rangle \rightarrow |5D_{5/2}; F'' = 4\rangle$ transition. Each laser beam is split into two beams which are sent on the rubidium cloud in counter-propagating configurations as shown in Fig. 1. Frequency and intensity control is done separately for each of the four laser beams with acoustic-optical modulators (AOM). After being frequency shifted, we couple the beams to single mode polarization maintaining fibres, which ensures linear polarization (in the z-direction) and uniform intensity.

All lasers used in the experiment are frequency stabilized by using modulation transfer spectroscopy. We modulate the laser diode current to modulate the laser frequency, which effectively increases the laser linewidth. This additional broadening is taken into account in theoretical calculations. The linewidths were checked by heterodyne beating of two stabilized lasers with similar locking parameters. For locking the laser tuned at 776 nm, we counter-propagate picked off beams from the 776 nm laser and the 780 nm laser through a heated^{87}Rb glass cell, where the 780 nm laser populates the $|5P_{3/2}; F' = 3\rangle$ level. We observe an absorption signal resulting from the $|5P_{3/2}; F' = 3\rangle \rightarrow |5D_{5/2}; F'' = 4\rangle$ transition. This signal is mixed with the modulation signal from the 780 nm laser to obtain a frequency locking error signal. Therefore, there is no need to modulate the 776 nm laser frequency to stabilize it.

Theoretical Calculation of the Radiation Pressure Force

To calculate the force plotted in Fig. 3, we first solve the optical Bloch equations (OBEs) to find the density matrix $\hat{\rho}$[18]:

$$\frac{d\rho_{nm}}{dt} = -\frac{i}{\hbar}[\hat{H}, \hat{\rho}]_{nm} - \gamma_{nm}\rho_{nm}, \quad n \neq m,$$

$$\frac{d\rho_{nn}}{dt} = -\frac{i}{\hbar}[\hat{H}, \hat{\rho}]_{nn} + \sum_{E_m > E_n} \Gamma_{nm}\rho_{mm} - \sum_{E_m < E_n} \Gamma_{mn}\rho_{nn}, \qquad (1)$$

(1)

where n, m = 0, 1, 2. The Hamiltonian $\hat{H} = \hat{H}^{(0)} + \hat{H}^{(1)}$ describes a three-level system interacting with laser fields. $\hat{H}^{(0)}$ is represented by a diagonal matrix with elements $H_{00}^{(0)} = 0$, $H_{11}^{(0)} = \hbar\omega_{01}$, and $H_{22}^{(0)} = \hbar(\omega_{01} + \omega_{12})$, whereas thc intcraction Hamiltonian is, in the dipole and the rotating wave approximation18,

$$\hat{H}_{12}^{*} = \hat{H}_{21} = -\frac{\hbar\Omega_{12}}{2}\left(e^{i(k_x x-\omega_{01}t-\delta_{\rightarrow}t)} + e^{i(-k_x x-\omega_{01}t-\delta_{\leftarrow}t)}\right),$$
$$\hat{H}_{23}^{*} = \hat{H}_{32} = -\frac{\hbar\Omega_{23}}{2}e^{i(k_y y-\omega_{12}t-\delta_{\uparrow}t)}, \quad (2)$$

(2)

and zero otherwise. The linewidths, $\Gamma_{12}=2\pi\times 6.1$ MHz, $\Gamma_{23}=2\pi\times 0.66$ MHz, and the transition frequencies, $\omega_{01}=k_x c$ and $\omega_{12}=k_y c$, where $k_x=2\pi/780\ \text{nm}^{-1}$ and $k_y=2\pi/776\ \text{nm}^{-1}$, correspond to the experiment; $\Gamma_{13}=0$; the coherences are $\gamma_{ij}=\frac{1}{2}(\Gamma_{ii}+\Gamma_{jj})$, where $\Gamma_{ii}=\sum_{j<i}\Gamma_{ji}$. Because transition $|0\rangle\rightarrow|1\rangle$ is driven with two lasers of different detuning values$\delta_{\leftarrow}=-\delta_{\rightarrow}$, there is no stationary solution, and the level populations oscillate with the frequency $|\delta_{\rightarrow}|\sim 6$ MHz (the phase of these oscillations depends on x, but this is not reflected in the force). The force on an atom is calculated by employing the Ehrenfest theorem[18], and averaging over the oscillations: $F'_y(\delta_{\rightarrow}, \delta_{\leftarrow}, \delta_{\uparrow}) = \langle -Tr[\hat{\rho}\partial\hat{H}/\partial y]\rangle_t$. The average over time $\langle\cdots\rangle_t$ is justified because the oscillations at frequency $|\delta_{\rightarrow}|\sim 6$ MHz are much faster than the characteristic scale for measuring the force,τ_F, which is on the order of a few milliseconds.

The outlined procedure for calculating the force assumes that the lasers are perfectly monochromatic. However, there is finite laser linewidth that should be taken into account to quantitatively describe the two-step two-photon resonances in Fig. 3, i.e., $F_y(\delta_{\uparrow})$. The spectral profile of the diode laser can be described by a Gaussian in the frequency domain, $\propto \exp((\omega-\omega_0)^2/2\sigma^2)$. In the experiment, the modulation transfer spectroscopy for laser locking is used, and the central laser frequency is modulated with the amplitude $d\approx 2\pi\times 1$ MHz and frequency $\eta=2\pi\times 14$ kHz: $\sim w0^{(t)}=\omega_0+d\sin\eta t$. Thus, the laser spectral profile, over time-scales larger than $1/\eta$, is given by

$$p(\omega;\omega_0) = \frac{p_0\eta d}{\sigma\sqrt{2\pi^3}}\int_{\omega_0-d}^{\omega_0+d}\frac{\exp(-(\omega-\tilde{\omega}_0)^2/2\sigma^2)}{\left|\frac{d\tilde{\omega}_0}{dt}\right|_{t(\tilde{\omega}_0)}}d\tilde{\omega}_0$$
$$= \frac{p_0}{\sigma\sqrt{2\pi^3}}\int_{\omega_0-d}^{\omega_0+d}\frac{\exp(-(\omega-\tilde{\omega}_0)^2/2\sigma^2)}{\sqrt{1-\left(\frac{\tilde{\omega}_0-\omega_0}{d}\right)^2}}d\tilde{\omega}_0, \quad (3)$$

where p_0 is the normalization factor.

The detuning values of 780 nm lasers oscillate as follows: $\tilde{\delta}$ $\rightarrow=\delta_{\rightarrow}+d\sin\eta t$ and $\tilde{\delta}=\delta_{\leftarrow}+d\sin\eta t$. Note that $\eta\ll|\delta|\rightarrow$, and at the same time $1/\eta$ is sufficiently smaller than τ_F. This means that we can employ the separation of scales to take into account finite laser linewidth. We calculate the resulting force $F_y(\delta_{\uparrow})$ on the cloud by averaging F'_y $(\delta_{\rightarrow}, \delta_{\leftarrow},\delta_{\uparrow})$ with the appropriate laser profile distribution in the frequency domain:

$$F_y(\delta_{\uparrow}) = \int_{-\infty}^{\infty} p(\omega;\omega_{01}+\delta_{\rightarrow})F'_y(\omega-\omega_{01},\omega-\omega_{01}+2|\delta_{\rightarrow}|,\delta_{\uparrow})d\omega. \quad (4)$$

The width σ was determined to be $2\pi\times 1.5$ MHz by fitting to the experimental profiles. The Rabi frequencies used in the calculation are discussed in the Results Section.

REFERENCES

1. Madison, K. W., Chevy, F., Wohlleben, W. & Dalibard, J. Vortex Formation in a Stirred Bose-Einstein Condensate, Phys. Rev. Lett.84, 806–809 (2000).
2. Abo-Shaeer, J. R., Raman, C., Vogels, J. M. & Ketterle, W.Observation of Vortex Lattices in Bose-Einstein Condensates,Science 292, 476–479 (2001).
3. Lin, Y.-J., Compton, R. L., Jimnez-Garca, K., Porto, J. V. &Spielman, I. B. Synthetic magnetic fields for ultracold neutral atoms, Nature 462, 628–632 (2009).
4. Le Blanc, L. J. et al. Observation of a superfluid Hall effect, PNAS109, 10811–10814 (2012).
5. Aidelsburger, M. et al. Experimental Realization of Strong Effective Magnetic Fields in an Optical Lattice, Phys. Rev. Lett. 107, 255301 (2011).
6. Struck, J. et al. Tunable Gauge Potential for Neutral and Spinless Particles in Driven Optical Lattices, Phys. Rev. Lett. 108 225304 (2012).
7. Miyake, H., Siviloglou, G. A., Kennedy, C. J., Burton, W. C. &Ketterle, W. Realizing the Harper Hamiltonian with Laser-Assisted Tunneling in Optical Lattices, Phys. Rev. Lett. 111, 185302 (2013).
8. Aidelsburger, M. et al. Realization of the Hofstadter Hamiltonian with Ultracold Atoms in Optical Lattices, Phys. Rev. Lett. 111, 185301 (2013).
9. Aidelsburger, M. et al. Measuring the Chern number of Hofstadter bands with ultracold bosonic atoms, Nature Physics 11, 162–166 (2015).
10. Jotzu, G. et al. Experimental realization of the topological Haldane model with ultracold fermions, Nature 515, 237–240 (2014).
11. Ray, M. W., Ruokokoski, E., Kandel, S., Mottonen, M. & Hall, D. S.Observation of Dirac monopoles in a synthetic magnetic field,Nature 505, 657–660 (2014).
12. Bloch, I., Dalibard, J. & Nascimbene, S. Quantum simulations with ultracold quantum gases, Nature Physics 8, 267–276 (2012).
13. Dalibard, J., Gerbier, F., Juzeliunas, G. & Öhberg, P. Colloquium: Artificial gauge potentials for neutral atoms, Rev. Mod. Phys. 83, 1523–1543 (2011).
14. Goldman, N., Juzeliunas, G., Öhberg, P. & Spielman, I. B. Light-induced gauge fields for ultracold atoms, Rep. Prog. Phys. 77, 126401 (2014).
15. Cooper, N. R. Rapidly rotating atomic gases, Adv. Phys. 57, 539–616 (2008).
16. Dum, R. & Olshanii, M. Gauge Structures in Atom-Laser Interaction: Bloch Oscillations in a Dark Lattice, Phys. Rev. Lett.76, 1788–1891 (1996).
17. Dubček, T. et al. Synthetic Lorentz force in classical atomic gases via Doppler effect and radiation pressure, Phys. Rev. A 89, 063415 (2014).

18. Metcalf, H. J. & Van Der Straten, P. Laser Cooling and Trapping, (Springer, New York, 1999).
19. Steck, D. A. Rubidium 87 D Line Data. Available online at:http://steck.us/alkalidata/rubidium87numbers.pdf (2010) Date of access: 1st June 2015.
20. Sheng, D., Perez Galvan, A., and Orozco, L. A. Lifetime measurements of the 5d states of rubidium, Phys. Rev. A 78, 062506 (2008).
21. Chabe, J. et al. Coherent and incoherent multiple scattering, Phys. Rev. A 89, 043833 (2014).
22. Cheneau, M. et al. Geometric potentials in quantum optics: A semi-classical interpretation, Europhys. Lett. 83, 60 001 (2008).

CHAPTER 10

CME DEVELOPMENT: THEORIES AND OBSERVATIONS

Jun Lin, Nicholas A. Murphy, Chengcai Shen, John C. Raymond Katharine K. Reeves, Jiayong Zhong, Ning Wu4 Yan Li

Keywords: Sun: flares Sun: coronal mass ejections (CMEs) Plasmas Magnetic reconnection Instabilities

INTRODUCTION

Magnetic reconnection is a ubiquitous process in magnetized plasmas, in which magnetic energy is converted into heat and kinetic energy of the plasma, and the kinetic energy of energetic particles. In the simplest picture of this process, two plasmas with magnetic fields of opposite polarity are brought together, producing a magnetically neutral region, or a neutral point between the two fields. Magnetic field lines quickly break and reconnect, and release the magnetic energy. When stretched or squeezed, the neutral region collapses to a current sheet (CS; see also Figs. 2.1 and 2.2 of Priest and Forbes 2000, which show the evolution in various magnetic configurations including an X-type neutral point, also known as the X-point, and/or a CS) where magnetic reconnection takes place.

Usually, in the solar and the other astrophysical environments, we may not be able to pin down mechanism for reconnection in a specific event due to the limited observations, and also due to the fact that even the detail in the reconnection region theoretically is still an open question. Different mechanisms could give rise to the same output. For example, Sweet-Parker CS with anomalous resistivity (Priest 1982; Biskamp 2000; Priest and Forbes 2000), Petschek-type reconnection (Petschek 1964; Biskamp 1986; Forbes and Priest 1987; Yokoyama and Shibata1994; Kulsrud 2001; Drake et al. 2006; Baty et al. 2009; Vršnak et al. 2009; Zweibel and Yamada2009; Ko et al. 2010; Liu et al. 2012), and turbulent reconnection (Strauss 1988; Bhattacharjee and Yuan 1995; Lazarian and Vishniac 1999; Lin et al. 2007, 2009; Lazarian et al. 2012, 2014). These different dissipation styles may even work together to yield a much more complex pattern and more efficient diffusion than that dominated only by a single mechanism (e.g., see also Mei et al. 2012). Therefore, in the present work, the term of "current sheet" should be understood in a more general sense than was used traditionally such that it refers to any region or structure in which magnetic diffusion occurs or has occurred rapidly.

In the solar eruptive process, the closed magnetic field is severely stretched, and the neutral region generally appears in the form of a CS (e.g., see also Priest and Forbes 2002; Lin et al.2003; Forbes et al. 2006). Magnetic reconnection through the CS helps the magnetized plasma ejected by the eruption propagate away smoothly and produces a flare near the solar surface. Usually, the eruption happens in an environment where the electrical conductivity of the plasma and the length scale of the magnetic field are fairly large, so rapid dissipation of the magnetic field is almost impossible (e.g., Priest 1982; Priest and Forbes 2000) unless the thickness of the CS or quasi-separatrix layers (e.g., Démoulin et al. 1996; Priest and Forbes2000) is small so that fast magnetic reconnection could occur easily.

The magnetic configuration in the solar atmosphere in which magnetic reconnection occurs and two-ribbon flares are produced was suggested by Carmichael (1964), and the famous Kopp-Pneuman model for the two-ribbon flare (Fig. 1) was developed on the basis of Carmichael's original idea (e.g., see Kopp and Pneuman 1976; Švestka 1976; Švestka and Cliver 1992). In this model, energy is stored in a force free magnetic arcade prior to the eruption. Eventually, the field erupts to form a fully open magnetic configuration including a CS due to instabilities in the system (Fig. 1a). The open configuration then relaxes into a closed, nearly potential field

Citation: Jun Lin Nicholas A. Murphy Chengcai Shen John C. Raymond Katharine K. Reeves Jiayong Zhong Ning Wu Yan Li, CME Development: Theories and Observations, Doi 10.1007/s11214-015-0209-0.

via magnetic reconnection in the CS, releasing the magnetic energy stored in the system beforehand, and returns to the initial closed structure. This implicitly suggests that the magnetic energy stored is mainly released by the post-eruption reconnection process (e.g., see also Forbes et al. 1994). The clear manifestation of this process in observations is a bright post-flare loop in the corona and a pair of separating bright ribbons on the solar disk, for which the "two-ribbon flare" was named. Several early versions of the two-ribbon flare configuration could be found in the work of Švestka and Cliver (1992). It is now known that the motions of the flare ribbons and loops are not due to mass motions of the plasma but rather due to the upward propagation of the energy source onto new magnetic field lines at higher altitudes (Schmieder et al. 1987); the hot X-ray loops are the newly reconnected closed loops, located at the outermost edge of the loop system, mapping to the outer edge of the flare ribbons; while the cool Hα loops are formed from the hot ones by conductive and radiative cooling (see Ko et al. 2003 and references therein for a brief review on this issue). (Note: The term "altitude" used in this paper means the distance measured from the surface of the Sun.)

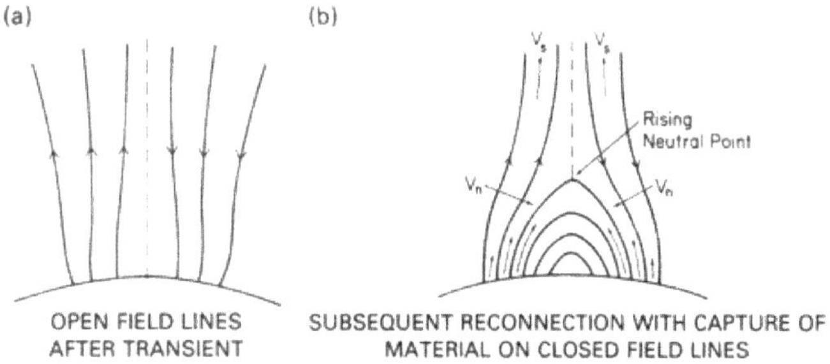

Figure 1: Postflare/CME loop model by Kopp and Pneuman (1976). (**a**) The magnetic field is pushed open by an eruption, and a current sheet separates two antiparallel magnetic field lines. (**b**) The opened configuration relaxes into a closed, nearly potential field via magnetic reconnection in the current sheet. This process produces two bright and separating flare ribbons on the solar disk and a continually growing flare loop system in the corona (reproduction of Fig. 1 of Kopp and Pneuman 1976)

Because of the approximate force-free environment in the low corona, the open or semi-open magnetic structure required for the Kopp-Pneuman model cannot exist unless disruption occurs and the closed magnetic field is quickly stretched outward. According to this basic physical properties of this process, Forbes and co-workers (e.g., see Forbes and Isenberg 1991; Isenberg et al. 1993; Forbes et al. 1994; Forbes and Priest 1995) investigated in detail how a closed coronal magnetic configuration could evolve to a Kopp-Pneuman-type configuration. They pointed out that the coronal magnetic field continues to evolve in response to the motion of plasma in the photosphere, and the kinetic energy of mass motions in the photosphere is thus converted into the magnetic energy that is stored in the coronal magnetic field. Usually, this process takes tens of hours, even days, and the coronal magnetic field evolves quasi-statically until the energy stored in the corona exceeds the threshold and the system evolves to the critical point.

Generally, the amount of energy stored in the coronal magnetic configuration is a critical parameter used for describing the equilibrium state of the system (see discussions of Forbes1991). Equivalently, on the other hand, other parameters for the property of the photospheric boundary conditions may also constitute the critical parameter of the catastrophe model (see detailed discussions by Forbes and Isenberg 1991 and Lin et al. 2003). These parameters include the total magnetic flux sent from the photosphere to the corona (Forbes and Isenberg1991), strength of the background field (Isenberg et al. 1993; Lin et al. 1998), distance between the source regions on the photosphere (e.g., see Forbes and Priest 1995), and so on.

As the system evolves to the critical point, further evolution in or a perturbation to the configuration causes the loss of equilibrium in the system, and the consequent evolution becomes dynamic. Then the magnetic energy stored in the system is quickly released, giving rise to the eruption. The energy release often lasts from several tens of minutes to many hours. The transition from the quasi-static to dynamic evolution constitutes the catastrophe, and the corresponding model for solar eruptions are known as the catastrophe model (see Lin et al.2003 for more details). Alternative models to the catastrophe one for triggering eruptions include the sheared arcade model (Mikić et al. 1988; Mikić and Linker 1994; Linker et al. 2003; Amari et al. 2005, 2010; Reeves et al. 2010), the breakout model (Antiochos et al. 1999; Lynch et al. 2010; Karpen et al. 2012), the ideal MHD model on the basis of the kink and torus instability (Titov and Démoulin 1999; Török and Kliem 2005; Kliem and Török 2006; Fan and Gibson 2007; Kliem et al. 2010; Karlický and Kliem 2010), the tether-cutting model (Moore et al. 2001), and so on (see also Shibata and Magara 2011; Yang et al. 2012; Schmieder et al. 2013; Yan et al. 2014).

In the sheared arcade model, a set of simply-connected force-free magnetic arcades are sheared, and the arcades expand outward gradually, a CS develops between two magnetic fields of opposite polarity, and the evolution in the system turns into dynamic as the diffusion in the CS is invoked. In the breakout model, the magnetic configuration has a quadrupolar geometry. Shearing the central arcade causes this part of magnetic field to expand outward and interact with the background field overhead, and magnetic reconnection eventually results in the loss of the equilibrium in the system. In the ideal MHD model on the basis of the kink and torus instability, a toroidal electric current channel could be in the tokamak equilibrium (or Shafranov equilibrium, Shafranov 1966) as the magnetic tension due to the background magnetic field and the magnetic compression due to the curvature of the current channel balance one another (see also Lin et al. 1998, 2002), and the equilibrium is lost as the balance breaks down. Recently, Kliem et al. (2014) confirmed the equivalent role of the kink and torus instability to the catastrophic loss of equilibrium in triggering the eruption, and further pointed out that in both frameworks, the equilibrium in system could be driven to non-equilibrium through the tether–cutting process; Longcope and Forbes (2014) have also found that a flux rope in quadrupolar external field could reach a catastrophe along various evolutionary paths, depending on the detailed form of the initial equilibrium.

Basically, all these models (theories) of solar eruptions are fundamentally similar, or even the same, such that the energy driving the eruption is stored in the coronal magnetic field beforehand, and the loss of equilibrium in the system triggers the eruption. The difference lies only on the fact that magnetic reconnection is not necessary in triggering the loss of equilibrium in the catastrophe (equivalently the kink and torus instability) model, and it is needed in the other models. The consequent evolutionary features in the disrupting configuration is the same in either model after the loss of equilibrium has occurred (e.g., see also detailed discussions of Lin et al. 2003). In the present work, we follow the catastrophe model for simplicity.

Because the coronal magnetic field is line-tied to the dense photosphere (see the definition and discussions given by Forbes 1991), the closed magnetic field in the lower corona is severely stretched during the catastrophe, creating a CS between two fields of opposite polarity and thrusting the upper part of the magnetic configuration (usually including a filament or current-carrying flux rope) outward. In the ideal magnetohydrodynamic (MHD) environment, the CS is infinitely thin and attached to the photosphere. The increasing magnetic tension due to the magnetic field lines passing over the flux rope and anchoring to the photosphere tries to halt the outward motion of the flux rope, and the evolution of the system will be eventually stopped after the flux rope (or the whole system) finds the new equilibrium state. This implies that the catastrophe cannot develop to a plausible eruption in the ideal MHD environment (see also Lin et al. 2003 for detailed discussions).

In the real coronal environment, on the other hand, the electric resistivity of the Spitzer type is very small, but not zero. So finite diffusion exists although it is indeed very tiny. Therefore, dissipation inevitably occurs, especially in the CS. Since the resistivity is small and the catastrophe takes place much faster than the dissipation at the beginning of the eruption (see discussions of Forbes and Lin 2000; Lin and Forbes 2000; Lin 2002), the CS is able to develop to a large scale, and various plasma instabilities would occur in the CS. Linear theory of the tearing mode instability indicates that the CS becomes unstable as its length exceeds 2π times its thickness (Furth et al. 1963; Priest and Forbes 2000), and non-linear investigations through numerical experiments suggest that this ratio could apparently exceed 10 and even up to 100, before the tearing mode is invoked (e.g., see Loureiro et al. 2007; Ni et al. 2010; Shen et al.2011; Bárta et al. 2011a, 2011b; Mei et al. 2012). Plasma instabilities yield turbulence inside the CS that may significantly enhance the dissipation of the magnetic field, resulting in an effective resistivity that is much higher than the traditional Spitzer value (Strauss 1986, 1988; Ambrosiano et al. 1988; Bhattacharjee and Yuan 1995; Shibata and Tanuma 2001; Skender and Lapenta 2010; Bárta et al. 2011a, 2011b; Shen et al. 2011; Mei et al. 2012; and references therein). The enhanced resistivity accelerates the dissipation in the CS in the form of fast reconnection, which causes the growth of the flare loops below the CS and the rapid expansion of the CME bubble above the CS (see Fig. 2, and Lin et al. 2004; Lin and Soon 2004). Hence, the disrupting magnetic field does not necessarily become fully open before the CME propagates outward from the relevant coronal magnetic configuration.

The scenario sketched by Fig. 2 suggests that the role played by reconnection in an eruptive process is twofold. First, reconnection breaks the magnetic field lines that pass over the flux rope and are anchored in the boundary surface at both ends.

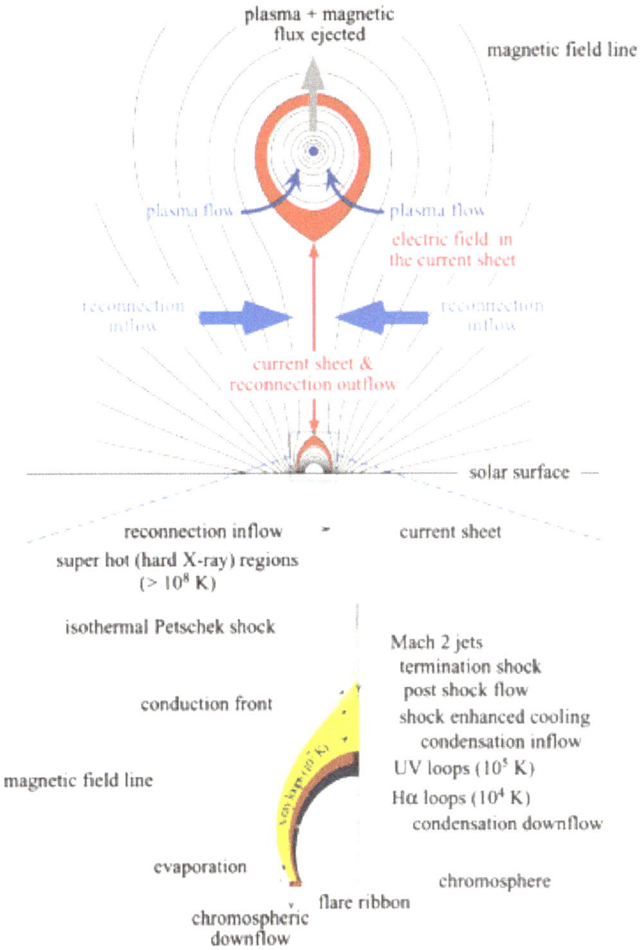

Figure 2: Upper part: Sketch of the flux rope/CME model of Lin and Forbes (2000) showing the eruption of the flux rope, the current sheet formed behind it, and the postflare/CME loops below, as well as the inflows and outflows associated with the reconnection. Lower part: Enlarged view of the postflare/CME loops (adopted from Forbes and Acton 1996). The upper tip of the cusp rises as reconnection happens continuously

These field lines produce strong magnetic tension in a stretched configuration that would otherwise prevent the flux rope from escaping to form a CME. [We note here that, in most cases, breaking magnetic field lines could either take place at the top of the magnetic arcade by reconnection between the arcade and the overlying background field as the break-out model described (e.g., see Antiochos et al.1999and Fig. 3), or take place by reconnection through the CS between the flux rope and the flare as described by the catastrophe model (e.g., see detailed described by Forbes and Lin2000; Lin and Forbes 2000; Lin 2002; Lin et al. 2003, as well as the two thick blue arrows in Fig. 2). On the other hand, works of Aurass et al. (2013) and Aurass (2014) have convincingly showed that both the reconnection sites could be observed simultaneously in the same event. Combinations of X-ray and metric radio spectroscopy and imaging of their data revealed structures and electron acceleration sites at locations beyond the SDO/AIA and SOHO/EIT FOVs but under the edge of the occulting disk of LASCO/C2. However, obtaining such observational data is not trivial, only when the event takes place near, but not at, the limb of the Sun, can it become possible. In the present work, we follow the case of the catastrophe as shown in Fig. 2].

Second, reconnection dumps large amounts of energy in the lower atmosphere of the Sun, yielding intense heating, which accounts for the traditional flare ribbons and loops. Figure 2incorporates the flare models summarized by Forbes and Acton (1996) and the CME model of Lin and Forbes (2000), and clearly illustrates the relationship among solar flares, eruptive prominences, and CMEs such that these three phenomena are different manifestations of a single physical process that involves a disruption of the coronal magnetic field (see also Martens and Kuin 1989; Priest and Forbes 2002). On the other hand, the scenario described by Fig. 2 is usually valid for the major eruptions that produce both solar flares and CMEs, and it may not be good for events that produce either a flare or CME alone.

The development of a long CS following the onset of the eruption is a very important prediction of the Lin and Forbes (2000) model. Because the timescale of reconnection is long compared to the timescale of the onset stage (Alfvén timescale), a long CS would be expected in the eruption even though reconnection is considered fast. The highest altitude flare loops reach and the distance that the rear part of the CME propagates during the typical eruptive event have indirectly confirmed the formation of the long CS (see discussions of Ciaravella et al. 2002; Lin2002; Ko et al. 2003; Lin et al. 2005).

The formation of a CS behind the CME was half-quantitatively predicted, for the first time, by Martens and Kuin (1989), and then quantitatively described by Lin and Forbes (2000) such that a long CS could be expected in the major eruption. The other models for the eruption, like the sheared-arcade model (Mikić et al. 1988; Mikić and Linker 1994) and the breakout model (Antiochos et al. 1999) in which reconnection creates the flux rope, did not explicitly address the development and the other related features/properties of the CS behind the CME probably because either the numerical resistivity was high and the CS is hard to form in the numerical experiments, or the main attention was paid to the triggering of the eruption. In the follow-ups of these works, on the other hand, with the improvement in the numerical techniques and more and more attention being paid on the CS structure, the related works also showed the appearance of a long CS behind the CME in these models, and detailed properties of the CS started to be investigated (e.g., see Linker et al. 2003; Riley et al. 2007; Reeves et al. 2010; Karpen et al. 2012 for details). Therefore, a CS behind the CME is a common feature present in all three models for solar eruptions, namely sheared-arcade, breakout, and catastrophe models.

In addition to these works that considered development of a single CS during the eruptive process assuming that the destabilization of the magnetic configuration is mainly caused by the evolution of the background field, another set of studies takes into account the role of the interaction between a destabilized structure in an active region with overlying closed magnetic fields via coronal magnetic breakout reconnection (e.g., see Antiochos 1998 for the theory, and Antiochos et al. 1999; Lynch et al. 2004, 2008, 2010; Zhang et al. 2005, 2006; Zhang and Wang2007 for simulations). They found that the interaction between two systems of magnetic configuration of different topology results in the formation of a CS parallel to the solar surface, and the disruption of the destabilized structure eventually yields the formation of another CS perpendicular to the surface (see Fig. 3 as an example). The latter CS is also known as the "flare CS" since magnetic reconnection occurring inside directly accounts for the solar flare. Therefore, two CS could be expected to develop in the eruption in the framework of this type of model, namely the "breakout CS", and the "flare CS" is expected to dominate the energetics. As shown by Fig. 3, the breakout CS always appears on the top of the central magnetic arcade above a certain height in the middle corona, and directly penetrates into the CME bubble.

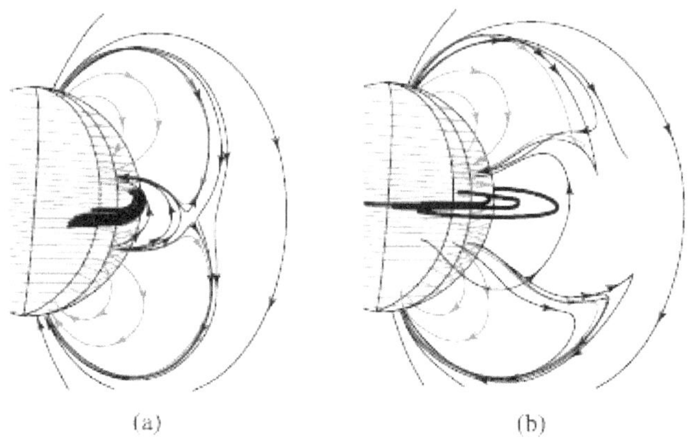

Figure 3: Evolution in the magnetic configuration of the breakout model (Antiochos et al. 1999). Because of the symmetry about the rotation axis, one side is shown. Shearing the arcade field (thick lines) creates a force-free current at the equator, and that the sheared region bulges outwards produces a toroidal current layer as well. Reconnection of the field lines in this layer allows the sheared field lines to open outward to infinity. (**a**) Magnetic configuration at early stage, and (**b**) magnetic configuration at late stage. From Forbes and Lin (2000)

Aurass et al. (2013) observationally verified that this process works from the very beginning of the impulsive flare phase. Using meter wave imaging and spectroscopy these authors have shown that the eruption started almost simultaneously with the formation of two CS perpendicular to each other: the flare CS at low heights, and an azimuthally extended breakout CS at about 0.4 $R_\odot$ in front of the flare CS. In a more recent work by Aurass (2014), the first observation is presented of the reverse coronal breakout (the closure of the coronal field behind the CME body). These two works also further confirmed that the breakout CS exists in a certain height range and is penetrating into the CME bubble.

For the first time, Yokoyama et al. (2001) identified the reconnection inflow above the top of the flare loop system, and Ciaravella et al. (2002) provided spectroscopic evidence of the CME/flare CS by displaying a narrow and hot region between a CME and the associated flare. This region was at the position where the CS is presumably located (see also Fig. 2). Webb and Cliver (1995) found evidence for magnetic disconnection in the form of concave-outward features in white-light images, and Webb et al. (2003) then analyzed the data from the Solar Maximum Mission (SMM) satellite, and noticed that many CMEs were followed by a long ray-like structure in white light, and they identified this structure with the CS. Ko et al. (2003) studied the emission features and properties of the CS in detail, identified several plasma blobs flowing away from the Sun along the CS and measured the speed of the reconnection outflow and the rate of magnetic reconnection as well as the magnetic field strength near the CS. Lin et al. (2005) performed direct measurement of the reconnection inflow speed near the CS, identified five moving plasma blobs to obtain the reconnection outflow speed and the rate of reconnection. McKenzie and Hudson (1999), Innes et al. (2003a, 2003b), Asai et al. (2004), Sheeley et al. (2004), Reeves et al. (2008a), and Savage and McKenzie (2011) studied the sunward reconnection outflow above the flare loops and recognized fine structures of the plasma flow, also known as the supra-arcade downflow (SAD).

The reconnection outflows play an important part in the structure of the CME. Numerical experiments performed by Riley et al. (2007) displayed bi-directional reconnection outflows along the CS in the solar eruptive events. Raymond et al. (2003), Lin et al. (2004), Lin and Soon (2004), and Bemporad et al. (2006) investigated the consequences of the reconnection process manifested by the thermal structures and kinetic behaviors of CMEs. They found that the CME bubble is surrounded by the hot reconnected plasma, and that the rapidly expanding CME bubbles are mainly due to the addition of magnetic flux brought by the reconnection outflow from the CS into the CME. Lin et al. (2004) and Lin and Soon (2004) further pointed out that it is the anti-sunward reconnection outflow that accounts for the three-component structure of the CME such that the mass and the associated magnetic field in the core of the CME come from the prominence before the eruption, and those surrounding the core are first brought into the CS by the reconnection inflow from the corona nearby, and then sent into the CME bubble around the core by the reconnection outflow. Observations of STEREO confirmed this scenario (e.g., see Davis et al. 2009), and indicated that the three-component structure of the CME could even be observed to a very large distance from the Sun (e.g., see Fig. 2 of Davies et al. (2009), which described mass flows from the Sun, including the CME).

Meanwhile, more and more attention has now been drawn to the CS, and studies on the topic have been continuously pushed forward. In this review, we introduce observational features of structures associated with the CS and related results of investigations, discussing studies and results in both theories and numerical experiments on the CS, and then pointing out the importance of these results to our understanding and knowledge of the CS connecting a CME to the associated flare, as well as properties of magnetic reconnection occurring in such a region. We display the first set of observational works on the CS, including identifications of the CS shortly after the prediction of Lin and Forbes (2000) and observational features of reconnection inside and around the CS in Sects. 2 and 3. Table 1 contains several highly valuable observations that are discussed in this work; the related events are listed in the order that they appear in the text for the reader's reference. Observational studies regarding the internal structure and the scale of the CS will be introduced in Sect. 4, and numerical experiments on the relevant topics and the results are unfolded in Sect. 5. We finally summarize this review in Sect. 6.

Table 1: Dates of the events discussed in this work and the section(s) where they are mentioned

Event date	**Section(s)**	**Event date**	**Section(s)**
23 Mar 1998	2, 4.1	03 Nov 2010	3.1
21 Apr 2002	2, 3.2	17 Aug 2011	2, 3.2
20 Nov 1988	2	20 Jan 1999	2, 3.2
23 Jul 2002	2, 3.2	08 Jan 2002	2, 3.3, 4.1
24 Aug 2002	2	09 Apr 2008	2, 3.4
26 Nov 2002	2	18 Aug 2010	3.4
28 Jul 2004	2	17 May 2002	2
01 Dec 2004	3.4	03 Nov 2003	2
26 Jun 2005	4.1	18 Mar 1999	3.1
07 Sep 2005	2, 4.1	18 Nov 2003	2, 3.1, 3.3, 4.1
04 Nov 2003	4.2	29 Jul 2004	3.1, 3.3
28 Oct 2003	6		

IDENTIFICATIONS OF THE CURRENT SHEET IN THE SOLAR ERUPTIVE EVENTS

Usually, direct observations of the current sheet are very hard, if not impossible. The high electrical conductivity and nearly force free environment confine the current sheet to a very local region in the direction perpendicular to its vertical extent (refer to Fig. 2). Thus, its thickness is usually small compared to the size of nearby large-scale structures. In addition, the low plasma density in the corona implies that the material inside the CS should be tenuous as well. Therefore, it is not easy to observe the current sheet in an eruptive process because both its size and emission are easily dominated by large-scale and bright structures nearby (see our detailed discussions below and those given by Forbes and Acton 1996) although the CS is indeed a common feature developed in the major eruption as we mentioned earlier. Using data from Yohkoh and SOHO, Švestka (1996), Švestka et al. (1998), McKenzie and Hudson (1999), and McKenzie (2000) studied behaviors of regions above the top of the flare loop system, and displayed typical characteristics of the reconnection region and revealed important information above the reconnection process.

Ciaravella et al. (2002) analyzed spectral data obtained from the Ultraviolet Coronagraph Spectrometer (UVCS; Kohl et al. 1995) on SOHO for an eruption on 1998 March 23. This event developed a slow CME propagating at speed of 100 kms^{-1} when the CME core was at the heliocentrical distance (namely the distance measured from the center of the Sun) of 3.5 $R_{\odot}$, and it reached a speed of 180 kms^{-1} as the core was at 15 $R_{\odot}$. It is usual for slow CMEs to accelerate to a distance of several solar radii (Sheeley et al. 1999). Meanwhile, flare loops in the low corona were observed by EIT in 304 Å, and emission from the high temperature coronal forbidden lines [Fe XVIII] (maximum emission occurs at temperature of 6×106 K) and [Ca XIV] (maximum emission occurs at temperature of 4×106 K) were identified by UVCS between the flare loops and the CME core.

Together with these two very high temperature lines, emission from some other active region spectral lines, such as Fe XVII and Ne IX were also observed. But the emission from the ordinary lines (namely the spectral lines that can usually be observed in the quiet corona) was dispersed along the UVCS slit, and the emission from [Fe XVIII] and [Ca XIV] appeared in a narrow and bright region. Figure 4 displays a composite of EIT (304 Å), LASCO/C2, and UVCS slit images. The UVCS slit image indicated by the arrow is the true location of the slit when [Fe XVIII] and [Ca XIV] emission was detected, and the other three slits are shown for comparison. The narrow and bright region in [Fe XVIII] and [Ca XIV] emission is located below the CME observed by LASCO/C2, and above the flare loop system observed mainly in Si XI emission by EIT in 304 Å. Figure 3 of Ciaravella et al. (2002) shows LASCO C1 images in Fe XIV λ5302.42 taken at different times on 1998 March 23, and displays the evolution of the magnetic configuration in the event; Figs. 8 and 9 of Ciaravella et al. (2002) reveal features of the resultant flare loop system seen in various wavelengths, respectively.

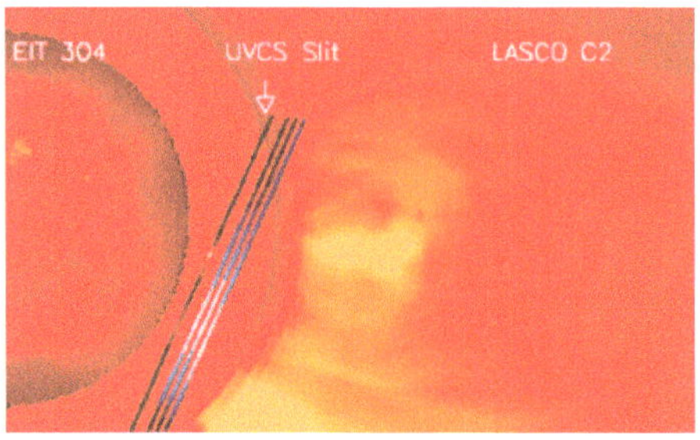

Figure 4: Composite image of EIT 304 Å (22:24 UT), LASCO C2 (12:33 UT), and line intensity distribution along the UVCS (16:56 UT) slit taken on 1998 March 23. The arrow indicates the true position of the UVCS entrance slit, and the intensity distribution along it is that of the [Fe XVIII] λ974 line. The other three slits were plotted for comparison and show (starting from the left) the line intensity distribution of Si XII λ499, O VI λ1032, and Lyβ. The UVCS data used in this figure were taken on March 23 with the slit pointed at 1.45 $R_\odot$ and PA=245°. The position of the peak in the [Fe XVIII] line is PA=257° with height 1.55 $R_\odot$. The different times were chosen to combine the data into a single image. At the time of the UVCS observation, the LASCO CME was much larger but similar in shape. From Ciaravella et al. (2002)

By comparing the relative locations of various important features developed in the eruptive process shown in Fig. 4 with the CME/flare magnetic configuration predicted by Lin and Forbes (2000), Ciaravella et al. (2002) suggested that the narrow and bright feature discussed above could be the region where the reconnecting current sheet between the CME and the associated flare was located. This is the first identification of the possible location of the CS via the spectroscopic approach. As we noted in a previous section, we use the term of CS in this paper in a general sense.

When studying the spectral data of the 2002 April 21 event, Innes et al. (2003a) noticed the emission of the high temperature (~107 K) spectral line, [Fe XXI] 1354Å, and SADs observed by both TRACE and SOHO/SUMER. This event was observed by LASCO and UVCS as well. It developed a fast CME (~2500 kms^{-1}) and an X-1.5 two-ribbon flare, and has been studied extensively (e.g., see also Innes et al. 2003b; Sheeley et al. 2004; Reeves et al. 2008a; and references therein). It is probably the first event developing a CS that was observed face-on [namely the line-of-sight (LOS) is perpendicular to the CS plane and along the two thick horizontal arrows shown in Fig. 2]. At the location where the [Fe XXI] emission was seen, the emission of low temperature spectral lines like C II did not appear. This, again, confirmed apparently plasma heating that took place as a result of magnetic reconnection in the CS above flare loops. Spectra of the reconnection outflow in the CS revealed flows of high speed up to 103 kms^{-1} (Innes et al. 2003b).

Webb et al. (2003) investigated 26 CME events observed in white light by the Solar Maximum Mission satellite. These events displayed magnetic disconnection features, basically rapidly evolving concave-outward bright regions behind CME fronts, and they are also known as the "candidate disconnection events (CDEs)". A set of ray-like structures were observed to form several hours after these CDEs, and they sometimes were seen as well to connect to the streamers above the limb. Two examples of SMM CMEs with CDEs, which were followed by trailing, transient ray-like structures, were chosen to demonstrate the approach for recognizing the structures and identifying the current sheet. Figure 5 displays the evolution of the 20–21 November 1988 CME over the southwest limb. It was typical of the events in several aspects. The pre-existing streamer swelled and brightened, then blew out as the CME, which had a broad, outer loop enveloping a dark cavity, within which appeared a bright, circular core. So this CME possesses the typical three-component structure, namely bright front, dark cavity, and bright core (see Lin et al. 2004; Lin and Soon 2004; and references therein).

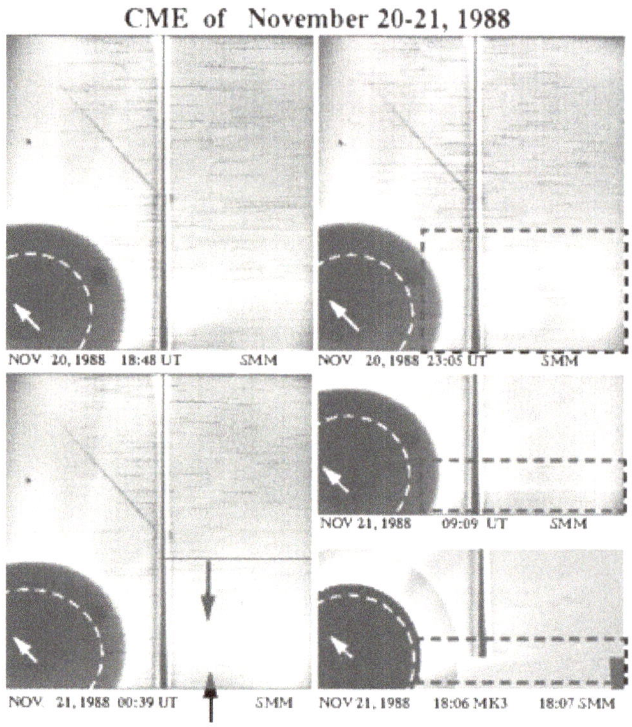

Figure 5: Evolution of the 20–21 November 1988 SMM CME in the southwest. The pre-existing streamer swelled and brightened (20, 18:48 UT), then blew out as the CME. The CME had a broad outer loop enveloping a dark cavity, within which appeared a bright, circular core (20, 2305 UT). The core later revealed a V-shaped "back" with a stalk extending below the occulter (21, 00:39 UT). Hours later, a very bright, narrow ray suddenly appeared a few degrees south of the axis of the previous CME and CDE (21, 09:09 UT). Below this panel is a Mauna Loa Solar Observatory (MLSO) image later on this day superposed within the SMM occulting disk, showing that the ray was part of a newly formed helmet streamer. The curved dashed lines indicate the limb of the Sun, and the arrows point to solar north. The vertical and diagonal lines and the small circular feature in each panel are detector artifacts. From Webb et al. (2003)

The bright core expanded outward, revealing a V-shaped "back" with a stalk extending below the occulter. Many hours later, a very bright and narrow ray structure suddenly appeared a few degrees south of the axis of the previous CME and CDE. As shown in the bottom right panels, the ray was also visible later in the day in the images obtained by the Mauna

Loa Solar Observatory's K-Coronameter, in which it appeared as an extension of a newly formed helmet streamer. After having compared with observations regarding emission and structure features of the CS (see Ciaravella et al. 2002), and with theoretical expectations of the CS dynamic behaviors given by both analytic solution (see Lin and Forbes 2000; Lin 2002) as well as numerical experiments (see, for example, Linker et al. 2003), Webb et al. (2003) regarded this kind of ray-like structures as evidence of the current sheet developed following a CME. This is the first time that a survey of ray-like structures has been performed and the results have been compared with CME models, and is also the first time that the CS was recognized by studying the white-light data. With the success of SDO, the CS associated structures right above flare loops have as well been observed in the AIA 131 Å and 94 Å channels in exquisite detail both as rays when viewed edge on (Reeves and Golub 2011), and as fans when viewed face on (Warren et al. 2011; Savage et al. 2012a; McKenzie 2013).

Raymond et al. (2003) studied the 2002 April 21 event and another two events occurring on July 23 and August 24, 2002, respectively, that all produced X-class (>1.5) flares with associated fast CMEs at speeds of 1700--2500 kms^{-1}. These events occurred in active regions targeted by the Max Millennium Campaign, and were well observed by many instruments both in space and on the ground. The UVCS slit was placed where both CMEs and the CSs behind successively passed through. The UVCS data manifested clear spectral signatures in terms of very rapid disruption of the pre-CME helmet streamer suggesting that all the eruptions were initiated in the magnetic structures below helmet streamers, very high Doppler shifts that indicating high speed motion component along the line-of-sight (LOS), and high-temperature plasma visible in the [Fe XVIII] emission line implying intensive heating taking place during the events.

Raymond et al. (2003) noticed that all three events displayed transient brightening's in the [Fe XVIII] line and little or none of the cool gas that usually dominates UVCS spectra of CMEs, and that the violent disruption of the pre-CME streamer was manifested by splitting of the low temperature (2.9×105 K) O VI line profiles into strongly red and blue shifted components, and by the rapid growth of the region of split profiles along the UVCS slit. Following splitting of the O VI line was transient brightening in the [Fe XVIII] line, placing it well inside the CME front. Consulting Fig. 2, we noticed that the hot plasma created in the current sheet by reconnection could enter the outer shell of the CME bubble (see the red component around the flux rope), and freezing of plasma to the magnetic field confines the hot plasma to the shell. Therefore, the plasma in the other components is cooler than those in the outer shell, which would lead to the transient brightening in the [Fe XVIII] line as the CME passes UVCS slit quickly. These results depicted the dynamic scenario of the fast CMEs via spectral signatures, which was also theoretically described by Lin and Soon (2004) and Lin et al. (2004). Raymond et al. (2003) pointed out that the features seen in [Fe XVIII] line were quite likely the post-CME CS.

Bemporad et al. (2006) studied an event on 2002 November 26, at the time of a SOHO-Ulysses quadrature campaign, which made their work unique because the SOHO-Ulysses data set allowed them to analyze a current sheet structure from its lowest coronal levels out to its in situ properties. The event occurred over the northwest limb and produced a CME with speed at the front increasing from 90 to 420 kms^{-1} and a flare that the Geostationary Operational Environmental Satellite (GOES) system did not class because the main part of the flare was behind the limb, but the growth of the post-CME loop system could be clearly seen. The orientation of the disrupting magnetic configuration resulted in a CS between the flare loop and the CME that had been continuously observed face-on for two days, which made it relatively easy to investigate its temporal behavior. The post-flare loop system continued to be observed in EIT Fe XII and He II for more than 24 hours, and its heights gradually rose from 0.06 to 0.34 $R_\odot$ above the solar surface (see also Fig. 2 of Bemporad et al. 2006, which displayed the images of the flare loop system at different times after the associated CME, as well as the change in the loop heights versus time). However, the typical cusp feature at the end of the CS did not appear in the EIT images, and it did not appear in the LASCO images, either.

About 1.5 hours after the CME left the field-of-view (FOV) of EIT, UVCS started to take spectral data with its slit located at a heliodistance of 1.7 $R_\odot$, covering emissions from both cool and hot plasmas. The emission of [Fe XVIII] indicated the high temperature of plasma due to magnetic reconnection in progress, and its eventual fading implied that the reconnection process was slowing down or stagnating because the free magnetic energy in the system ran out. At the time of the quadrature, Ulysses was directly above the location of the CME and intercepted the ejecta. High ionization state Fe ions (up to Fe XVII) were detected by the Ulysses SWICS throughout the magnetic cloud associated with the CME, compared to the normal charge state of Fe in the solar wind being about 10.

Poletto et al. (2004) extrapolated Ulysses measurements back to the Sun using the in situ measurement of the flow speed of the solar wind plasma, which gave a rough estimate of the origin time for the plasma, and yielded a positive identification of the plasma resulting from the 2002 November 26 CME by comparing features in the solar wind with phenomena directed toward Ulysses in the corona. Therefore, Bemporad et al. (2006) concluded that a relation of the ICME detected by Ulyssees to the CME observed near the Sun by several other instruments could be set up, and the model shown in Fig. 2 fits the scenario manifested by this event very well. Bemporad (2008) further strengthened this conclusion.

To our knowledge, this event is the only one so far whose eruption and consequences were observed (detected) by both remote sensing and in situ instruments. Detecting the high ionization state Fe ions by the Ulysses SWICS implies that magnetic reconnection was still occurring in a very elongated CS a couple of days after the initiation of the event, that the CS was still connecting to the rear part of the CME, and the reconnected hot plasma was continuously sent into the CME bubble as suggested by the theoretical scenarios described by Fig. 2 (e.g., see also detailed discussions of Lin et al. 2004 and Lin and Soon 2004). It is not easy to determine where the CS terminates at the far end although we know that the distance between the CME core and the CS end depends on the rate of magnetic reconnection in the CS and the local Alfvén speed around the CS (e.g., see also Forbes and Lin 2000; Lin and Forbes2000; Lin 2002). In fact, in the Petschek reconnection picture, there may no real distinction, or there may be a shock (e.g., see Shiota et al. 2005). This leaves an open question to our works in the future.

As for the lower tip of a CS and the associated typical cusp structure below it, on the other hand, there are many observations and studies that have been reported (see discussions in Sect. 3.2below for details). Usually, these observations were made not far from the surface of the Sun because the cusp structure above the flare loop system cannot get very high in the corona. Forbes and Lin (2000), Lin and Forbes (2000), and Lin (2002) pointed out that the kinematic behavior of the lower tip of CS, namely the cusp point over the flare loop system, is determined by the rate of magnetic reconnection in the CS itself. Both theoretical calculations (Lin 2002) and numerical experiments (Linker et al. 2003) indicated that the highest altitude to which the CS lower tip could reach in the coronal environment within the typical life time of the major flare (tens of hours) is around 2.5×105 km (0.36 $R_\odot$).

Observations suggested that this value varies from case to case, but is not very different from the theoretical value. We list results for this value deduced from observations of the typical events in various eras in Table 2 for comparison. We notice that the cusp structure above the flare loop system are usually observed below 0.5 $R_\odot$ from the solar surface. This height is obviously below the edges of the occulting disks of the LASCO/C2 and C3 that are located at 1 $R_\odot$ and 2 $R_\odot$ above the solar surface, respectively. Therefore, the cusp region in a flare event cannot appear in the FOV of either C2 or C3. LASCO C1 might be able to observe the structure, but it failed to work soon after the launch of SOHO. Hence, to our knowledge, no study of the CS lower tip on the basis of the LASCO observations has ever been reported.

Table 2: Heights of cusp structures observed in various events

Event date	Cusp heights ($R_\odot$)	Instruments	Reference
21 Feb 1992	0.54	Yohkoh/SXT	Forbes and Acton (1996)
06 Nov 1980	0.26	SMM	Švestka (1996)
20 Jan 1999	0.14	Yohkoh/SXT	McKenzie and Hudson (1999)
20 Apr 1998	0.22	Yohkoh/SXT	McKenzie (2000)
23 Apr 1998	0.40	Yohkoh/SXT	McKenzie (2000)
23 Mar 1998	0.25	SOHO/EIT	Ciaravella et al. (2002)
08 Jan 2002	0.40	SOHO/EIT	Ko et al. (2003)
23 Jul 2002	0.10	TRACE	Asai et al. (2004)
18 Nov 2003	0.25	SOHO/EIT	Lin et al. (2005)
26 Nov 2002	0.35	SOHO/EIT	Bemporad et al. (2006)
21 Apr 2002	0.17	TRACE	Reeves et al. (2008a)
02 May 2007	0.20	Hinode/XRT	Reeves et al. (2008b)
09 Apr 2008	0.25	Hinode/XRT	Savage et al. (2010)
22 Oct 2011	0.22	SDO/AIA	Savage et al. (2012a)
19 Jul 2012	0.32	SDO/AIA	Liu et al. (2013)
03 Nov 2003	0.18	NRH	Aurass et al. (2013)
19 Jan 2012	0.18	SDO/AIA	Innes et al. (2014)
07 Sep 2005	0.30	MLSO/MK4	Ling et al. (2014)

The SXR emission above the solar limb at altitudes ranging from 0.1 to 0.3 $R_\odot$ from the solar surface during the 2002 November 26 event was also observed by RHESSI (e.g., see Saint-Hilaire et al. 2009). This high-altitude coronal X-ray source (HACXS) lasted for 12 hours, and the corresponding temperature peaked around (1.0~1.1)×10^7 K with its emission measure increasing throughout this time interval. It was also observed to progress gradually outward at speed up to 14 kms^{-1}. Saint-Hilaire et al. (2009) noticed that the emission was dominated by the thermal component and lacked a contribution from nonthermal hard X-rays (HXRs). They interpreted the hot plasma as the signature of the material inside a CS in the wake of the CME, and suggested that heating occurring inside the CS could be directly due to the heat conduction from the HACXS and reconnection itself, and that the energetic particles were unlikely to account for the plasma heating. However, the information they collected could not help determine whether the CS and the HACXS were co-located in space. On the other hand, locations and temporal behavior of the HACXS and the EUV loop systems they observed supported the picture that the reconnection site above the loop top continuously moves upward, heating the local plasma to the temperatures of X-ray emission, before it cooled down and was later seen in the EUV. This implied that the EUV loops trailed the X-ray source in both space and time, and that the higher energies were located higher than the lower energies (see also Gallagher 2002), constituting the standard Kopp-Pneuman two-ribbon flare configuration.

Susino et al. (2013) studied EUV/UV data from UVCS, HXR data from RHESSI, and SXR data from GOES of the 2004 July 28 event that developed a partial halo CME with maximum velocity of ~754 kms^{-1} in the online LASCO CME catalog (Gopalswamy et al. 2009) and a small flare. A narrow emission feature in [Fe XVIII] λ974 Å was observed to stand for about 2.5 days by UVCS after the CME crossed the UVCS slit. Analyzing the UVCS data indicated apparent non-thermal broadening due to the turbulent motion inside the CS, and a large density gradient across the CS region, and studying the X-ray data showed that SXR and HXR sources evolved differently from the EUV/UV sources. They concluded that the reconnection process was very likely to be turbulent in the CS, but reconnection occurring above the post-eruption arcades was unlikely to be directly responsible for the high-temperature plasma sampled higher up by UVCS, instead the HXR could be the direct source of high temperature plasma sampled higher up by UVCS as suggested by Saint-Hilaire et al. (2009).

However, we also need to point out here that, according to the standard model of the two-ribbon flare (see Forbes and Acton 1996; Priest and Forbes 2002; and references therein), the HXR source on the top of the post-flare loop is basically created by the collision of the energetic particles and the downward reconnection outflow from the CS with the closed magnetic field lines where flare loops are believed to lie on (see also the lower part of Fig. 2). This is the so-called Masuda scenario of solar flare (e.g., Masuda et al. 1994; Forbes and Acton 1996; Krucker et al. 2008). The lower part of Fig. 2, which is a duplicate of Fig. 1 of Forbes and Acton (1996), describes clearly where the interaction takes place. As we can see from this figure, the interaction region is already detached from the CS although it is still located on the top of the flare loop system, and the hot plasma responsible for the HXR emission is confined in the closed field lines that are not magnetically connected to the CS (see also Fig. 4 of Lin 2004 that describes the evolution of magnetic field lines when the flare loop system is produced by reconnection). Therefore, the thermal energy of the HXR source region within flare loops is not able to propagate to the reconnection region via thermal conduction, and thus cannot account for observational results obtained by UVCS for the CS. Studies of Caspi and Lin (2010) clearly showed that the super-hot (~45 MK) plasma was observed high in the corona, co-locating with a non-thermal coronal HXR source, before the HXR emission could be detected in the lower altitudes. So Caspi and Lin (2010) concluded that this super-hot plasma originated in the corona.

In the framework of Petschek-type reconnection (Petschek 1964), on the other hand, the diffusion region is tiny, and most of the energy dissipation occurs in the shocks that bound the exhaust flow, which includes an upward component and a downward component (see also Vršnak et al. 2009; Ko et al. 2010). Numerical experiments of Mei et al. (2012) showed that a CS in the CME/flare process actually possesses complex structures, in which three types of magnetic reconnection: Sweet-Parker, Petschek, and turbulence, could even take place simultaneously. So UVCS is observing the reconnection region, though much of the emitting gas might not be heated at the position observed, but somewhere between this position and the X-line. As we mentioned earlier in this work as well as in our previous work (e.g., see Lin et al.2009), the reconnection region discussed here is, in fact, the CS in a more general sense.

Studies of Aurass et al. (2009) demonstrated as well heating of the plasma inside the CS by magnetic reconnection during the eruptive event occurring on May 17, 2002. This event was observed to take place on the east limb by many instruments. It produced an M-1.5 flare (see also Karlický et al. 2004) and a CME at an average speed of 665 kms^{-1} according to the SOHO/LASCO CME Catalog.[1] LASCO/C2 images enhanced by the wavelet method showed a bright "stem" feature behind the CME core at the position where a CS is presumably located according to the Lin and Forbes (2000) model. The EUV spectral data from UVCS allowed them to deduce the plasma temperature and density of the CS, and the results of Aurass et al. (2009) indicated that the CS extended from, at least, the heliocentrical distance of 1.5 to 2.1 $R_\odot$. Distributions of brightness in O VI and in Si XII images along the UVCS slit showed that a dimming in the O VI image is co-located with a brightening in the Si XII image in space, and analyzing the intensity ratio of these two spectral lines in that region gave a temperature up to 5×106 K, which indicated intensive heating occurring inside the CS.

Radio data obtained by AIP and Ondřejov radiospectrographs showed simultaneously several groups of drifting pulsating structures (DPSs) in both high (>2 GHz) and low (40--170 MHz) frequencies, and type III radio bursts in the low frequency. The DPS at low frequency and the type III burst took place during the decay of HXR emission observed by RHESSI. Karlický et al. (2004) noticed that the HXR source is located where the rising EIT loops appeared in the impulsive phase of the associated flare. The frequency range of the type III radio burst and the low frequency DPSs indicated the energetic particles were mainly accelerated in the middle and high coronal region where the plasma density corresponds to the frequency range between 40 and 170 MHz. Aurass et al. (2009) and Karlický et al. (2004) did not provide the information of the location of this region in space, but we can estimate the location and the extension of this region approximately by relating the above frequencies to the plasma density if the distribution of the plasma density in the corona is given.

Using the empirical model of Sittler and Guhathakurta (1999) for distributions of the plasma density in the corona, we deduced that the above frequencies correspond to a height range in the corona from the heliocentrical distance of 1.43 to 2.19 $R_\odot$ where energetic particles responsible for the DPS and the type III bursts were accelerated. This is consistent with that of the CS deduced from the UVCS data. So we conclude that those energetic particles were accelerated inside the CS by magnetic reconnection, which also accounts for heating plasma as indicated by the anti-correlation of the Si XII emission to the O VI emission in the CS (see also discussions by Aurass et al. 2009), and that meter wave radio imaging and meter wave radio spectroscopy provide us another access to look into the reconnection region based on the plasma emission due to energetic electrons accelerated in the CS (see discussions below as well).

Observations discussed above were mainly focusing on the CS below the flux rope and roughly perpendicular to the solar surface. Recently, radio observations, together with observations in the other wavelengths, suggested the formation and development of a CS parallel to the solar surface during the eruption (e.g., see Aurass et al. 2013; Aurass 2014). Aurass et al. (2013) studied an eruptive event taking place on November 3, 2003, which developed a major flare of X-3.9 class and a fast CME, focused on the impulsive phase, and looked into key details of the complementary radio data of high time cadence. They studied the radio data from the Leibniz-Institut für Astrophysik Potsdam (AIP), the simultaneous radio images were from the Nançay Multi-frequency Radio Heliograph (NRH), and the single frequency polarimeter data from the Trieste Astronomical Observatory (OAT). The radio emission was apparently non-thermal plasma radiation with a complex structure in both frequency and time. Veronig et al. (2006) studied the motion of the HXR flare loops and noticed the shrinkage of the loop system during the impulsive phase of the event, and the radio data analyzed by Aurass et al. (2013) were obtained right in the time interval when the top of the flare loop system reached its lowest location in height (see also Fig. 2 of Aurass et al. 2013, which displays the impulsive phase synopsis of the event with the AIP dynamic spectrum, radio and X-ray data time line, as well as RHESSI HXR counts).

Two stationary radio sources at meter wavelengths were observed by Aurass et al. (2013) radially aligned at 0.18 and 0.4 $R_\odot$ from the solar surface above the active region and the HXR sources, respectively. The lower source was believed to be the outflow region just above the X-line in the CS, while the upper one was related to that occurring in the CS above the top of the expanding arcade, and the radio emission at lower frequencies surrounded the upper source at the expected locations as displayed in the breakout scenario (see also Fig. 3). Aurass et al. (2013) thus concluded that the upper radio source was the most compelling evidence to date for the onset of breakout reconnection during an eruption. The height stationarity of the breakout sources and their dynamic radio spectrum discriminate them from propagating disturbances. Timing and location arguments revealed for the first time that both the earlier described above the flare loop top HXR source and the lower radio source were emission from the reconnection outflow above the vertical flare (tail) CS. Furthermore, Aurass (2014) investigated the radio emission of the same event in the late gradual phase. In addition to the data from AIP, NRH, and OAT, the data from RHESSI, SOHO/LASCO, and SOHO/EIT were also used. The results indicated that the late gradual phase started about 6 min and 15 s after the end of the impulsive phase, and that the late phase consisted of two stages. At the first stage, a radio emission re-occurred at the breakout reconnection site, the maximum of this source shifted northeast in the following several mins, then another source appeared aligned with the motion azimuthally. In this process, the breakout reconnection site decayed and then re-appeared. Aurass (2014) found that the projected speed of the motion is about 1200 kms^{-1}, compared with 1100 kms^{-1} deduced by Vršnak et al. (2006) from the drift rate of the type II radio burst observed in the same event. Aurass (2014) also noticed that the disturbance caused by these motions to the corona was also observed by SOHO/EIT in 171 Å, and that the emission at high frequency (236.6 MHz) occurred at outermost location from the Sun, and that at low frequency (150 MHz) was closer to the Sun. This means that the local density distribution behaves in a way opposite to the normal coronal density distribution.

The reappearance of the breakout reconnection site marked the beginning of the stage 2, in which radio emission sources at frequencies lower than 236.6 MHz were seen to move at velocity ~400 kms^{-1}, apparently lower than the speed observed earlier. The feature moving at this speed was also observed by SOHO/LASCO C3 behind the CME four hours later. At this stage, the radio sources at higher frequencies appeared right above the active region with growing intensity. This scenario was explained as the process in which the CME bubble went through the breakout CS above the disrupting magnetic configuration, the global coronal magnetic field reconnected with the field around the tail CS that connected CME to the associated flare, and the second stage is the phase of the eruption in which the coronal breakout relaxation took place after the CME had left. Aurass (2014) believed that this was the first report of the coronal breakout recovery, and that their interpretation of the observation confirmed some predictions given by the numerical experiments of Lynch et al. (2008).

PLASMA FLOWS INSIDE AND AROUND THE RECONNECTION CURRENT SHEET IN THE ERUPTION

In addition to high temperatures inside the current sheet, there exist other important observational features. As shown by Fig. 2, two reconnection outflows move along the current sheet in opposite directions. The sunward (downward) flow contributes to the formation of flare loops, and the anti-sunward (upward) flow governs the evolution of the structures of the CME bubble (see also Lin and Soon 2004; Lin et al. 2004; Mei et al. 2012 for more details). Therefore, observations of the reconnection outflows above the flare loop system and below the associated CME could be expected. In addition, Fig. 2 also indicates the reconnection inflow near the CS, which could be considered as another observational consequence of the model. In this section, we will discuss the reconnection inflows, the sunward (downward) outflows, the anti-sunward (upward) outflows, bidirectional outflows, and finally outflows observed during laser driven reconnection.

The Reconnection Inflow near the Current Sheet

The first aspect of reconnection that we will discuss is observations of inflowing plasma into the CS region. As reconnection proceeds, the reconnection inflow continuously brings the magnetized plasma into the CS, and the magnetic energy is converted into heat and kinetic energy of the plasma and energetic particles. The amount of the magnetic energy entering the CS is balanced by the amount of energy dissipated by reconnection, otherwise either the magnetic field or the plasma pile up outside the CS if the dissipation rate is slower than the input rate, or reconnection ceases if the dissipation is faster than the input. Thus a fast reconnection process requires a fast reconnection inflow. Therefore, the reconnection inflow speed vi is a proxy to describe the rate of energy conversion in the CS. Usually, we use dimensionless parameters to describe the processes/phenomena of interest in order to deduce the related scaling laws, and MA=vi/vA is often used as the dimensionless measure of vi, where vA is the local Alfvén speed in the same region. In the magnetized plasma, the flow speed of plasma is lower than vA, and so is vi, in the absence of the fast mode shock. The advantage of using MA is that it ranges from 0 to 1, and can be easily related to other parameters via various scaling laws. Hence, MA is frequently used in studies of magnetic reconnection. Usually, fast reconnection has MA of around 0.01~0.1 (see discussions of Petschek 1964; Priest and Forbes2000).

As we noted before, identifying the current sheet is not trivial, so it is difficult to measure either vi or vA near the current sheet. To our knowledge, Yokoyama et al. (2001) performed the tentative measurement of vi, and then MA for the first time. They studied images of the cusp region above the flare loop system of the event of 1999 March 18 obtained by SOHO/EIT and Yohkoh/SXT, and noticed a well-recognized cusp structure and a plasma blob in the SXT image (or movie) and a blob-like cavity in the EIT 195 Å image (or movie). The cavity was dark in EUV wavelength and bright in SXR, which suggested that the temperature of the plasma in the cavity was about 4×106 K. As the cavity moved away from the solar limb, an X-like structure appeared behind it, and some thin filament-like structures were seen to merge together from the two sides of the X-like structure. This merging pattern was identified as the trajectories of the inflow near the reconnection region. A speed of the reconnection inflow between 1.0 and 4.7 kms^{-1} was deduced by measuring the speed of merging motion seen in EIT 195 Å images. A disadvantage of this measurement exists though since the structure seen in EIT 195 Å images is usually diffuse, so it is difficult to determine its velocity. Chen et al. (2004) re-analyzed the data of Yokoyama and Shibata (2001), and realized that the merging motion toward the X-point was, in fact, caused by the change in the location of the reconnection region, and was not the true inflow.

In the spirit of works by Yokoyama et al. (2001) and Ko et al. (2003), Lin et al. (2005) looked into various aspects of an event taking place on November 18, 2003. The event started with a rapid expansion of a few magnetic arcades located over the east limb of the Sun and developed an energetic partial-halo CME, a long CS, and a group of bright flare loops in the wake of the CME. It was observed by several instruments, both in space and on the ground, including EIT, UVCS, and LASCO on board the SOHO, the RHESSI, and the MLSO MK-4 coronameter. Lin et al. (2005) combined the data from these instruments to investigate various properties of the eruptive process, including those around the CS. The maximum velocities of the CME leading edge and the core were 1939 and 1484 kms^{-1}, respectively. The average reconnection inflow velocities near the current sheet over different time intervals ranged from 8 to 85 kms^{-1}, and the average outflow velocities ranged from 460 to 1075 kms^{-1}, which brought MA to the range from 0.007 to 0.18. According to discussions above, the values of MA in this range indicate the fast reconnection, which was expected since the event studied produced a fast CME associated with a group of post-flare loops. In the work of Lin et al. (2005), the inflow speed was determined directly according to the change in Lyα images of the UVCS slit that described the pattern of the reconnection inflow in a simple and straightforward fashion, so the problem that Chen et al. (2004) found with Yokoyama et al. (2001) was avoided.

As for the local Alfvén speed v_A in the inflow region, Lin et al. (2005) took an approximation to first order that the magnetic pressure $B_i^2/8\pi$, in the inflow region balances the dynamic pressure, $\rho_{out} v_{out}^2/2$, in the outflow region as the magnetic reconnection takes place continuously. Here v_{out} and ρ_{out} are the plasma density and velocity in the outflow region, respectively. To the first order of approximation again, $\rho_{out}=\rho_i$, where ρ_i is the plasma density in the inflow region. Therefore, equating the magnetic pressure in the inflow region to the dynamic pressure in the outflow region gives, $v_{out} = v_A = B_i/\sqrt{4\pi\rho_i}$, which is the Alfvén speed in the magnetic reconnection inflow region. This means that the local Alfvén speed v_A in the inflow region can be approximated with the reconnection outflow speed vout to the first order.

In principle, the approach taken by Lin et al. (2005) to determining the inflow speed near the CS was not different from that taken by Yokoyama et al. (2001) and then by Chen et al. (2004), as well as by the other authors who focused on the similar topic. The only difference lies on the manifestation of the same process observed by different instruments and/or at different wavelengths, and such manifestations vary from case to case as well. When performing the measurement, we chose those features of the motion pattern that could be recognized easily. In the work of Yokoyama et al. (2001) and Chen et al. (2004), EIT data was used to determine the inflow speed because the motion of the relevant structures seen in EIT 195 Å was the most apparent manifestation of the reconnection inflow; and in the work of Lin et al. (2005), on the other hand, the Lyα data were used because the manifestation of interest could be seen most clearly in this wavelength although the same manifestation was also shown in EIT 195 Å (see Fig. 2 of Lin et al. 2005 that displays a set of EIT 195 Å images showing the process that is characterized by the magnetic arcades' disruption, the driven magnetic reconnection, and the formation of flare loop system with a cusp structure on the top).

It could be seen clearly in the EIT 195 Å movies that the eruption commenced with severe stretching of the closed magnetic arcade followed by approach of the two legs of the stretched arcade to one another, and a group of flare loops topped with a cusp area subsequently developed at a lower region (see Fig. 2 of Lin et al. 2005 for details). Figure 6 displays a composite of LASCO/C2, UVCS/Lyα slit, and EIT 195 Å images taken roughly at the same time in this process. If we ignore the error in co-locating the images due to the fact that these images were not taken exactly at the same time, Fig. 6 displays the main aspects of the magnetic configuration that may appear in the major eruption, which usually produces both flare and CME (cf. Fig. 2). The image of the UVCS slit is actually a composite of 5 filtergrams in Lyα taken at 5 times successively. The distribution of the brightness along the slit indicates the change in the intensity of Lyα emission in that direction, and the horizontal scale of the UVCS images does not represent the extension in space, but in time.

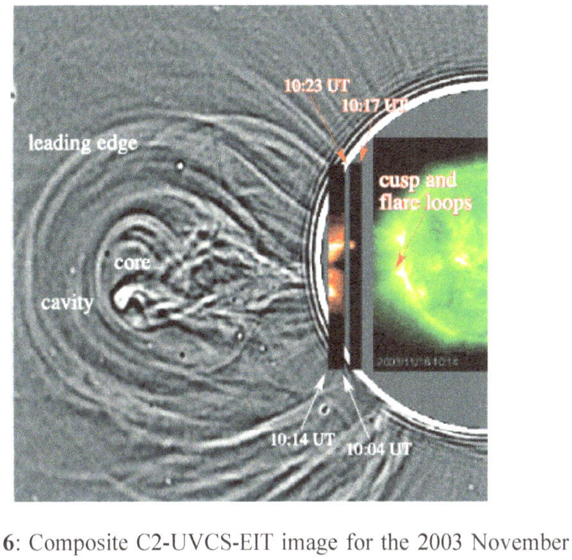

Figure 6: Composite C2-UVCS-EIT image for the 2003 November 18 event. The C2 image was taken at 10:26 UT and has been enhanced

by the wavelet technique in order to emphasize the morphological features at various positions of the CME, the EIT 195 Å image was taken at 10:14 UT, the left UVCS image was made by stacking five UVCS exposures in Lyαthat were taken at an altitude of 1.70 $R_\odot$ and cover the time interval from 10:04 UT (right edge) to 10:14 UT (left edge), and the right UVCS image was made by stacking four UVCS exposures in Lyα that were taken at an altitude of 1.52 $R_\odot$ and cover the time interval from 10:17 UT (right edge) to 10:23 UT (left edge). From Lin et al. (2005)

Therefore, from the right to the left of the UVCS slit images in Lyα, changes in the shape of both the dark gap and bright spots indicate the decrease in the gap width with time (see also Fig. 11a of Lin et al. 2005, which showed five Lyα intensity profiles along the UVCS slit taken at 1.70 $R_\odot$ from 10:04 to 10:14 UT during the event. The dip in each profile is suggestive of the dark gap shown in Fig. 6). Comparing with the images taken at the corresponding times that display motions of the eruptive arcade legs (see Fig. 2 of Lin et al. 2005), we can see that the dark gap coincided with the motions of the two legs of the eruptive arcades in both space and time, and that the gradual disappearance of the gap could be ascribed to the reconnection inflow near the CS. Measuring the gap widths at different times allows us to look into the evolution in this region. We plot the width of the gap at five different times in Fig. 7, and obtain correspondingly four speeds of the gap shrinkage as indicated in the figure. Since this event occurred on the limb and the orientation of the disrupted magnetic configuration led to a CS just along LOS (i.e., edge-on) as illustrated in Fig. 2, these speeds were approximately the speed of the reconnection inflow near the CS.

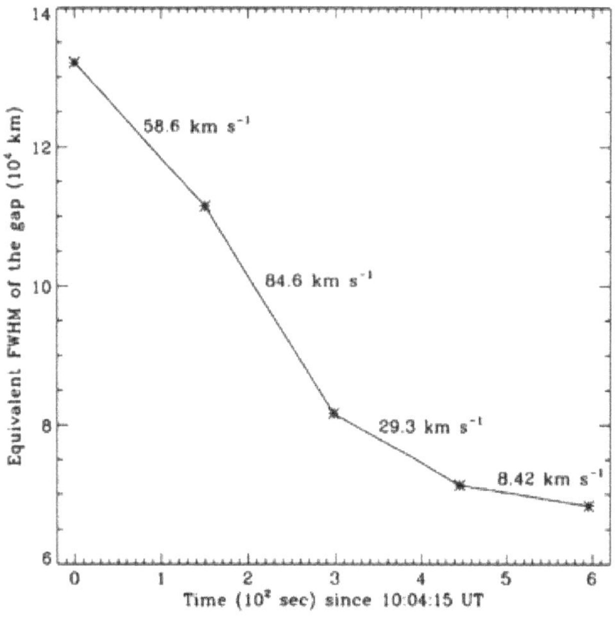

Figure 7: Width variations of the Lyα gap on the UVCS slit image for the 2003 November 18 event as seen in Fig. 6 versus time and the corresponding speeds of the reconnection inflow, vinear the CS (see also Figs. 5 and 11a of Lin et al. 2005). From Lin et al. (2007)

We note here that the dark gap seen in Lyα image could be due to either high temperature or high speed of plasma flow perpendicular to LOS or both in that region. At the location shown in Fig. 6, on one hand, the emission of Lyα comes from the resonant scattering of the photospheric emission by the neutral hydrogen atoms (H I), whose intensity is proportional to the density of H I. So Lyα emission may disappear if there are few H I atoms due to high temperature. On the other hand, the intensity of Lyα emission also depends on the speed of H I perpendicular to LOS. Usually, if this speed exceeds 370 kms^{-1}, the Lyα scattering profile is Doppler shifted away from the chromospheric emission profile, and the emission disappears in Lyα. This phenomenon is known as the Doppler dimming (e.g., see Noci et al. 1987 and references therein). Generally, the Doppler dimming becomes apparent as the speed of H I atoms exceeds 200 kms^{-1}.

Therefore, the bright spots in the image of the UVCS slit corresponded to the region where the H I density is high, the LOS speed is low (apparently <370 kms^{-1}), and temperature is in the range of (0.7~3) $\times 10^5$ K. The dark gap between the two spots resulted from either low density or high speed of H I atoms, or both. A low density of H I atoms means that the temperature is so high that H I atoms have mostly ionized. Further investigation suggested that the temperature in that area may reach up to 6.7×106 K, but it is also quite possible that the plasma speed is high there (see also discussions of Lin et al. 2005). No matter what happens, it always implies that the dark gap embeds the reconnecting CS connecting the CME to the associated flare. In other words, the dark Lyα gap marked the reconnection outflow that could result from the Petschek-type reconnection. Because the reconnection inflow and outflow regions are in close proximity to each other, the change in the size of the outflow region indicated by the decreasing width of the gap must be related to the reconnection inflow near the sheet. In another word, the decreasing width of the Lyα gap suggests squeezing of the outflow region, and indicates the reconnection inflow as well. We shall see later that the asymptotic behavior of the curve in Fig. 7reveals more important information on the reconnection current sheet.

One more issue related to the work of Yokoyama et al. (2001) needs extra attention. They use $MA=v_i/v_A$ to calculate MA, but it is usually impossible to measure v_A directly in reality. So they estimated it through an indirect approach from the soft X-ray emission measure by assuming that all the magnetic energy released during the eruption went to heating via reconnection, and that the depth L of the source region along the LOS was on the same order of flare loop length, so that the source region of the SXR emission fills the whole volume of L^3 when calculating the electron density. These two assumptions introduce some uncertainty in their results: The first assumption obviously causes the magnetic field near the reconnection site to be underestimated because much (perhaps as much as half) of the released energy goes to kinetic energy of the reconnected plasma flows, and the second assumption leads the total thermal energy to being overestimated because the source region of soft X-ray could never fill the whole volume of L^3.

To avoid these uncertainties, Ko et al. (2003) and Lin et al. (2005) estimated the local Alfvén speed around the reconnection region by measuring the speed of the reconnection outflow directly. The standard theory of magnetic reconnection indicates that these two speeds are equal to one another (Priest and Forbes 2000). Because the current sheet developed by the event on November 18, 2003 was also observed edge-on, the features of the reconnection outflow could be seen easily. Lin et al. (2005) identified several blobs in the outflow and speeds of five of them could be measured, and uncertainties in estimating the local Alfvén speed near the CS could be more or less avoided.

Later, Narukage and Shibata (2006) reported observations of reconnection inflows near the reconnection site after analyzing the EIT data for a set of events. They surveyed the events observed by EIT from 1996 to 2000 and recognized six events that apparently manifested reconnection inflow. The approach used by Yokoyama et al. (2001) to deduce the speed of the reconnection inflow was improved, and the inflow speed measured through the new method was larger than previous ones by a factor of 2.5. Correspondingly, Narukage and Shibata (2006) found that the rate of magnetic reconnection for the events studied, including that investigated by Yokoyama et al. (2001), varied from 0.001 to 0.07. This result is consistent with those obtained by other authors (e.g., see Table 4 of Narukage and Shibata 2006).

In a work focusing on the reconnecting current sheet above the solar flare, Liu et al. (2010) studied the eruption of a trans-equatorial arcade on 2004 July 29, which resulted in a halo CME and a C2.1 flare. A bright elongated feature extending from the tip of the cusp structure to the CME was observed in the impulsive phase, and was identified with the reconnecting current sheet as predicted by Lin and Forbes (2000). The sheet was seen in the high temperature spectral line Fe XXIV whose emission peaks at 2×107 K. Estimates of the plasma density ranged from 106 to 109 cm−3 between the upper and the lower tip of the sheet. These values are consistent with those found by Ko et al. (2003). Furthermore, a merging motion of the disrupting arcade legs was also noticed, and the corresponding inflow speed was about 6 kms^{-1}near the sheet, which is consistent with that obtained by Lin et al. (2005).

Similar to the event studied by Lin et al. (2005), the reconnection inflow was observed in the form of sweeping of the magnetized plasma into the CS region at the beginning of an event occurring on 2010 November 3 started with an erupting flux rope (Savage et al. 2012b). The reconnection outflow was soon observed no more than a minute after the inflow was detected (Fig. 8a). This event was observed by both SDO/AIA in several passbands and RHESSI simultaneously (see also Reeves and Golub 2011; Cheng et al. 2011; Bain et al. 2012; Glesener et al. 2013), and a bright and elongated linear feature, as marked in Fig. 5c of Savage et al. (2012b), that was referred to as the candidate CS was recognized within the first few minutes of the eruption. The RHESSI data indicated that the non-thermal component of X-ray emission (≥25 keV) peaked first and the thermal-dominated total emission (≤25 keV) reached maximum 3 minutes after. The reconnection inflow was first seen in high temperature line AIA 131 Å at speeds ranging from 660 to 690 kms^{-1}, and that seen in a low temperature line, say 171 Å, did not appear until 4~7 min at apparently lower speed (~150--260 kms^{-1}). The local Alfvén speed near the CS is most likely in the range from 500 to 3000 kms^{-1}, which allows for deviations due to magnetic field and density variations. Using this information, they found that, on average, MA decreased from 0.4 at the beginning to 0.1 about 14 min later, and that the reconnection process is more of the Petschek-type than Sweet-Parker, which indicated the occurrence of fast reconnection.

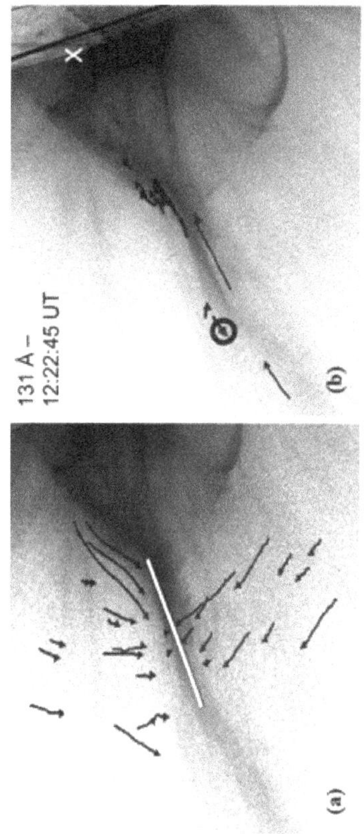

Figure 8: (**a**) Composite image showing the inflow tracks from all four selected AIA filters (131 Å, 211 Å, 193 Å, and 171 Å) overlaid onto a 131 Å image (see also Fig. 9 of Savage et al.2012b). (**b**) The sunward reconnection outflow tracks seen in 131 Å, with an anti-sunward flow track highlighted by a circle. The white "X" marks the footpoint of the magnetic configuration used to determine outflow heights. From Savage et al. (2012b)

Features of the reconnection outflow inside the CS were also detected almost at the same time as the inflow was detected in the event. Unlike features of the inflow, however, only those observed in 131 Å and 94 Å channels were more likely to be the outflow along the CS (Fig. 8b), those seen in the 211 and 193 Å band passes were not in the typical outflow (SAD or SAD loop [SADL]) morphology, but appeared more like density enhancements traveling down along the legs of newly formed loops. When overlaid on the 131 Å image sequence, the flows in 211 and 193 Å could be seen indeed to move along the loop legs. Because of less clarity, the information about the outflow seen in 94 Å was not included in Fig. 8b. In total 12 outflows were recognized in each of the three AIA wavelengths, 131 Å, 211 Å, and 193 Å. They estimated outflow speeds averaged over 12 outflows (11 sunward and 1 anti-sunward; the anti-sunward speed ~150 kms^{-1}) on the plane-of-sky for the SADs/SADLs tracked in 131 Å of ~240 kms^{-1} versus a median value of the speed of ~103 kms^{-1} for the 12 211 and 193 Å outflows. It is interesting to note that the retracting structures like SADs and SADLs moved apparently slower than the reconnected plasma flows in the CS.

Recently, both the reconnection inflow and outflow were also reported in a compact flare. Su et al. (2013) studied a flare on 17 August 2011 that was created by collision of two coronal loop systems and observed in both EUV and X-ray emissions. The multi-wavelength EUV observations from SDO/AIA showed inflowing cool loops and newly formed, outflowing hot loops. RHESSI X-ray spectra and images simultaneously displayed the appearance of plasma heated to >107 K at the expected locations. The velocity of two loop systems approaching to one another varied from 10 to 70 kms^{-1}. In this process, magnetic reconnection between two systems yielded the ejection of the newly formed loops at speed between 90 and 440 kms^{-1}. Ignoring the projection effect and assuming the outflow speed to be identified with the local Alfvén speed, Su et al. (2013) eventually ended up with that MA is between 0.05 and 0.5. These results are similar to those obtained by Takasao et al. (2012) who noticed that the inflow speed was between 12 and 90 kms^{-1}, the outflow speed between 220 and 460 kms^{-1}, and MA between 0.055 and 0.2 accordingly.

Downward Reconnection Outflows

Observations of plasma flowing out of the reconnection towards the Sun have also been made, and we discuss these results here. McKenzie and Hudson (1999) studied a long-lived event that created a group of growing flare loops and a CME. This event was earlier observed by Yohkoh on 20 January 1999. For the first time, they recognized the apparent motions above the post-flare loops, and identified the motion with the reconnection outflow in the CS. Similar mass motions were later observed as well in the 2002 April 21 event by TRACE (e.g., see Innes et al. 2003a,2003b; Sheeley et al. 2004; Reeves et al. 2008a; and detailed discussions below). Comparing the magnetic configurations around and/or related to these mass motions with that in Fig. 2, we believe that McKenzie and Hudson (1999) had observed a CS above the flare loops roughly face-on although observing a CS face-on in the Yohkoh era was not a trivial job. On the other hand, nowadays, with the observations from the SDO/AIA, currently the highest resolution solar observatory capable of coronal measurements, identifying a CS face-on has become less difficult (e.g., see Warren et al. 2011; Savage et al. 2012a, 2012b; McKenzie 2013; Hanneman and Reeves 2014; Doschek et al. 2014).

In the event observed by McKenzie and Hudson (1999), the CS manifested as a fan of spikelike "rays" above the loops. Morphological features of the similar type had been seen in many events by Yohkoh (Švestka et al. 1998), but in this case there was a much clearer view of mass motions in the region above the arcade. McKenzie and Hudson (1999) noticed that the motion indicated the field line retraction but did not see cusp structure on the top of flare loops. This is probably because they observed the CS face-on and the cusp feature could be seen only when the sheet is observed roughly edge-on (see also detailed discussions on this issue by Forbes and Acton1996). In the late phase of the event, McKenzie and Hudson (1999) found that the downward motion was represented by a dark void in soft X-ray (SXR) moving at speed ranging from 100 to 200 kms^{-1}, and that the temperature in the void reached up to 9×106 K, which was slightly hotter than the surrounding fan-shaped supra-arcade plasma, and the density was about 108 cm−3. Later, McKenzie (2000) reported 11 long-life CME-associated events observed by Yohkoh/SXT that displayed similar sunward mass motions at speed ranging from 50 to 500 kms^{-1}, which were also observed recently by the SDO/AIA (Savage et al. 2012a). Because of the weak signal of the X-ray data in the void, the temperature deduced for the dark void by McKenzie and Hudson (1999) was somewhat suspect.

In the work of Reeves et al. (2008a), these dark voids were also named as SADs. Now terms "SADs", "dark voids" and "dark lanes" are usually used in parallel. Spectroscopic observations indicated for the first time SADs to be voids (Innes et al. 2003a), and by analyzing the data from SDO/AIA with high signal level and increased temporal cadence, Savage et al. (2012a) argued that SADs were actually density depletions left in the wake of thin flux tubes retracting from a reconnection site in the supra-arcade region, and should no longer be considered cross-sections of newly reconnected, large evacuated flux tubes. Hanneman and Reeves (2014) further found that SADs usually had temperatures very similar to or lower than that of the surrounding flare arcades, which ranged from 3.1×106 to 1.3×107 K, and that their emission measure was typically about an order of magnitude lower than the nearby arcade. Studies by McKenzie (2013) also displayed evidence of vorticities that varied in time, stagnation points (S-points), and numerous velocity shears in the flows observed in the hot (~107 K) plasma above the flare arcades. By comparing observations with numerical experiments, Innes et al. (2014) suggested that SADs resulted from secondary instabilities of the Rayleigh-Taylor type at the head of reconnection jets.

We need to note here that S. Savage (2014, private communications) keeps pointing out that SADs and plasmoids usually appearing in numerical experiments (e.g., see Bhattacharjee et al.2009; Huang and Bhattacharjee 2010; Bárta et al. 2011a; Shen et al. 2011; Mei et al. 2012; and references therein) are not the same thing. SADs are low-density regions, while plasmoids show up in simulations as high density structures, and plasmoids appear when we view CSs edge-on, while SADs appear when we view CSs face-on. There are a few papers that discuss theories behind SADs (Cécere et al. 2012; Cassak et al. 2013; Scott et al. 2013; Guo et al. 2014), though these are still not fully accepted.

Following McKenzie and Hudson (1999) and McKenzie (2000), Sheeley and Wang (2002) identified downward motions of the coronal mass at the heliocentrical distance from 2 to 6 $R_{\odot}$, with each flow at speed varying from 50 to 100 kms^{-1}. The event occurring above the west limb on April 21, 2002 provided us a good opportunity to study the sunward reconnection outflows (see Fig. 5 of Reeves et al. 2008a, which displayed an excellent example of the sunward outflow seen in the 195 Å channel with TRACE. The high-temperature response of TRACE at this wavelength provided important information of the hot plasma above the flare loops). These reconnection outflows in the CS were observed in the early phase of the event, so both the temperature and the speed of the plasma flow were high and reached up to 103 kms^{-1} and 107 K, respectively.

In addition to the flow itself, the deceleration of flow was also observed as a result of interaction with the closed flare loops below. Figure 9 displays the time profiles of the void location inside the current sheet, which shows the void motion and its rapid deceleration (see the dark void specified by the white arrow). This void moved at speed of about 600 kms^{-1}, and then underwent a significant deceleration at up to 1.5×103 ms−2, which suggested a violent interaction between the reconnection outflow and the closed flare loops, and might further imply the formation of the termination shock on the top of the loop system (Aurass and Mann2004; Aurass et al. 2006, 2011). So far, on the other hand, whether a termination shock could form as a result of interaction of the reconnection outflow with the flare loop is still an open question, and more detailed investigations are surely needed (e.g., see also discussions of Gao et al. 2014). We note further here that the deceleration of an outflow and the formation of a termination shock might also occur without the loop top obstacle, the increase in the plasma density in the related region may play the role of the obstacle.

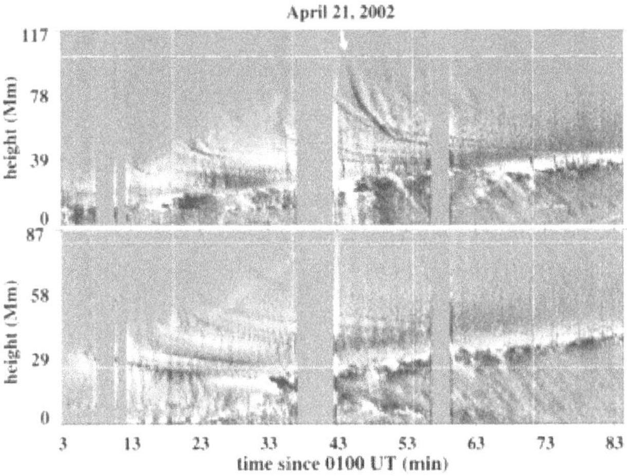

Figure 9: Height-time maps of the plasma blobs produced in the event of 21 April 2002. The toppanel was obtained for the path of the "tadpole" (indicated by arrow in Fig. 1 of Sheeley et al. 2004) and shows steep tracks rapidly decelerating to form new layers of the slowly descending cloud. The bottom panel shows a neighboring region without "tadpoles" and emphasizes the relation between the descending tracks of the plasma cloud and the rising ramp of post-flare loops. From Sheeley et al. (2004)

Asai et al. (2004) presented a detailed examination of downward motions of the plasma above flare loops observed in the 2002 July 23 event. This event produced an X-4.8 flare and a fast CME at speed of 2.6×103 kms^{-1}, which were observed by several instruments both in space and on the ground, including TRACE, RHESSI, NoRH, UVCS and LASCO. The EUV images obtained by TRACE showed dark downward flows (sunward motions) above the postflare loops, not only in the decay phase but also in the impulsive and main phases. They also found that the times when the downward flows started to be seen corresponded to the times when bursts of nonthermal emission in HXRs and microwaves were produced. This implies that the downward motions of the plasma occurred when magnetic energy was quickly released as a result of magnetic reconnection, and it is worth noting that the downflows in this flare was observed in the impulsive phase. To our knowledge, the downflows occurring in the impulsive phase are not often reported.

Similar downward flows over the top of flare loops were also observed in both HXR (25~50 keV and 12~25 keV) and SXR (6~12 keV) by RHESSI in 3 events studied by Sui and Holman (2003) and Sui et al. (2004). In these events, high temperature loops or the loops observed in high energy bands (25~50 keV or 12~25 keV) were seen above the low temperature loops or the loops observed in low energy bands (6~12 keV). Meanwhile, the loop system observed in both bands manifested downward motion in the HXR impulsive phase at speed ranging from 8 to 32 kms^{-1} with the loops seen in high energy band moving faster than those in the low energy band. The flare loops observed in the radio bands with frequencies ranging from 17 to 32 GHz as well as those observed in 193 Å by TRACE confirmed such a downward motion of the flare loop system in the early stage of the eruption (Li and Gan 2005, 2006). Veronig et al. (2006) saw the same feature when studying an X-3.9 flare such that the mean downward velocities range from 14 kms^{-1} in the RHESSI 10--15 keV energy band to 45 kms^{-1} in the 25–30 keV band, and that observed by the SXI instrument was 12 kms^{-1}. Reeves et al. (2008b) studied a B-class and a C-class flare, and found that individual loops moved downward at speeds ranging from 5 to 48 kms^{-1} and from 2 to 4 kms^{-1}, respectively. Their estimates showed that measurements of the shrinkage in these two events were between 17 % and 27 %. Recent work by Liu et al. (2013) on an M7.7 flare showed that the speed of such a descent motion of the overall X-ray and EUV loop-top varied from 7 to 23 kms^{-1}.

The downward motion observed in the above events could be a combination of the plasma flow and the shrinkage of flare loop system created by magnetic reconnection in the CS (e.g., see also Švestka et al. 1987; Lin et al. 1995; Forbes and Acton 1996; Lin 2004; Reeves et al. 2008a,2008b; Liu 2013; Liu et al. 2013). Both of them are the products of magnetic reconnection, but the former is the motion of plasma and the latter is the motion of magnetic structures. Observationally, they mixed together and could hardly be distinguished from one another previously due to the low spatial resolution and time cadence of the instruments. SDO/AIA allows for unprecedented scrutiny of the SADs. Savage et al. (2012a) showed that SADs, which once were thought to be identifiable as the shrinking loops themselves, appeared to be "side effects" of the passage of the loops through the supra-arcade plasma, and demonstrated the observational difference of SADs and loop shrinkage.

The shrinkage of the loop system was observed by Sui et al. (2004) for the first time in HXR. It usually occurs within a short period at the beginning of the impulsive phase, and then the loop system displays apparent upward motion as usually seen in two-ribbon flares. Sui et al. (2004) suggested that this motion is due to the fact that the current sheet develops as a result of the collapse of the X-type neutral point. This scenario was first introduced by Dungey (1953) such that the X-type neutral point is locally unstable if the source region of the relevant configuration could move freely, and the X-point would consequently collapse and develop to a current sheet with a Y-type neutral point (or a reversed current and singularity) at each end (see also Figs. 2.1 and 2.2 of Priest and Forbes 2000). In this scenario, the formation of the reversed current may push the flare loop system to move downward at the very beginning of the event.

Considering the fact that the magnetic configuration is severely stretched in the eruptive process, we believe that the current sheet should be produced as the local magnetic structure of the X-point is stretched (see Fig. 2.10b of Priest and Forbes 2000, which demonstrated the bifurcation of an X-point in response to motions of sources that stretch the magnetic structure around the X-point) instead of collapses (see Fig. 2.10c of Priest and Forbes 2000 for the bifurcation of the X-point due to squeezing of the related structure). Analytic studies (Forbes and Isenberg 1991; Isenberg et al. 1993; Forbes and Priest 1995; Lin and Forbes 2000) indicated that the catastrophic loss of equilibrium in the coronal magnetic configuration that includes a current-carrying flux rope thrusts the flux rope outward, severely stretching the magnetic structure around the flux rope and resulting in the formation of an X-point and then a CS with a Y-point at each end. Numerical experiments of Mei et al. (2012) demonstrate clearly this process occurring in a disrupting magnetic configuration. Because stretching directly produces a low pressure region near the CS, and both magnetic field and plasma around the CS between the two Y-points are pushed toward the CS, forming the reconnection inflow all along the CS, by the pressure gradient. Since the pressure gradient is not just confined to the region near the Y-point, but exists all along the CS, the reconnection inflow is expected all along the CS. Because the CS is created by stretching, a CS with reversed current and singularity may not be produced, and the downward motion of the flare loop system at the beginning of the impulsive phase should be due to the other reasons. Lin (2004) ascribed this motion to the standard shrinkage of the newly formed closed flare loops.

In a two-ribbon flare process, individual flare loops do not stay at the height when they are newly formed, but decrease to a lower altitude at which their shape is more potential, and the earlier the loops are formed, the faster the loops shrink (Lin et al. 1995; Lin 2004). In the impulsive phase of HXR, the shrinkage of individual loops is much more significant than the growth of the loop system resulted from accumulations of individual loops such that the speed of the former could be close to 200 kms^{-1}, and that of the latter is less than 100 kms^{-1}. In reality, manifestations of the two motions are mixed with one another and the highly dynamic behavior of loops in the early stage of the flare makes distinguishing them from one another very difficult, so the component that is moving faster dominates the observational consequence. But this question is still open and more investigations in detail are necessary.

After having studied all the flare events collected, Savage and McKenzie (2011) identified 62 cases of sunward plasma flows above the flare loops. These events had been observed by Yohkoh, TRACE, SOHO/LASCO, SOHO/SUMER, and Hinode/XRT over almost 20 years. They looked into 35 events among which 10 displayed the mixture of the sunward flow and the loop shrinkage, and 25 showed the pure plasma flow in the current sheet. In individual events, the largest number of pure plasmoid motions identified in the sheet was 25, and the largest number of the mixture of plasmoid motions and shrinkages was 60. Overall, the average velocity of these motions was about 150 kms^{-1}. Furthermore, Hara et al. (2011) found that the speed of the reconnection outflow near the top of the flare loop system was around 200 kms^{-1} by analyzing the data from Hinode/EIS, and Warren et al. (2011) obtained a speed close to 150 kms^{-1} from SDO/AIA data. Compared with the anti-sunward flow, the sunward flow is apparently slow. Shen et al. (2011) pointed out that this is generally true because the sunward flow is always slowed down by its interaction with the closed flare loops below. Studying the SDO/AIA data, Takasao et al. (2012) brought the speed of the anti-sunward flow to the range from 220 to 460 kms^{-1}, and that of the sunward flow to the range from 250 to 280 kms^{-1}.

Upward Reconnection Outflows

In this part of work, we discuss the upward (anti-sunward) outflows in the CS. After Ciaravella et al. (2002) confirmed observationally for the first time the existence of the reconnection region behind the CME where a CS should be located as predicted by Lin and Forbes (2000), Ko et al. (2003) performed the first quantitative study of the reconnection outflow leaving the Sun in the 2002 January 8 event although they were not the first to observe the anti-sunward reconnection outflow (e.g., see discussions by Webb et al. 2003; Simnett 2004 for details). The event studied by Ko et al. (2003) occurred on the east limb and produced a fast CME associated with a growing SXR flare loop system (the main part of the flare was behind the limb so it is hard to determine the class of the flare). The advantage of studying this event is that it occurred on the limb and the current sheet above the flare loop system could be observed edge on, so that enough emission measure from the sheet would be collected and reliable diagnostics of plasma properties of the sheet could be performed.

Combining the data from the UVCS, LASCO, EIT, and CDS on board SOHO, as well as those from the MLSO MK4 coronameter (Elmore et al. 2003), Ko et al. (2003) investigated the morphological and dynamical properties of this event, and the physical properties of the current sheet. The velocity and acceleration of the CME reached up to 1.8×103 kms^{-1} and 1 kms−2, respectively. The acceleration is found to occur mainly at the lower corona (<2.76 R⊙). The post-CME loop systems showed behaviors of both postflare loops (upward motion with decreasing speed) and soft X-ray giant arches (upward motion with constant speed, or acceleration) according to the definition of Švestka et al. (1987). In the current sheet, the presence of highly ionized ions, such as [Fe XVIII] and [Ca XIV], suggested temperature as high as (3--6)×106 K, and the plasma outflows had speeds ranging from 300 to 650 kms^{-1}. Absolute elemental abundances in the CS showed a strong first ionization potential effect similar to that found in the active region streamers. The magnetic field strength in the vicinity of the CS was found to be of the order of 1 G.

In the images (movies) of LASCO, the anti-sunward plasma flows could be recognized by tracking a group of plasma blobs (or plasmoids) moving along the current sheet. Figure 18 of Ko et al. (2003) displays two examples, one showed accelerating motion pattern, and another one showed tendency of slowing down. Selecting 5 blobs with sharp edge that could be easily recognized and following their motions, Lin et al. (2007) plotted their heights at different times as shown in Fig. 10, and further deduced the velocity of each blob, which is between 300 and 650 kms^{-1}. The distance of every two adjacent blobs is also marked in Fig. 10, and the importance of these distances will be discussed later.

In the event of 18 November 2003, many blobs were also recognized in the LASCO data (see Fig. 11), and the speeds of the plasma blobs moving away from the Sun along the current sheet were between 460 and 1075 kms^{-1} [see also Fig. 8 of Lin et al. 2005 that is a composite of LASCO C2 (at 17:50 UT), MK4 (17:49 UT), and EIT 195 Å (17:48 UT) images of the event, clearly illustrating the positions of the post-flare/ CME loops, cusp point, and extended current sheet, which resembles the typical Kopp-Pneuman configuration for major flares]. Figure 11 indicates that the CS could extend up to 20 $R_\odot$. According to Forbes and Lin (2000), Lin and Forbes (2000), and Lin (2002), the upper (outer) tip of the CS is right behind the CME rear edge, and the distance between these two locations depends on both the rate of magnetic reconnection inside the CS as well as the local Alfvén speed near the CS. Because of the observational limitations, the CS became too diffuse to be recognized far away from the Sun. Therefore, it is hard to tell where the CS terminates at the far end although it is now well known the lower tip or the near end of the CS terminates at the top of the associated flare loop system. Figure 11shows that the far end of the CS disappeared at around 20 $R_{\odot}$, but it does not necessarily mean that it terminated there. More advanced techniques and higher sensitivity detectors are needed in order to resolve this problem in the future.

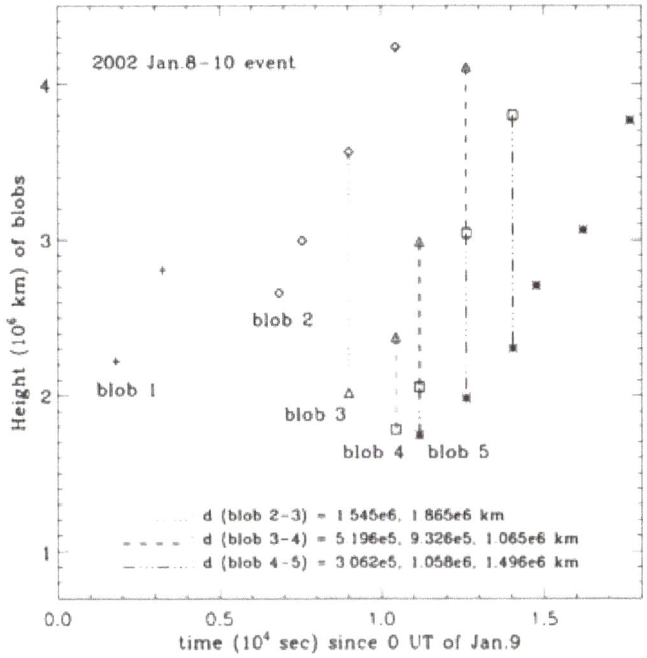

Figure 10: Heights of five well-recognized blobs in the 2002 January 8 event versus time. The corresponding distances of two successive blobs are also indicated. From Lin et al. (2007)

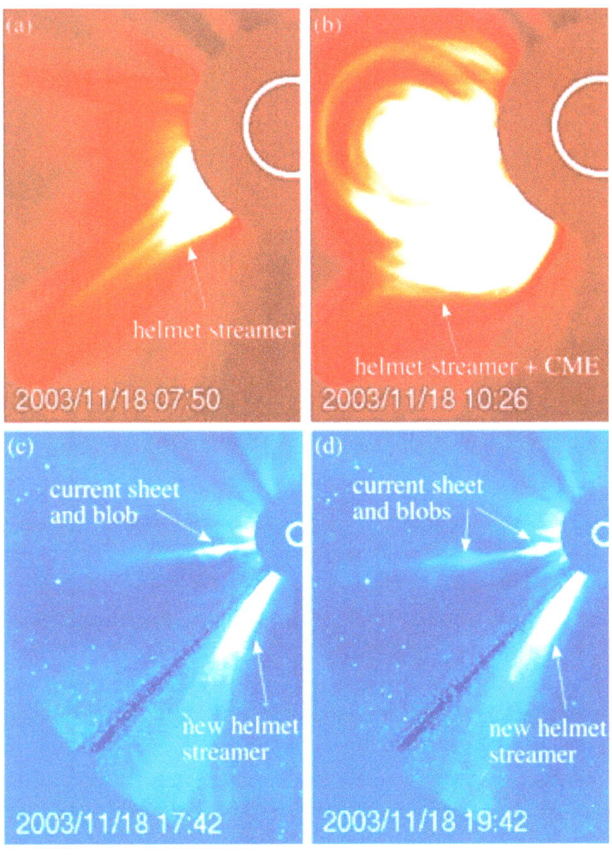

Figure 11: LASCO C2 and C3 images. (**a**) A helmet streamer exists in the region that the CME will go through. With rapid expansion of the CME, (**b**) the helmet streamer is severely deflected from outside, and (**c**, **d**) the southern leg of the CME (disrupting arcade) merges with the helmet streamer, forming a new helmet streamer. A couple of blobs of the reconnected plasma flowing along the current sheet can also be recognized in (**c**) and (**d**). All times are UT. From Lin et al. (2007)

After having analyzed the LASCO data obtained in the time interval between 1996 and 2009, Song et al. (2012) found more plasma blobs flowing inside the post-CME CS. The speeds of these blobs spread over the range from 200 to 103 kms^{-1}. In recent works by Liu (2013) and by Liu et al. (2013) on an event that produced an M7.7 flare and a fast CME at speed >103 kms^{-1} (see also Patsourakos et al. 2013), the reconnection outflow was reported to leave the Sun along the current sheet over the flare loop system at speed up to 1.4×103 kms^{-1}. These results are consistent with those of Ko et al. (2003) and Lin et al. (2005), and suggest a positive correlation of the outflow speed with the class of the associated flare. In addition, Song et al. (2012) also studied the plasma blobs flowing inside the helmet streamer, and noticed that these blobs are apparently slower than those inside the CME current sheet. They concluded that the difference in velocities of plasma blobs could be ascribed to the reconnection processes occurring in the different environments. Similar phenomenon was also observed in numerical experiments on the reconnection process taking place in two-ribbon flares (see also Forbes and Malherbe 1991; Riley et al. 2007; Shen et al. 2011).

We need to note here that the CS inside the helmet streamer is very different from the post-CME CS in their dynamic property although they are the same in the intrinsic physical property. The former is nearly stationary and could exist for a long time, but the latter is highly dynamic and is produced in the eruptive process. We note here that "stationary" means that the helmet streamer and the CS inside already existed before the CME and continued to exist after the CME [see the structure located at the polar angle (PA) of around 140o in all the panels of Fig. 11], and that "dynamic" means that the post-CME CS did not exist beforehand, but was created and developed during the eruption (see the structure located around PA=100o in Figs. 11c and 11d, which did not appear in Figs. 11a and 11b). Panels in Fig. 12 display LASCO C2 images of the same process that had been enhanced by the wavelet technique, which further indicates that discriminating a post-CME CS from a streamer CS is not difficult, although in both cases the tearing mode instability could take place in the same fashion (e.g., see also Einaudi et al. 2001) since the occurrence of the tearing mode depends on the length to thickness ratio of the CS (e.g., see Furth et al. 1963; Loureiro et al. 2007; Ni et al. 2010; Shen et al. 2011; Bárta et al. 2011a,2011b; Mei et al. 2012) no matter whether the CS is stretched by the expanding solar wind, or by the disrupting coronal magnetic field.

In addition, observations in the radio band revealed dynamical features of plasma blobs inside the CS. When studying the trans-equatorial coronal loop eruption occurring on 29 July 2004, Liu et al. (2010) found that a current sheet was produced at the early stage of the event, and that radio observations made in space and on the ground revealed important information on the CS and dynamic features of plasma inside the sheet. First of all, a set of DPSs at metric frequencies was detected (see Fig. 2 of Liu et al. 2010). Compared with those usually observed at the decimetric frequency (Kliem et al. 2000), the DPS observed in this event was obviously produced at higher altitudes and displayed the pattern of the anti-sunward motion in group, which was suggestive of tearing of the sheet and the resultant magnetic islands moving away from the Sun along the sheet. The fine details included in DPSs indicated interactions of energetic electrons trapped in the magnetic islands (or plasma blobs). In the associated LASCO data, the motion of plasma blobs was also recognized at speed of ~360 kms^{-1}. Second, a type III radio burst was detected at frequencies ranging from 1.08 to 13.83 MHz, which implied the escape of some energetic electrons from trapping to the high corona and interplanetary space.

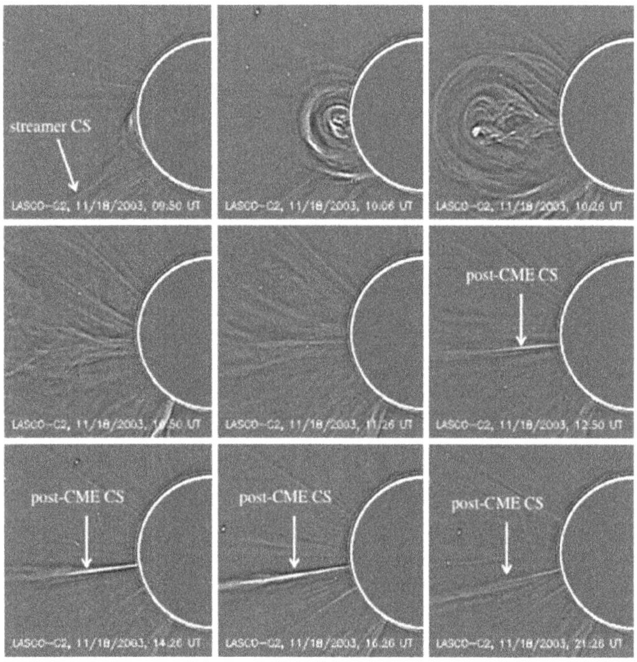

Figure 12: LASCO C2 images of the 2003 November 18 event that have been enhanced by the wavelet technique, which is good at emphasizing the fine features of large-scale structures. The leading edge, cavity, and core of the CME are easily recognizable. The post-CME and streamer CSs have been apparently specified. From Lin et al. (2005)

Bi-Directional Reconnection Outflows Observed in a Single Event

Conservation of the momentum of the plasma flow determines that the bi-directional outflow is the ubiquitous consequence of the magnetic reconnection process no matter in what type of reconnection. As we clarified earlier, the term "current sheet" used in this and our previous works possesses a general meaning such that it refers to the region where fast dissipation/reconnection of the magnetic field takes place. In most cases, we are able to observe the reconnection outflow in one direction, and it is usually difficult to identify bi-directional flows simultaneously in a single event. This is because the bifurcation point of the reconnection outflow usually appears at low altitude, say lower than 1 $R_\odot$ as suggested by Shen et al. (2011). So the sunward flow in the eruptive event usually cannot be observed by a coronagraph with a large occulting disk like LASCO/C2 and C3, while instruments with FOV near the solar surface can only recognize the anti-sunward motions. In the streamer detachment event, on the other hand, the bi-directional reconnection outflows have indeed been reported with LASCO data and defined as "in/out pairs" by Sheeley and Wang (2007). This indicates that, unlike that in the eruptive event, the bifurcation point of the reconnection outflows in the streamer detachment event is located high in the corona.

To our knowledge, the eruptive event on April 9, 2008 is the first one in which both sunward and anti-sunward reconnection outflows were observed simultaneously (e.g., see Savage et al.2010). This event took place over the active region AR 10989, and produced an X-class flare and a CME. The active region was located about 23° behind the west limb when the event occurred, and several instruments in space observed both the flare and the associated CME. These instruments include XRT on Hinode, TRACE, SECCHI on STEREO A, and LASCO on SOHO. GOES did not classify the flare because most of the flare was located behind the limb. Occulting of the brightest structures of the flare by the solar limb allows the less bright structures above the active region to be recognized more easily. As the eruption was in progress, the CME moved away from the Sun like a cart wheel, so this CME is also known as the "cart-wheel CME".

Focusing on the data from XRT and LASCO, and using the data from SECCHI for reference, Savage et al. (2010) found a large amount of mass seen in EUV and SXR to be expelled from the Sun at speed ranging from 80 to 180 kms^{-1} in the time interval from 09:16 and 10:11 UT. Structures of the associated flare that could be seen were about 7×104 km above the limb. The resultant white-light CME entered the FOV of LASCO at 11:06 UT and swept the FOV of LASCO C2 at speed of ~450 kms^{-1}. Landi et al. (2010) studied the initial acceleration phase of the CME in the same event, analyzed the physical properties of the CME core, and their data set included all other manifestations associated a CME: flare loops, CS, and coronal dimming. Landi et al. (2012) extended the work of Landi et al. (2010) to the diagnoses of plasma behind the CME, including that in the CS, to complement the works by Savage et al. (2010) and Patsourakos and Vourlidas (2011) on the same event.

After the CME left the FOV of XRT, a set of the enhanced images (or movies) displayed sunward and anti-sunward plasma flows, together with the shrinkage of flare loops. Savage et al. (2010) noticed that the plasma flows were along a thin bright feature (see Fig. 13), which became more apparent after 11:00 UT and displaced 25° southward within 6.5 hours at speed of about 4° h^{-1}. The rotation of the Sun at the corresponding latitude is at speed of 0.6° h^{-1}, which could not account for the displacement, so the motion of the thin structure should be due to the self-adjustment of the disrupting magnetic configuration as suggested by Ko et al. (2003).

In the XRT FOV, this structure was thin and located between the CME and the flare. Comparing it with those similar ones observed previously (Ciaravella et al. 2002; Ko et al. 2003; Webb et al.2003; Lin et al. 2005; Bemporad et al. 2006), we are able to identify this structure with the CME/flare current sheet. What is especially interesting was that a set of plasma blobs flowing in the sheet were seen between 12:18 and 12:35 UT. These blobs appeared to move bi-directionally. The blobs that were selected for further analyses were recognized by naked eyes because of the lower signal-noise ratio of the data. From 11:11 to 17:24 UT, totally 16 plasma blobs were identified with 3 moving sunward and 13 anti-sunward.

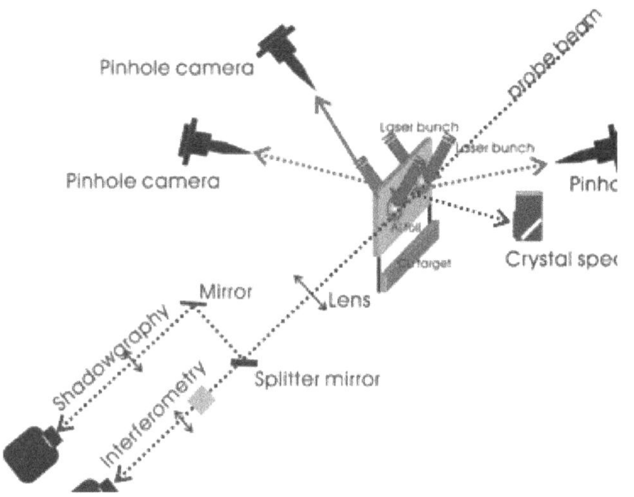

Figure 13: The CS in the region between two arrows developed in the "cart-wheel" event on April 9, 2008. The figure in the left panel was taken at 11:08:42 UT, and that in the right panel was taken at 17:31:55 UT. The change in the orientation of the current sheet resulted from the self-adjustment of the associated magnetic configuration in the eruption at speed of around 4° h^{-1}. From Savage et al. (2010)

Analyzing trajectories and motions of these 16 blobs moving along the current sheet, Savage et al. (2010) found that the velocity of these blobs ranged from 21 to 165 kms^{-1} with the mean value of 120 kms^{-1} and an average of 109 kms^{-1} for both sunward and anti-sunward motions after correcting the projection effects. Overall, the anti-sunward motion was faster than the sunward one, and all the sunward motions were slower than the free-fall speed, which indicated that the motion of the plasma blob was somehow blocked. As we mentioned earlier in this work, slowing down of the downward reconnecting outflow could be due to interactions of the flow with the forward plasma of increasing density, and the turbulence produced by such an interaction might impede the flow as well.

On the other hand, the observed downward motion may as well include a component that comes from the shrinkage of flare loops below the current sheet. Savage et al. (2010) pointed out that the SAD motion at the slowest velocity of 21 kms^{-1} in the event they studied could be of the shrinkage of the flare loops below the current sheet (see also Lin et al. 1995; Lin 2004) rather than the downward reconnection outflows. (This issue seems more complex than expected. In another flare event studied by Savage et al. 2012a, SADs were found to be the "side effects" of the passage of loops through the supra-arcade plasma. More discussions on this issue can be found in Savage et al. 2012b.) Similar events were analyzed by Sui et al. (2005) and Saint-Hilaire et al. (2009), in which the X-ray emission sources in the corona appeared to move both downwards and outwards with time. The outgoing blobs with HXR emission, indicating a very hot source of ~107--108 K, were identified with the plasma flows, and those moving downward were ascribed to the shrinkage of the newly formed flare loops. Recent work by Liu et al. (2013) indicated that the speed of the loop shrinkage could reach up to 300 kms^{-1}, which might be an important issue that is worth extra attention when we study the supra-arcade down flows as a result of magnetic reconnection.

For the blobs moving away from the Sun, tracking their motion becomes difficult after they have left the FOV of XRT, but is still doable by combining the data from LASCO C2 and C3. Savage et al. (2010) performed an extrapolation from the XRT images, estimated roughly where a specific blob might appear in LASCO images, and then located that blob in LASCO images. Therefore, the information from XRT and LASCO could be put together, giving a comprehensive scenario of the blob motion along the current sheet in a long period. Savage et al. (2010) noticed that velocities of blobs entering LASCO FOV from XRT FOV remained nearly unchanged.

Similar plasma flows due to reconnection in the eruption were continuously observed later. Takasao et al. (2012) reported observations and studies of a C4.2 flare occurring on 2010 August 18 on the northwest limb. Flare loops developed in this event could be seen in both hot (>7×106 K) and cool (4×105 K) temperature spectral lines, and a sheet structure could be recognized to extend from the top of flare loops to the outermost corona, and bi-directional plasma flows were identified in both low and high temperatures images, which suggested the wide range of the plasma temperature inside the sheet as observed in previous events (e.g., see McKenzie and Hudson 1999; Innes et al. 2003a; Asai et al. 2004). Speeds of the anti-sunward flow were found between 220 and 480 kms^{-1}, those of sunward flow were between 250 and 280 kms^{-1}, and significant deceleration of the sunward flow was also detected as the flow approached to flare loops. Interactions of the flow with flare loops eventually slowed the flow to the speed of 70--80 kms^{-1}. Observations by different instruments in various wavelengths, for example Yohkoh/SXT (Kolomański and Karlický 2007) and RHESSI (Milligan et al. 2010), as well as numerical experiments revealed the same information (see also Shen et al. 2011; Mei et al. 2012).

In addition to the above observations at optical, EUV, and X-ray wavelengths, together with the associated numerical experiments, it is worthy to note that Kliem et al. (2000) suggested that plasma blobs inside a CS could be "seen" in the radio band via some specific radio characteristics. Theses characteristics include impulsive structures with slowly drifting property on the radio dynamic spectrum, which results from energetic particles trapped in the magnetic islands, or plasma blobs, inside CS. Recent studies by Aurass et al. (2011) and Gao et al. (2014) confirmed the idea of Kliem et al. (2000). Karlický et al. (2010) noticed that the coalescence of plasmoids could even account for the occurrence of DPSs observed in solar flares. Gao et al. (2014) investigated the spectral fine structure of a solar radio burst in a CME/flare event on 2004 December 1. The data displayed zebra-like emission patterns with weak polarization, including positive, negative and zero frequency drifting stripes, during the impulsive phase of the event. They noticed that the process of the formation and the motion of plasma blobs inside reconnecting CS was similar to what occurred in the event observed by Aurass et al. (2011).

Outflows Observed During Laser-Driven Reconnection

Before ending our discussions on the reconnection outflow, we turn to the topic of the laser-driven reconnection process taking place in the laboratory to look into detailed features and physical properties of the reconnection region. Since performing in situ measurements for various parameters and the energetic particles in the reconnection region of the solar flare is impossible (at least in the foreseeable future) although performing the similar measurements in the Earth's magnetosphere environment has been done for decades (see Eastwood et al. 2007; Retinó et al. 2007, 2008; Huang et al. 2012; and references therein), conducting alternative laboratory experiments becomes a very viable approach.

Recently, the field of laboratory astrophysics has quickly developed (Remington et al. 2006). In this field, experiments are performed in laboratories, simulating various astrophysical processes, probing important features and key properties of these processes, to allow us to examine the physics of these processes. Although systems in different environments possess very different scales that can vary over a huge range, scaling laws of plasma and MHD processes allow us to relate them to the same process in many cases (Ryutov et al. 2000). Laser-driven magnetic reconnection (LDMR, Nilson et al. 2008; Li et al. 2007; Zhong et al. 2010) is one of the most significant topics of the laboratory astrophysics, booms and draws extensive attention lately.

Referring to solar X-ray observations, some very similar images were also found in a recent LDMR experiment. Faraday rotation method (Stamper and Ripin 1975) measured the magnetic field in laser-produced plasmas, which can be up to a megaGauss. At that time, the structure of magnetic field was not very clear until the proton radiography technique became available. With this technique, the magnetic field is produced in the plasma bubble due to the Biermann battery effect (see Biermann 1950; Roxburgh 1966; Widrow 2002) can be measured/detected. Such a magnetic field resulting from the anisotropic pressure in the plasma when a solid target is intensively irradiated by the long pulse laser and evaporates (nanosecond laser, with intensity of around 1015 Wcm−2). Regarding the Biermann battery effect, interested readers are referred to a review by Widrow (2002) for more details.

Yates et al. (1982) realized the occurrence of magnetic reconnection when an unexpected X-ray emission was observed between two laser spots in a multi-beams experiment. As the two bubbles expanded laterally and encountered each other with oppositely directed magnetic fields, reconnection took place and the field lines were topologically rearranged in the diffusion region. Later, with the proton radiography technique, Nilson et al. (2006) and Li et al. (2007) diagnosed the LDMR, and some striking features were found, such as the collimated jets and magnetic null point in the diffusion region.

With long pulse lasers, Zhong et al. (2010) constructed LDMR to model the loop-top X-ray source and outflows, which are often observed in solar flares. In their experiments, two aluminum (Al) foil targets were placed on the same plane, and a copper (Cu) target was placed right below the Al targets with its plane perpendicular to the Al plane (Fig. 14). Two plasma blobs were created as the Al targets were irradiated by two intense laser beams. Magnetic fields of opposite polarity associated with the plasma blobs moved toward each other as the blobs expanded, and magnetic reconnection took place between them. Two outflows were observed in the experiments. Downward outflow directly collided the Cu target and produced a very hot X-ray spot, which was used to model the loop-top source of the X-ray emission observed in the two-ribbon flare (Fig. 15). The flow velocity was directly measured to be 400 kms^{-1}, which agrees well with the typical Alfvén speed in the lab environment.

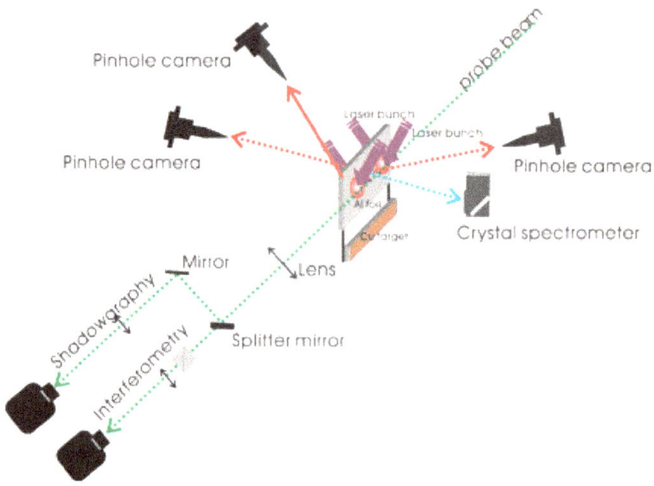

Figure 14: Experimental set-up for laser-driven magnetic reconnection performed by Zhong et al. (2010). Four bunches of long pulse (1 ns) lasers are focused on both sides of a thin Al foil target. Shadowgraphy and interferometry are used to diagnose the plasma evolution with a short pulse (120 ps) probe beam (shown as a green dotted line). The MR occurs between the two laser focus spots, and is detected by three X-ray pinhole cameras. The reconnection outflow/jet can thus interact with a pre-set Cu target. X-ray spectra from the heated plasma are recorded by a crystal spectrometer set in front of the target. From Zhong et al. (2010)

However, it is not straightforward to extrapolate directly the phenomena occurring in laboratory to those observed on the Sun because of the huge difference in scales between two environments. By applying the MHD scaling laws, on the other hand, they found that the physical parameters in solar flare and laser plasma systems are very similar, such that the characteristic length scales change from thousands of kilometers to hundreds of micrometers. Therefore, the output of their experiments as shown in Fig. 15 is considered a laboratory counterpart of the solar flare. Another important point drawn from experiments was that the reconnection process could be easily controlled by adjusting the experimental parameters. Namely, they produced inclined outflows with two asymmetrically arranged lasers, which could accordingly produce asymmetrical magnetic fields. Therefore, a "bench-top solar flare" could always be created in the laboratory according to the requirement and duplicate various important processes and/or phenomena that may take place in the solar eruption, and even those occurring in the deep universe.

In addition to helping study the solar flare, the experiment of the laser-driven magnetic reconnection may help us look into details of coalescence of plasmoids inside the reconnecting CS as well because the hot X-ray spot appearing in the experiment of Zhong et al. (2010) could be treated as a plasmoid. Although a hot X-ray spot and a plasmoid are produced in different ways in different environments, they are both the magnetized plasma blobs, and magnetic reconnection could take place when a hot spot (plasmoid) approaches to another spot (plasmoid). This indicates that the results of Zhong et al. (2010) confirmed from the point of view of the laboratory experiment that the coalescence of plasmoids in the CS causes the secondary magnetic diffusion, which partly explains why fast magnetic reconnection could still occur in a thickness CS during the major eruption.

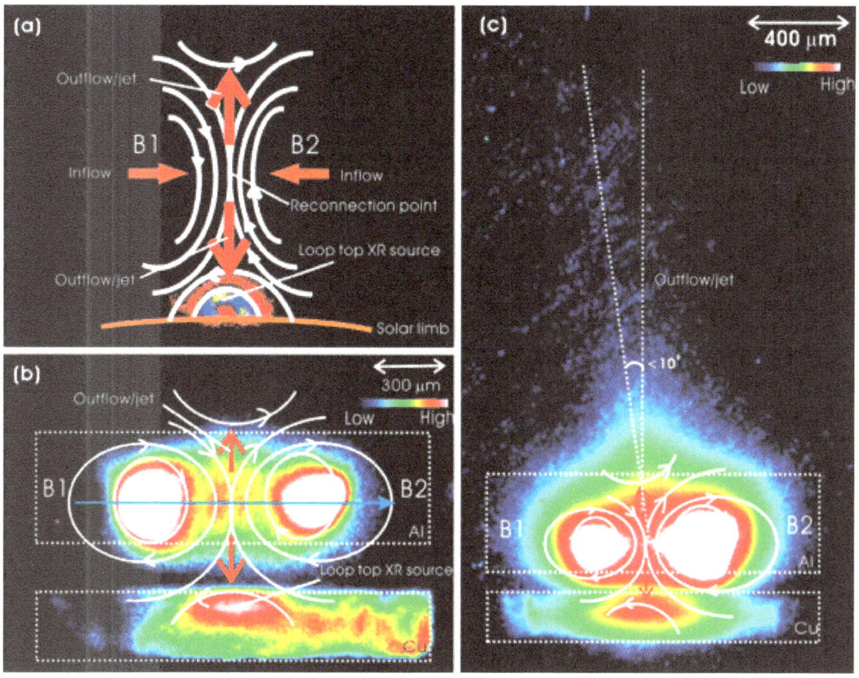

Figure 15: Loop-top-like X-ray source and outflows observed in the laboratory. (**a**) Magnetic reconnection model for the loop-top X-ray source in an eruptive solar flare, with a sketch depicting the X-ray observation scheme of Masuda et al. (1994). (**b**) The pinhole X-ray image observed forward of the Al foil target. Magnetic field lines are illustrated based on the flux surface of the plasma bubbles. The Al and Cu targets are the rectangles enclosed by white dotted lines. The red arrows indicate outflow/jet directions. (**c**) X-ray image with two laser spots separated by 400 μm400 μm and with a foil thickness of 10 μm. The asymmetry of the laser intensity on the Al target causes an imbalance of the laser spots as well as of the magnetic fields B1 and B2, and further induces the inclination of the upward outflow. The downward outflow impinges on the Cu target and results in a hot X-ray source. From Zhong et al. (2010)

Figure 15b also shows the existence of the outflow jets with an opening angle of 10∘10∘, leading to a feature becoming broader with increasing distance from the center of the CS, which is similar to what is observed for CS on the Sun. We note that the scale of the system in lab is only a few mm, so we may not necessarily be able to expect a very long CS in such a system. On the other hand, regarding the scales of the reconnection process occurring in different environments, the magnetic configuration shown in Fig. 15 has a scale of about 2 mm, and the CS is about 0.3 mm in length. According to Table 1 of Zhong et al. (2010), which displays the similarity of solar flares to laser-produced plasmas, 1 mm in lab could be equivalent to up to 1010 cm1010 cm on the Sun, therefore the CS of 0.3 mm long in lab could be scaled to 3×109 cm3×109 cm on the Sun, which may not be considered short.

INTERNAL STRUCTURES AND THICKNESS OF CURRENT SHEETS CREATED IN SOLAR ERUPTIONS

The high electrical conductivity and force-free environment in the corona confine the reconnection region to a thin layer, so a long and thin current sheet is expected to develop in major eruptive processes. Such features of the reconnecting current sheet allow us to treat the current sheet as an infinitely thin line when the main purpose is to study the global behavior of the eruption (e.g., Lin and Forbes 2000). Although indirect evidence indicates the existence of the current sheet in the eruption (e.g., see Forbes and Acton 1996 for a brief review), direct observation of a thin current sheet is difficult. This is because both size and emission of the current sheet are easily dominated by other large-scale structures nearby (e.g., see Ko et al.2003 for detailed discussions). It is often assumed that the current sheet is too thin to be observable since its thickness,d, is believed to be limited by the proton Larmor radius (Litvinenko 1996; Wood and Neukirch 2005; and references therein), which is tens of meters in the coronal environment. This view is based on information about small-scale magnetic reconnection in laboratory (with the size of meters). When the sheet forms in the solar eruption, on the other hand, it could develop in length at speed of a few hundred km s^{-1} (Forbes and Lin 2000; Lin and Forbes 2000; Lin 2002). In such a highly dynamical process, the Larmor radius of particles cannot govern the scale of the magnetic structure. Instead various plasma instabilities will likely occur and play an important role in diffusing magnetic field and in governing the scale of the sheet (Strauss 1986, 1988).

Lin et al. (2007) analyzed a set of unique data for three eruptions observed by UVCS and LASCO. They found that, in some circumstances, the current sheets are observable, and their thickness in real events may range from 10^4 to 10^5 km! This led to a very large effective electric resistivity, η_e , of the plasma inside the current sheet, which could reach up to 5×10^5 ohms m, about 12–13 orders of magnitude larger than Spitzer's classical resistivity in the quiet corona, and 4–5 orders of magnitude larger than the anomalous resistivity that is due to interactions between electrons and low frequency ion-acoustic turbulence (e.g., see Priest 1982, pp. 80–81).

These results imply that large-scale plasma turbulence operates within the sheets during the eruption, and poses a serious challenge to the existing understanding and theories of reconnection and particle acceleration in the sheet. In sections below, we discuss observations and consequent results of the thickness of the CS, and relations of these results to the theories mentioned above.

Observed Apparent Thickness of Current Sheets

As we mentioned earlier, the dark Lyα gap observed in the 2003 November 18 event marks the reconnection outflow region, which is close to the inflow region. So the decrease in the width of the dark gap in Lyα images displayed by Fig. 7 means the shrinkage of the outflow region in size, and depicts the reconnection inflow outside the CS as well. In addition, the plot also shows the asymptotic feature, which suggests the ceasing of the reconnection in- flow at the edge of the current sheet. Therefore, we can deduce the apparent thickness, d, of the sheet according to the tendency of the curve, and obtain d = 6.4×10^4 km. Because of the projection effects, this value is considered the upper limit to the actual value of d (see also discussions of Lin et al. 2007 in detail).

Similarly, we are also capable of deducing the apparent value of d via studying the image of the UVCS slit in [Fe XVIII]. As shown by Fig. 15 of Ko et al. (2003), Figs. 1 and 2 of Ciaravella and Raymond (2008), and Fig. 4 of this work, the bright spots in [Fe XVIII] on the UVCS slit approximately showed the location and the extent of the CS in space because the occurrence of [Fe XVIII] emission indicates heating as a result of reconnection that is taking place or has taken place in the CS. Plot the intensity of the [Fe XVIII] bright spot along the slit, we obtain the full-width at half-maximum (FWHM) of the intensity distribution, which in fact gives the apparent value of d. Applying this method to the intensity of [Fe XVIII] emission shown in Fig. 4, Lin et al. (2009) obtained d = 1.1×10^5 km for the event of 21 March 1998. Figure 16 showed two distributions of the [Fe XVIII] intensity along the UVCS slit in two time intervals in the event occurring in the period between January 8, 2002 and January 11, 2002. Histograms in both panels are the observational data, and the dotted curves are Gaussian profiles fitted to the data. The apparent value of d is determined according to the FWHM of the Gaussian profile, which gives d = 1.3×10^5 km.

We may need to note here that the possibility exists that the apparent value of d seen in [Fe XVIII] might contain the component contributed by a so-called "thermal halo" around the CS. Numerical experiments performed by Yokoyama and Shibata (1998, 2001) indicated that if the heat conduction was included in the simulation, the thermal energy inside the CS could spread out of the CS because of the difference in temperature between the CS and the inflow region, heating the plasma outside the CS and forming a thermal halo around the CS. Seaton and Forbes (2009) and Reeves et al. (2010) confirmed this result in the numerical experiments with higher resolution and in the spherically symmetric configuration. Spreading of the thermal energy could result in a CS embedded in the thermal halo, and what we would observe directly is actually the halo and the apparent value of d is in fact the scale of the halo instead of the CS. We shall discuss this issue in detail later.

When studying the event of 26 June 2005 and performing the relevant plasma diagnoses, Vršnak et al. (2009) also re-investigated the LASCO white-light data of the events of 8 January 2002 and 18 November 2003. They noticed that the current sheet could be wavy and be of multi-ray-like features, which implied complex internal structures of the CME/flare current sheet. Their results confirmed that the apparent value of d could be between a few 10^4 and a few 10^5 km

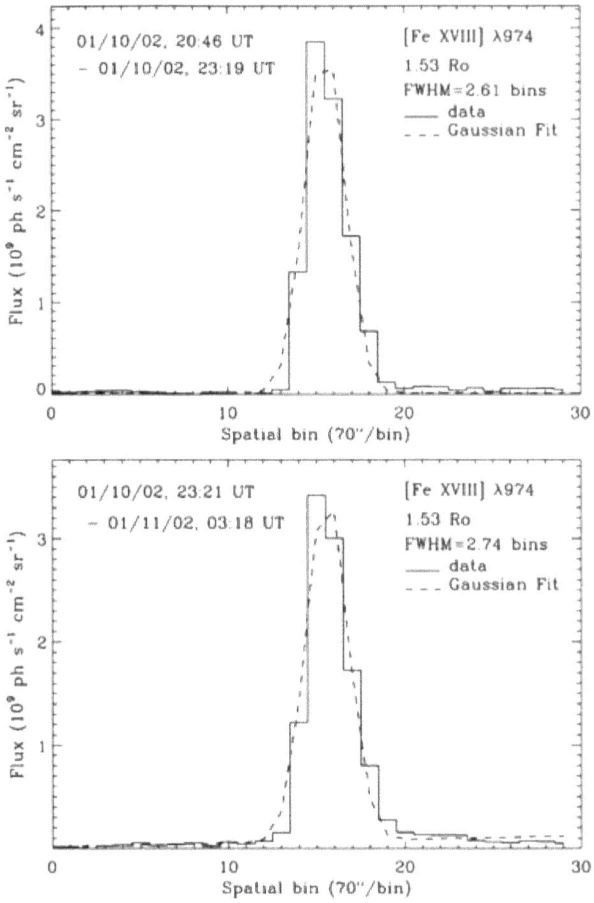

Figure 16: Distributions of the [Fe XVIII] intensity along the UVCS slit in two time intervals during the event occurring in the period from January 8, 2002 to January 11, 2002. Upperpanel: 20:46 UT to 23:19 UT on January 10, and lower panel: 23:21 UT on January 10 to 03:18 UT on January 11. Histograms in both panels are the observational data, and thedotted curves are Gaussian profiles fitted to the data. The apparent value of d is determined by the FWHM of the Gaussian profile. From Lin et al. (2009)

The CS is also often observed by LASCO on SOHO, studying the LASCO white-light images may also reveal the information of apparent d. In addition to analyzing the UVCS data, Lin et al. (2009) also investigated the size of the current sheet developed in the above two events via studying the LASCO/C2 and C3 data by an approach known as limb synoptic maps (LSMs; Li et al. 2000). An LSM is made by extracting a strip around the solar limb at a fixed altitude from an image, and then aligning the strips in time sequences. It displays the position and/or displacement of an object in polar angle (PA) at various altitudes and times, helps determine the scale (especially d) of some smaller features like the current sheets in a simple and straightforward fashion, and makes it easy to look into the variation of d in space. In this approach, the intensity distribution of the current sheet along the strip is first displayed, and then the apparent value of d is determined by the FWHM of the Gaussian profile used to fit the observational distribution. Lin et al. (2009) found that analyzing the LASCO data by LSMs still brought the apparent value of d to the range from 10^4 to 10^5 km. In addition to plasma flows inside the current sheet in the wake of the cart-wheel CME, Savage et al. (2010) measured the apparent thickness of the current sheet (see Fig. 13), and found that d ≤ 4 × 10^3 km. Re-studying the same event, Landi et al. (2012) later measured d again, and obtained that d was at least 1.3×10^5 km. It might be difficult in judging which result is more reliable than another one, but it is clear that both values are huge compared to the skin depth of the coronal plasma. This means the properties of the CME/flare CS that always involves in a highly dynamic fashion are governed by the global behaviors of the large scale structures nearby and the associated dynamical evolution. We may also note that Savage et al. (2010) and Landi et al. (2012) used different data to study the scale of the CS, the former used the HINODE/XRT data and the latter used HINODE/EIS data. Therefore, the thermal halo effect could probably play a role in governing the apparent values of d.

Most recently, Ling et al. (2014) studied the current sheet formed in the eruption of 7 September 2005. The source region of this event was active region AR 10808, which was likely associated with a number of fast halo CMEs from late August to September 5. When it was on the visible disk, this region produced 25 M-class, and 10 X-class flares over ten days. The eruption of 7 September 2005 produced the last extremely energetic flare of solar cycle 23 (optical class 3B, X-ray class > 17, saturating GOES instrument), which led to a severe radio blackout affecting communication and navigation systems, and a fast CME at speed of about 2500 km s^{-1}, which was the fastest ever observed by MK4. As expected, the event developed a long current sheet connecting the CME to the associated flare.

This event is probably the first one of which only the data obtained by the ground-based instrument were available for studying the CME/flare CS because SOHO was not operating during the event. What is nice about the work of Ling et al. (2014) is that they used data obtained by the MLSO K-coronameters, which have been continuously improved in resolution at low altitudes, and the radius of the occulting disk has been continuously decreased; therefore the FOV of the present MK4 extends from 1.08 to 2.86 $R_\odot$ in heliocentrical distance, which allows us to observe the objects in the very low corona in white-light. Usually, the region below 1.5 $R_\odot$ is very important for investigating the connection of the current sheet to the flare loop system as well as for looking into the lower part of the sheet itself since the high temperature flare loops and the associated cusp region in a typical eruptive process often appear above the altitudes ≥ 4 × 10^4 km (e.g., see Lin 2002 and references therein).

Observations of the event on September 7, 2005 by MK4 provided a very good opportunity for studying the early evolution of the current sheet at low altitudes (Fig. 17), including the formation and development as well as physical properties of the current sheet although only white-light data are available. Furthermore, the small radius of the occulting disk of MK4 even allows us to see the lower tip of the CS in the FOV of the coronagraph. Consistent with indications by Table 2 that the flare loops in the major eruption could usually exceed the height from the solar surface up to 0.14 $R_\odot$, the lower tip of the CS developed by this event indeed appeared in the MK4 FOV.

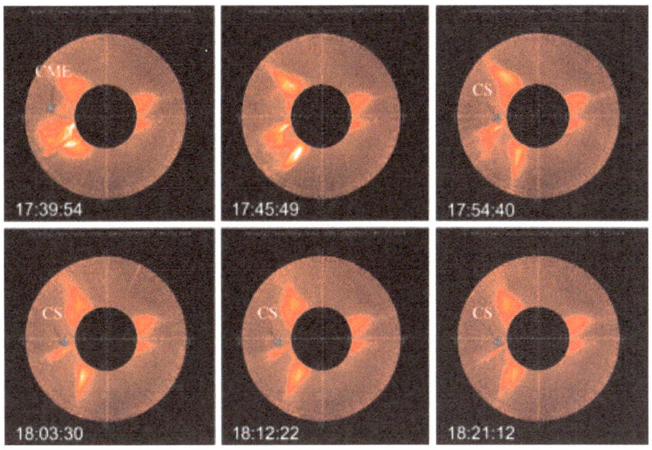

Figure 17: MK4 white light vignetted data in rectangular coordinates showing the progression of the CME and its trailing ray on 2005 September 7. The observation times (UT) are indicated on the images. CME and CS are specified in each panel. From Ling et al. (2014)

From the MK4 data, it is straightforward to determine the apparent value of d by directly measuring the FWHM of the brightness distribution of the sheet in the direction perpendicular to the sheet extension, and d as a function of the distance from the solar center could be further obtained. Results of Ling et al. (2014) showed that d varied with both distance and time such that the smallest value of d was obtained near the sheet base, and widened with height and time. After the measurement became possible, the narrowest region of the sheet had the width of ~ 7 × 10^3 km, which was measured near the base of the sheet, and d had the typical value of 2.8 × 10^4 km at most locations on the CS. This result is consistent with those deduced from the LASCO data that the apparent width of the CS becomes larger with height and time. We need also note here that, on the other hand, in the framework of the traditional Petschek-type reconnection, the jet of the reconnected plasma from the diffusion region may become wider at higher altitudes (cf. Fig. 15). In this sense, the feature in the region that was observed to increase in the width might not be the CS, but a collimated plasma jet.

Before we end this section, it is interesting to note that the value of d in different cases observed by various instruments does not differ from one another very much. Observations by LASCO and UV spectroscopy gave the similar results for d in the lower corona ranging from 6.4×10^4 to 2.1×105 km6.4×104 to 2.1×10^5 km, and this value almost doubled at higher altitudes of 4--5 R⊙4--5 R⊙. At even lower altitudes, say between 1.1 and 1.2 R⊙1.1 and 1.2 R⊙, it is remarkable that observations of the ground-based instrument (for example, MK4) and those of the space-borne instruments (for example, Hinode/XRT) indicated that the width of the white-light current sheet (7.0×103 km7.0×103 km) and that of the X-ray current sheet (4--5×103 km4--5×103 km, see also Savage et al. 2010) possessed almost the same apparent values at similar or slightly different altitudes, but the associated CMEs had quite different speeds (2.5×103 kms−12.5×103 kms^{-1} compared to 450 kms−1450 kms^{-1}).

Impact of Projection Effects on Apparent Values of d

Because of projection effects and possible complex

morphology of the current sheet and the associated magnetic field, the above values for d are upper limits of the true values of d. Suppose the current sheet is a plate of true thickness dd and true depth D, and LOS is in the xx-direction (Fig. 18). Generally, the emission measure (EM) of the CS observed in a given spectral line is related to the density inside the CS, nene, and the path length in CS along LOS, D′, in the way of n2eD′ne2D′; and the brightness of white-light images or that of the scattered Lyαα lines, Isc is proportional to neD′. As a result of the modest EM of the current sheets in eruptive events compared to that of the nearby corona, on the other hand, the current sheet may be faint and difficult to detect if it is not observed roughly edge-on. Below, we discuss how projection effects impact measurements of d (see also Lin et al. 2009)

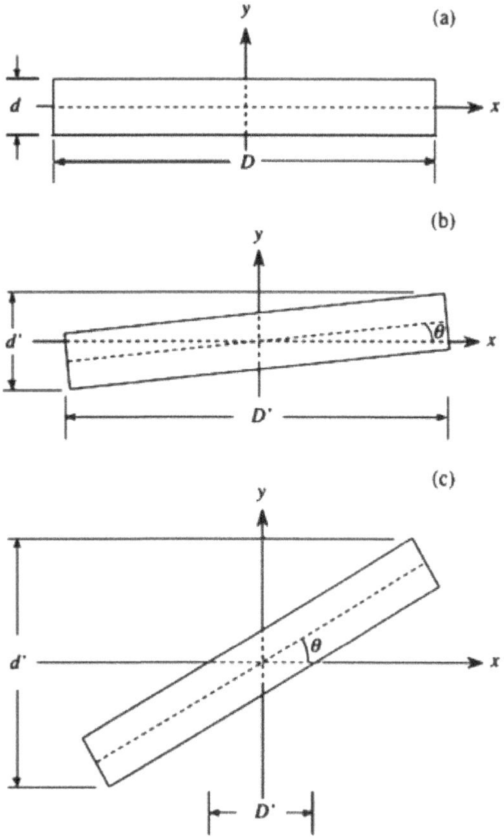

Figure 18: Schematic diagram of orientations of the sheet and the corresponding parameters for the sheet properties. (**a**) The sheet is observed edge-on; (**b**) the sheet is observed at a small angle; and (**c**) the sheet is observed at a large angle. From Lin et al. (2009)

All the images we analyzed above are projections of the emission from an optically thin three-dimensional structure onto the two-dimensional plane of a detector. Since the intensity recorded by the detector is an integration of the total emission along LOS, the intensity level is related to both the density and the column depth along LOS. Therefore, viewing a bright object at various angles may yield different impressions (see detailed discussions of Forbes and Acton1996). For a current sheet of given d and extent in other dimensions, seeing it edge-on results largest integral path and thus the largest EM, and seeing it face-on, on the other hand, results in the shortest path and the lowest EM.

In the case of a bright sheet appearing in the dark background, as in the LASCO images, the brightness, b, of the sheet is linearly dependent on the length of LOS inside the sheet. For simplicity, the plasma density in the background is set to be zero. If the current sheet is observed edge-on (Fig. 18a) and LOS is parallel to the long axis of the sheet, the observed thickness of the sheet is its true thickness, d, and its brightness (or the EM) b is proportional to the path along LOS in the sheet s = D, namely b ∝ s. If the current sheet is tilted by an angle θ, the apparent sheet thickness becomes d and the path along LOS becomes s = D' (Figs. 18b and 18c).

Lin et al. (2009) noticed that, suppose the uniform distribution of the mass inside the sheet, in the case of small θθ such that tanθ≪d/D (Figs. 18b), d′ is related to d in the way of d′/d≤2, which indicates that the increase in the apparent thickness caused by the projection effects is within the factor of 2. With increase in θθ, the apparent brightness of the sheet, b′, is related to b,d, and θ such as $b'/b=(d'/d-\cos\theta)^{-1}b'/b=(d'/d-\cos\theta)^{-1}$. Here, d′/d is always larger than unity, and cosθ is between 0 and 1. So the impact of cosθ could be ignored as d′/d gets big enough, and the brightness of the sheet is roughly inversely proportional to the observed thickness of the sheet. This implies only a limited impact of projection effects on measuring dd.

In reality, direct observations of the current sheet are also affected in an important way by the signal-to-noise ratio of instruments themselves in addition to being affected by brighter structures nearby. Combining these effects leads to that the current sheet becomes hard to be identified as d' /d exceeds 10 (see also Forbes and Acton 1996). According to the relation of EM and I_{cs} to n_e and D': $EM \sim n_e^2 D'$, $I_{cs} \sim n_e D'$, we have $n_e \sim EM/I_{sc}$ and $D' \sim I_{sc}^2/EM$. With D' deduced this way, d' could be measured directly from observations. Comparing D' and d' could help us evaluate the impact of the projection effects. Consulting Fig. 18, if the projection effects could be ignored, d' should be small compared to D' (Figs. 18a and 18b); and a serious consideration of projection effects is needed if d' and D' are comparable to each other (Fig. 18c). Using this method, Ciaravella and Raymond (2008) found that d' /d is between 2.5 and 5 for the current sheet developed in the eruption of 2003 November 4.

In addition to the case of planar CS as shown in Fig. 18, the CS in reality may be very likely to have a wavy configuration (see Fig. 2 of Vršnak et al. 2009, which depicted a CS that looks very like a curtain) as well. In this case, the complex internal structure could account for "multi-ray" features of the post-CME CS that were also often observed (see Ko et al. 2003; Vršnak et al.2009; Ciaravella and Raymond 2008; Ling et al. 2014), and brings more uncertainties in determining the geometric size of the CS in a specific case since it is hard to tell how many folders a wavy CS could have. So it is worth investing more effort in investigating this issue in the future. Such effort is two-fold: enhancing the resolution of observations in multiple wavelengths so that fine structures in the CS could be well recognized and studied, and performing the corresponding numerical experiments so that we could understand the physics governing manifestations of the CS.

Reasonableness of Large Values of CS Thickness and Resistivity

Results and discussions given above indicated that a CS developed in the solar eruptive events is recognizable and the apparent value of d could be easily obtained when the plane in which the CS extends is roughly parallel to LOS, namely it is observed edge-on. Because it is difficult to recognize a CS and to measure its thickness as the CS is observed face-on, most of the CSs we have recognized and studied were observed edge-on although recognizing a face-on CS is much less difficult in the SDO era (e.g., see Warren et al. 2011; Savage et al. 2012a, 2012b; McKenzie2013; Doschek et al. 2014; Hanneman and Reeves 2014).

The existing observations and the related results confirm that the apparent values of d could be up to 10^3–10^5 km although these observations were made for different events by different instruments, and impact of the projection effects on the measurements turned out to be limited. But our expectation is to completely remove this impact in order to get the true values of various observables. For the time being, on the other hand, it is impossible to remove technically the impact of the projection effects completely on the apparent values of observables although observations by STEREO could help partly resolve this problem (e.g., see Isavnin et al.2013 and references therein). To reach a more meaningful conclusion about the true value of the CS thickness, the corresponding lower limits need to be figured out by combining the well-known theory and observations.

Forbes and Malherbe (1991), Y. Fan (2005, private communication), and Riley et al. (2007) suggested approaching this problem by interpreting the plasma blobs flowing along a CS as the magnetic islands resulting from the tearing mode. According to our previous discussions, we realized that those blobs of various sizes appearing in the reconnecting CS are very likely to be the products of different types of plasma instabilities and the consequent turbulence, as well as the resultant small structures (see Kliem et al. 2000; Shibata and Tanuma 2001; Bárta et al.2011a; Shen et al. 2011; and references therein). The existence of internal small structures of the CS could be indirectly indicated by the location of the HXR source region above the flare loop (Sui et al. 2006), or by the sub-second pulse of the HXR emission (see Aschwanden 2004and references therein) as well as the time-profile of the quasi-periodic reversals of radio polarization at 17 GHz (Huang and Lin 2006). These results suggest that particle acceleration occurs inside reconnecting CS. Plasma instabilities and the associated consequences could be one source of these structures, and the inhomogeneities of plasma and magnetic field in both space and time during the reconnection process could be another one. Interested readers are referred to Kliem et al. (2000), Shibata and Tanuma (2001), Lin et al. (2006), Liu et al. (2010), and so on.

The tearing mode instability is probably the mode of the most importance of the various possible instabilities that can act in the CS (see discussions of Priest 1988; Priest and Forbes2000, pp. 177–185). The other possible instabilities that may occur in a structure like CS are gravitational and rippling modes instabilities. The former exists when a gravitational (or equivalent) force acts transverse the CS to produce a density stratification, and the latter takes place as the magnetic diffusivity is non-uniform across the CS that is usually caused by the temperature structure in the CS. As showed by Furth et al. (1963), the tearing mode could be excited as long as the resistivity exists and the aspect ratio (namely the length divided by the thickness) of the CS exceeds $\sim 2\pi$, without requiring gravitational and resistivity gradients. The tearing mode drives turbulence leading to the release of magnetic energy and reorganization of the magnetic field after topology changes. Because reconnection in the corona is often patchy (e.g., Guidoni and Longcope 2011), the turbulence is likely to be intermittent. The tearing mode and the resulting turbulence largely govern both the geometric features of the CS and many of the important properties of magnetic reconnection. We are able to deduce the lower limit of the CS thickness by relating observational consequences of the turbulence due to the tearing mode to the corresponding theories.

Long current sheets can easily become unstable to the tearing mode. This is shown by both theories (e.g., Furth et al. 1963) and numerical experiments (Ambrosiano et al. 1988). In two dimensions, the current sheet is torn into many small-scale magnetic islands (see Figs. 6.1 and 6.3 of Priest and Forbes 2000, which described possible modes of instabilities that may develop in the CS, and more detailed features of a torn CS, respectively). In three dimensions, the current sheet develops flux ropes which are the counterpart of two-dimensional magnetic islands (compare to Daughton et al. 2011). The turbulence developing from the tearing mode enhances diffusion of the large-scale magnetic field. The turbulence converts part of the magnetic energy into the kinetic energy of the perturbation, which in turn creates perturbed small-scale fields via the dynamo process (Bhattacharjee and Yuan 1995). The growth of the perturbed field is also governed by the tearing mode (see Strauss 1986, 1988), and small-scale field stochasticity and braiding produced in this process yields an effective perpendicular momentum transport, which leads to an anomalous electron viscosity (see also discussions of Biskamp 1993, p. 22). This results in a double diffusive process and causes more efficient diffusion in the reconnecting current sheet. If the process operates in a CME CS, then there is likely to be reasonably fast reconnection even in a fairly thick CS.

When the tearing mode develops (Fig. 19) with the growth rate slower than the hydromagnetic rate but faster than the resistive diffusion rate, its wavenumber k is related to l such that $S^{-1/4} < kl < 1$ (e.g., see Furth et al. 1963), where l is the half thickness of the CS (namely d = 2l, referring to Fig. 19), $S = \tau_d/\tau_A$ is the Lundquist number of the CS, and $\tau_A = l/v_A$ and $\tau d = l^2/\eta$ are the times at which the Alfvén wave and the resistive diffusion traverse the CS, respectively. We note here that S is the local Lundquist number of the CS, not the global one of the whole system, and V_A is the local Alfvén speed near the CS. Therefore, $S = \tau_d/\tau_A = vA_l/\eta = v_A/v_i = 1/M_A$ according to the standard theory of magnetic reconnection (see Priest and Forbes 2000).

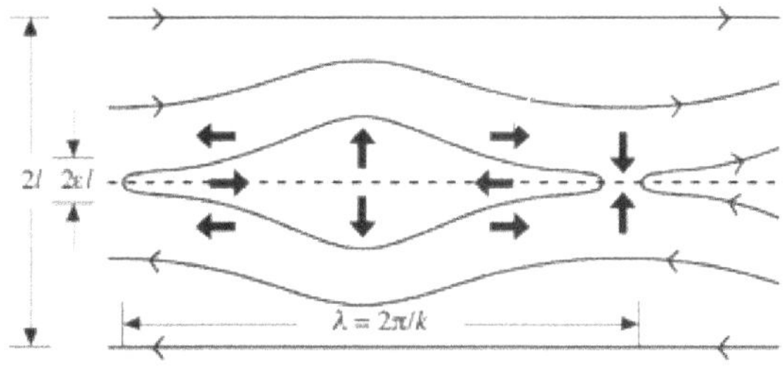

Figure 19: Interior structure of the current sheet in which the tearing mode instability develops.Thick arrows show plasma flow, and thin arrows are for magnetic field lines. (Courtesy of E.R. Priest)

If the plasma in the CS has sufficiently high conductivity such that both S and $S^{1/4}$ are large compared to unity, magnetic dissipation or reconnection will be confined to a very thin CS and the tearing mode develops with very long wavelengths (e.g., see Furth et al. 1963; Priest and Forbes 2000). In this case, kl<< 1 holds so that the turbulence could grow at the rate between $1/\tau_d$ (the resistive diffusion rate) and $1/\tau_A$ (the hydromagnetic rate). For the events so far observed to take place in the solar atmosphere, on the other hand, MA (S) varies from 10^{-3} (10^3) to 10^{-1} (10) (e.g., Yokoyama et al. 2001; Ko et al. 2003; Lin et al. 2005; Liu et al. 2010; Savage et al. 2010). So although S>>1 did hold for these events, $S^{1/4}$>> 1 did not. Therefore, kl possesses a lower limit of finite value, providing us an approach to determining the lower limit to the apparent value of d such that

$$S^{-1/4} < kl, \qquad l > M_A^{1/4}/k, \qquad l_{\min} = \lambda M_A^{1/4}/(2\pi), \tag{1}$$

where $\lambda = 2\pi/k$ is the wavelength of the turbulence.
We may also note here that Loureiro et al. (2007) analyzed the relation of l to the Lundquist number in a different way. They investigated the dependence of kl on the global Lundquist number, S_L, for the whole system instead of the local one, S, for the CS, as the rate of reconnection reached maximum. They started with a CS configuration described by the Dawson function, and found that at the maximum reconnection rate, $kl \sim S_L^{-1/8}$. If the plasma inside the CS undergoes turbulence in a more general sense, the minimum rate of reconnection would depend on SL in a little more significant way, $M_A \sim S_L^{-3/16}$ (see Lazarian and Vishniac 1999), and kl is roughly proportional to the speed of plasma blobs compared to the local vA (e.g., see also Lazarian et al. 2014 and references therein), which is a factor of around 3 less than MA according to Shen et al. (2013). This means that $kl \sim S_L^{-3/16}$ in the turbulent reconnection process. We see that SL could be as large as 10^{12}, but $S_L^{1/8} \approx 31.6$ and $S_L^{3/16} \approx 178.$, which are not very large, and the corresponding value of l may not be very different from that of l determined by (1). Later, Bhattacharjee et al. (2009) studied the reconnection process starting from a Harris sheet, and brought the dependence of kl on S to $kl \sim S^{-1/4}$ as the fastest reconnection took place. Therefore, the possible values of l governed by scaling laws for the turbulent reconnection have roughly the same order of magnitude no matter how the process is initiated. More discussions about the dependence of features and properties on S and/or SL manifested by the turbulent reconnection can be found in the introduction of Murphy et al. (2013) as a brief review

With M_A and λ given, we are able to estimate l_{min} according to (1). As suggested by Ko et al. (2003) and Lin et al. (2005), the plasma blobs observed to flow along the CS were identified with the magnetic islands (turbulent eddies) resulted from the tearing mode instability and the consequent turbulence, so the value of λ in Eq. (1) could be identified with the distance between the two adjacent blobs (see also Fig. 10). Usually, the value of M_A could be obtained via different approaches (see also Yokoyama et al. 2001; Ciaravella et al. 2002; Ko et al. 2003; Lin et al. 2005; Bemporad et al. 2006; Ciaravella and Raymond 2008; Savage et al. 2010; Liu et al. 2010, 2013; Liu 2013; Su et al. 2013), and then estimate lmin according to (1) is straightforward. For example, the minimum of the distances of every two adjacent blobs observed in the 2002 January 8 event is 3.062×10^5 km (see Fig. 10), namely $\lambda = 3.062 \times 10^5$ km, and M_A ranged from 0.01 and 0.03 (Ko et al. 2003). Taking M_A = 0.01 gives $l_{min} = 1.7 \times 10^4$ km and $d_{min} = 3.4 \times 10^4$ km, which is different from $d_{max} = 1.5 \times 10^5$ km by a factor of 4. This is consistent with the result of Ciaravella and Raymond (2008).

Further investigations indicated that a more general expression relating l_{min} to M_A and λ exists such that

$$l_{\min} = M_A^{\alpha}\lambda/(2\pi), \tag{2}$$

for different boundary conditions and different modes of the tearing (for instance, double or even multiple tearing, Pritchett et al. 1980; Priest 1985). Equation (2) is generally valid for different cases with α being a constant between 1/7 and 1 (see also Loureiro et al. 2007; Lin et al. 2009). Taking α = 1/7 for the instability in a CS with a periodic internal structure (see also Bobrova and Syrovatskii 1980 for more details), we have larger l_{min} than the above results; taking α = 1/4 for the "constant-ψ" approximation (namely fluctuations perpendicular to the CS are small) results in the scaling law given in Eq. (1); and taking α = 1 for the case that the "constant-ψ" approximation breaks down (see also Pritchett et al. 1980; Priest 1985) brings l_{min} to the range from 7.4×10^2 to 4.55×10^3 km, which is 10~100 times smaller than the apparent values of l. This difference is surely larger than 4 that we have just mentioned shortly, but apparently much less than the difference between 10^4 km and a few meters or a few km! This indicates that the true value of d should not differ very much from the apparent value as long as the CS could be recognized in observations.

Before ending this part of work, we point out that many authors have inferred observationally the presence of turbulence inside the post-CME CS, which could account for the large d and η_e, from the detection of significant non-thermal line broadening in EUV/UV spectral lines. For instance, Ciaravella et al. (2002) deduced that the turbulent motion had speed less than 60 km s^{-1} in the CS (see also discussions of Kohl et al. 2006); Ciaravella and Raymond (2008) measured [Fe XVIII] line widths in a CS that were 100 km s^{-1} larger than the thermal widths as a result of a combination of bulk motions and turbulence. The Petschek-type and turbulent reconnection processes must work together to account for observations (e.g., see also discussions of Mei et al. 2012 from a theoretical view point). Susino et al. (2013) noticed significant non-thermal broadening in the CS as a result turbulent motions when analyzing the EUV data from UVCS. Bemporad (2008) further pointed out that the post-CME current sheet is an assembly of multiple small-scale structures, which result from the turbulence and could be responsible for both high temperature and large thickness of the CS. Doschek et al. (2014) found 20–60 km s^{-1} non-thermal motions with Fe XXIII and Fe XXIV lines above flare loop in EIS spectra.

CS AND ITS INTERNAL FINE STRUCTURES OBSERVED IN NUMERICAL EXPERIMENTS

Numerical simulations and experiments provide us a very important and practical approach to studying these instabilities and their impact on the magnetic reconnection. There have been a large number of works and results of magnetic reconnection in the fields of plasma physics, geophysics and space sciences, stellar and interstellar physics, and interested readers are referred to the books by Priest and Forbes (2000) and by Biskamp (2000), or works of Shay et al. (2011) and Opher et al. (2011), or reviews by Cassak and Shay (2012) and Lazarian et al. (2014). We are focusing in the present work on reconnection occurring in the CS during solar eruptions, as well as the physical properties of magnetic reconnection and the CS revealed by these results.

An inevitable question related to our discussions in previous sections is: How can fast magnetic reconnection take place in such a thick CS? A leading explanation is that the occurrence of the tearing mode in the CS results in turbulence and that leads to small scale structure that enhances the diffusion of the magnetic field, and thus allows the magnetic energy to be converted into heat and kinetic energy at a reasonably fast rate in a thick CS. The question following this is then in a real solar eruptive process, what is the smallest size of these fine structures so that the energy conversion could be considered fast? Ichimaru (1975) noted long ago that a basic property of magnetic reconnection is that it is turbulent, and a reconnecting CS should contain a wide spectrum of structures at various sizes. Shibata and Tanuma (2001) argued that, in an environment of high magnetic Reynolds number, like the solar corona and elsewhere in astrophysics, magnetic reconnection must take place in a fractal CS, which is produced by the tearing mode instability. Recent observations from SDO/AIA have also shown clear evidence for turbulence in supra-arcade plasma sheets (Reeves and Golub 2011; McKenzie2013; Doschek et al. 2014). In such a CS, island-like magnetic structures of various scales relate the large scale features of the CS itself and those out of the CS to the micro-scale structures inside the CS in a self-consistent way.

For a CS with large aspect ratio, say larger than 2π, the tearing mode is linearly unstable (e.g., see Furth et al. 1963; Priest and Forbes 2000). In this stage, the large CS fragments into many structures (namely islands) of intermediate sizes. Continuous evolution in the system leads to the non-linear development of the instability. Smaller islands form in between islands that appeared earlier and are moving apart, which manifests the scenario of filamentation of the CS and has also been discussed by Shibata and Tanuma (2001) and Loureiro et al. (2007). [The instability with flow was considered by Loureiro et al. 2007 as well.] This process yields the successive decrease in the size of magnetic islands until the size of islands approaches the kinetic size of the system so that fast dissipation or fast reconnection is allowed to take place. In this process, the magnetic energy in large scale structures continuously cascades to small scales at which the magnetic field is quickly diffused.

Shibata and Tanuma (2001) found that, in the typical coronal environment, fast reconnection may occur as the number of cascading steps reaches 6. They noticed that the tearing mode develops slowly in the linear stage, it takes around $3 \times 10^4 \sim 10^6$ s for the first island to form after the instability commences. The system then evolves non-linearly, and the formation of magnetic islands accelerates such that the second island appears 100~1000 s after the first one, and the CS quickly fragments. On the other hand, recent numerical experiments indicated that these times depend on the background of the system and the initial configuration of the reconnecting magnetic field. Shen et al. (2011) used an initial configuration commonly applied to model two-ribbon flares with the magnetic field line-tied to the lower boundary and with the other boundaries open. They found that the first island appears 1.6×10^3 s after the system starts to evolve as a result of perturbation equivalent to $0.03V_A$ at the open boundary. This time shortens to 900 s if the perturbation is enhanced to $0.1V_A$ in a similar initial configuration (L. Ni 2014, private communication).

In the simulation of an eruption invoked by the catastrophic loss of equilibrium that produces both CME and flare, the first blob appeared 290 s after the disruption took place in the relevant magnetic configuration, the second one was created 79 s after the first one and the third one appeared 21 s following the second (Mei et al. 2012). In these and other similar numerical simulations (e.g., see Skender and Lapenta 2010; Bárta et al. 2011a; and references therein), the reconnection process always accelerates after the formation of the first plasmoid, which manifests the non-linear property of the tearing mode in the later stage and provides strong support of the result of the analytic studies by Shibata and Tanuma (2001). The non-linear stage of the tearing mode is also known as the plasmoid instability (e.g. see also Bhattacharjee et al.2009; Huang et al. 2010; Huang and Bhattacharjee 2010; Shen et al. 2011; Murphy et al. 2013; Shen et al. 2013 and references therein).

Forbes and Priest (1982) started working on the magnetic reconnection process taking place in the CS of two-ribbon flares, confirmed the kinematic features of post-flare loops predicted by the Kopp-Pneuman model (Kopp and Pneuman 1976), and investigated the rising of the post-flare loop and the interaction between the sunward reconnection outflow and the flare loop top. They noticed the occurrence of the slow mode shock extending downward from the principal reconnecting region in the CS (see also recent works by Shen et al. 2011; Mei et al. 2012) and the special upside-down Y-shape structure of the two-ribbon flare, and that it took a while for the fast reconnection to commence in the CS.

Forbes and Priest (1983) further studied the non-linear evolution of the reconnecting magnetic configuration. Two new features, which had not been considered by the existing models of the flare loop, appeared in their numerical experiments. These two features were the fast mode shock located between the top of flare loops and a pair of slow mode shocks extending from the lower tip of the CS, and the formation and the annihilation of magnetic islands in the CS above the flare loop (see also discussions of Bárta et al. 2011a; Shen et al. 2011; Mei et al. 2012). In the process of magnetic reconnection, some islands move upward and some downward, and they pointed out that the annihilation resulted from the coalescence instability.

We now know that the first phenomenon noticed by Forbes and Priest (1983) is the so-called termination shock on the top of flare loops, and that it results from the interaction of the reconnection outflow with flare loops, namely the product by the high speed flow when meeting an obstacle. It is the same in both the origin and physical properties as the bow shock on the top of the magnetosphere (see discussions of Priest and Forbes 2000, pp. 322–324; Lin et al. 2008), and as the termination shock near the edge of the heliosphere. In the reconnection process, the speed of the outflow is roughly the local Alfvén speed near the principal X-point (e.g., see detailed discussions of Lin et al. 2008; Shen et al. 2011), at which the reconnection rate reaches its maximum. On the other hand, the sunward outflow is usually slowed down by the closed flare loops (see Table 2 of Shen et al. 2011), which could result in the termination shock, and the stationary type II radio burst of nearly zero frequency drift reported by Aurass et al. (2002) constituted a possible evidence of the termination shock. Forbes and Malherbe (1986) further studied the properties of the termination shock, and realized that the termination shock may cause the accumulation of the thermal energy on the flare loop top, enhancing the formation of flare loops below the CS. On the basis of the works by Forbes and Priest (1983) and Forbes and Malherbe (1986), Yokoyama and Shibata (1997, 2001) investigated the impact of anisotropic heat conduction and chromospheric evaporation on magnetic reconnection and the consequent formation of flare loops, revealing much information that is more realistic and enriching our knowledge about the relevant issues.

The second feature noticed by Forbes and Priest (1983) has continuously been displayed and duplicated in recent numerical experiments with higher and higher resolutions (Riley et al.2007; Bárta et al. 2008, 2010, 2011a, 2011b; Shen et al. 2011, 2013; Murphy et al. 2013). These experiments also confirmed that it is inevitable for magnetic islands to form in the CS developed in the coronal environment of high S, and that small islands have a tendency to merge into bigger islands. Theoretical and numerical studies by Karlický and Bárta (2011) further indicated that collisions or interactions between two adjacent islands may produce energetic particles (see also Nishizuka and Shibata 2013), and contribute to the HXR and the radio emissions as well.

Subsequent works turned to pay more attention to the small scale structures inside the reconnecting CS as well as the possible observational consequences. In the spirit of Ko et al. (2003) and Lin et al. (2005) regarding the motion of plasma blobs moving in the CS observed in white-light, Riley et al. (2007) simulated evolutionary features of the CS and kinematic behaviors of magnetic islands in the CME/flare CS for the first time, specifying that the motion of the plasma blobs bifurcates somewhere in the CS, and that one group of blobs moved sunward and another one moved anti-sunward. They noticed that the location where the bifurcation took place was not far from the solar surface, and was usually within one solar radius from the solar surface. All the blobs, together with the associated reconnection outflows, observed above this location move anti-sunward, and those below this location move sunward (see also the movies attached to the papers by Ko et al. 2003; Lin et al. 2005; Savage et al. 2010). This explained why the plasma blobs appearing in FOVs of the LASCO C2 and C3 were always observed to leave the Sun, and why those appearing at lower altitudes were always observed to move sunward (e.g., see the events and the associated references listed in Table 2 of Shen et al.2011).

In addition to the above behaviors, the results of Shen et al. (2011) revealed that the occurrence of the tearing mode, and thus the formation of magnetic islands, depends on the value of S: magnetic reconnection undergoes smoothly and the CS is dissipated gradually without any small structure appearing in the CS at $S < 900$, and the situation changes apparently with the formation of magnetic islands at $S > 1000$. As a follow-up of Forbes and Malherbe (1986), Shen et al. (2011) re-investigated the reconnection process taking place in the two-ribbon flare CS in the same magnetic configuration, even using the same code, but with much higher grid resolution and even adaptive mesh refinement (AMR) because of the significant improvement in calculation techniques and resources. So not only did Shen et al. (2011) duplicate all the phenomena observed by Forbes and Malherbe (1986), but they obtained much richer information than before on the reconnection process in the flare CS.

First of all, evolution in the system with asymmetric boundary conditions (namely, the line-tied boundary condition on the bottom side and the open boundary on the other three sides) does not immediately cause the tearing mode to occur although the aspect ratio of the CS exceeds 2π greatly at the very beginning, which is the criterion deduced from the linear theory of the tearing mode (e.g., see also Furth et al. 1963; Priest and Forbes 2000). This is consistent with the result of the simulations with symmetric boundary conditions (e.g., see Loureiro et al. 2007; Ni et al. 2010; Mei et al. 2012). Two reasons lead to the difference between analytic solutions and numerical experiments: One, it turned out that the most significant consequence of the tearing mode will not appear until non-linear effects become dominant; and two, comparing with the magnetic configuration studied by Furth et al. (1963), those studied in the numerical experiment contain a reasonably large amount of mass, and could be as close to realistic as possible.

In these numerical experiments, the system commences to evolve as the two magnetic fields of opposite polarity and plasmas at either side of the CS approach one another, most of mass in the CS is expelled, and the CS gets thinner and thinner yielding a larger and larger aspect ratio. Loureiro et al. (2007) found that the first magnetic island, thus the tearing mode instability, occurs after the aspect ratio exceeds 50, and Ni et al. (2010) noticed that the threshold value of this ratio could be even larger than 100. The threshold value of the ratio in the case of Shen et al. (2011) was not available since they investigated the KoppPneuman configuration (e.g., see Kopp and Pneuman 1976) in which the CS extended from the solar surface to infinity. However, their results clearly showed that the tearing mode did not appear at the very beginning although the aspect ratio of the CS was infinity until the CS was squeezed to be thin enough. Figure 20 displayed space distributions of plasma density ρ, electric current density J , x-component of velocity v_x , and z-component of velocity v_z, respectively, at two times of the system in evolution. As expected, many blobs followed the first one to appear in the CS successively with the tearing mode commencing to take place, and then the CS thickness gradually approached a constant (see solid curve in Fig. 21 with L_0 being the scale of the simulation domain)

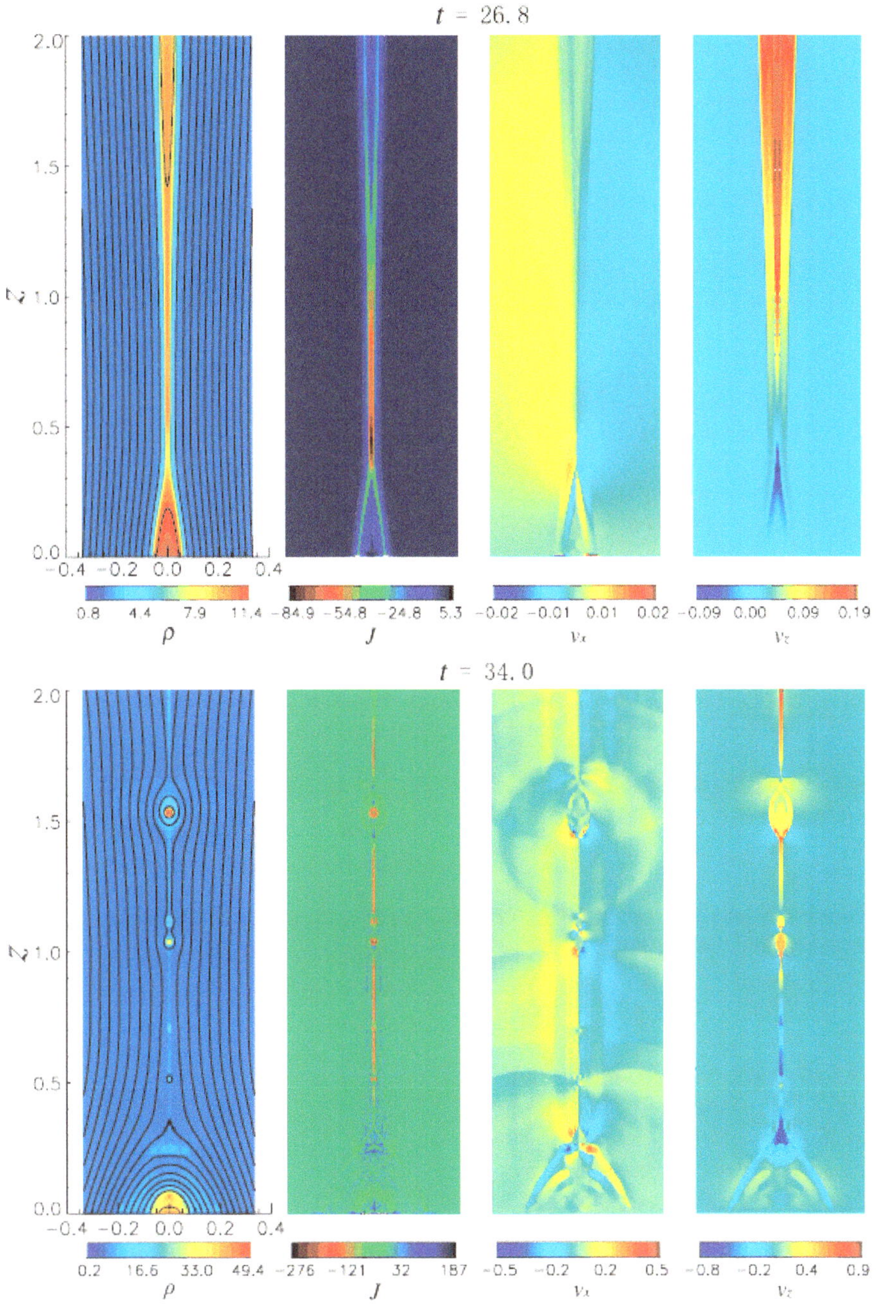

Figure 20: Evolution of various parameters (plasma density ρ, electric current density J, x- and z-components of velocity) in the magnetic configuration including a semi-infinitely long CS as magnetic reconnection progresses. Upper row: reconnection is in the Sweet-Parker stage; lower row: reconnection is in the turbulent stage. From Shen et al. (2011)

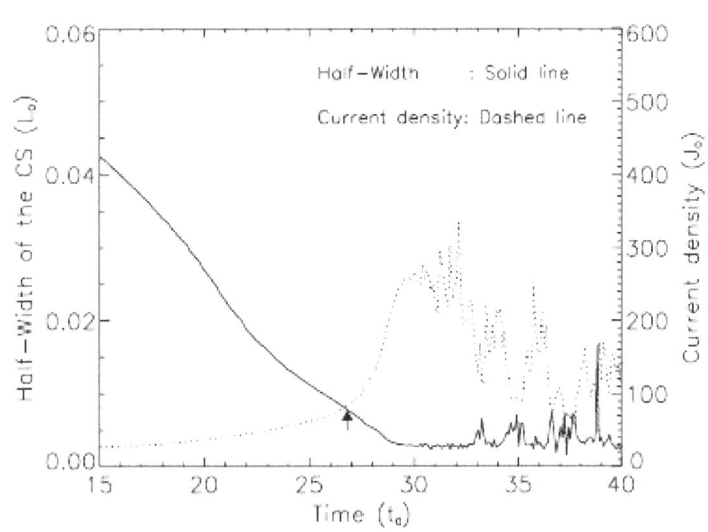

Figure 21: Variations of the half-width w (solid curve) and the electric current density J (dotted curve) of the CS near the PX-

point versus time. The arrow specifies the time t = 26.8 when the first magnetic island (O-point) starts to form. Here L_0 is the length scale of the simulation domain, and J0 is the characteristic value of the electric current density. From Shen et al. (2011)

Second, the electric current density in the CS gradually increases before the appearance of the first plasma blob, and a significant increment took place following the formation of the first blob. After it reached maximum, the current density roughly stayed constant for a while until a large amount of magnetic field and the associated energy had been dissipated (see dashed curve in Fig. 21 with J° being the characteristic value of the electric current density).

The third important result of Shen et al. (2011) is that a special X-point exists inside the CS as indicated by observations in the CME/flare CS and in the magnetotail CS (see discussions of Lin et al. 2008 and references therein). The reconnection rate, MA, is higher at this special X-point than at the other X-points inside the CS, and the sunward and antisunward reconnection outflows bifurcate around this point. This special X-point is known as the principal X-point (or PX-point). As shown by Fig. 20, at an earlier time, t = 26.8, the reconnection process is at the Sweet-Parker stage, we see a narrowest location at the CS, where the plasma flows bifurcate (see the right panel in the first row); at a later time, t = 34.0, the process evolves to the turbulent stage, many magnetic islands (blobs) associated with many X-points appear, and we see again that motions of this blobs, together with the reconnection outflow, bifurcate at a special point (see the right panel in the second row). This figure shows how the PX-point looks and how it distinguishes from the other X-points. Interested readers are referred to Fig. 6 of Shen et al. (2011), which clearly demonstrates the time-variations of the heights of different X-points. Further studies of the value of M_A at the PX-point showed that M_A remained small but kept increasing gradually after the first plasma blob was produced in the CS. The consequent evolution in M_A manifested an apparent growth and the oscillating behavior, with the highest value close to 0.1 (see the dashed curve in Fig. 22) following the formation of a group of magnetic islands and the associated X-points.

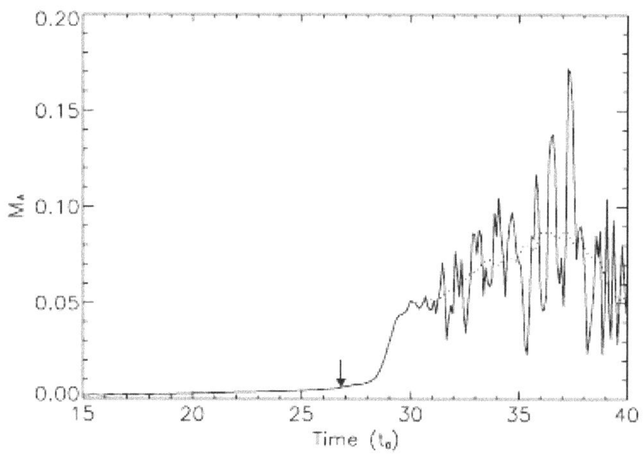

Figure 22: Rate of magnetic reconnection M_A near the PX-point as a function of time. The solid line shows the instant value and the dotted line is for the corresponding average value. The arrow indicates time = 26.8 when the first magnetic island forms. From Shen et al. (2011)

Furthermore, Shen et al. (2011) noticed that a fluid stagnation point (S-point), at which the fluid speed was roughly zero, existed near the PX-point. They were very close to one another, but not co-located with one another in space. They were moving anti-sunward slowly and their relative positions switch back and forth as reconnection progresses. Looking into detailed features of plasma blobs and these two special points reveals that the direction in which a plasma blob moves is related to the relative positions of the S- and PX-points. When the S-point is above the PX-point, newly formed blobs move upward. When the S-point is below the PX-point, new blobs move downward. No blob was observed to appear between the S-point and PX-point, this implies that plasma blobs never go through either of the two points.

Plasmoids are not limited to the CS between the CME and the flare arcade. Studies involving the breakout reconnection configuration have shown that X- and O-type nulls and associated plasmoids also appear in the CS that forms ahead of the eruption, referred to as the breakout current sheet. Karpen et al. (2012) found that the shape of the X-points early in the reconnection is very flat and sheet-like, and that they change shape to X-points with broader opening angles soon after the appearance of reconnection jets in the breakout sheet. Such a change in shape indicates a transition from a form of slow tearing in the CS to a faster and strongly driven reconnection process. This metamorphosis in null-point shape is mirrored in the X-point nulls that appear in the flare CS later in the eruption, indicating a similar evolution in the reconnection there. Karpen et al. (2012) also noticed that the number of X- and O-type nulls in the breakout CS are strongly but not perfectly correlated with the number of nulls in the flare CS during the explosive eruption phase, indicating that there is feedback between the two locations where reconnection is occurring.

Recently, Guo et al. (2013) investigated plasmoid instability and its consequences by comparing the distribution of plasmoids obtained from LASCO data in a CME event by Ko et al. (2003) with those deduced from a resistive MHD numerical experiment. They found that the plasmoid scale distribution of their experiment as well as the theoretical and numerical results of Huang and Bhattacharjee (2012) were consistent with observations of Ko et al. (2003). In addition, based on the nonlinear scaling law of the plasmoid instability, they inferred a lower bound on the CS width, assuming the underlying mechanism of CS broadening is resistive diffusion. They noticed that the lower limit to the CS width deduced from the non-linear scaling law was around 3×10^3 km, which is apparently lower than those obtained by Lin et al. (2007, 2009). The difference was ascribed to the fact that Lin et al. (2007, 2009) used the linear scaling law and they used non-linear scaling law that seemed more appropriate for studying a turbulent CS (see also Huang and Bhattacharjee 2010).

From an alternative point of view, Bárta et al. (2008) performed a simple statistical analysis for the existing observational results, and found the hypothesis of Kliem et al. (2000) that internal features could be visualized in the radio spectra supported by observations. By using a 2.5-D MHD model, they simulated reconnection in a Kopp-Pneuman configuration with an inhomogeneous and gravitationally stratified atmosphere, and demonstrated the formation and subsequent ejection of plasmoids inside the CS. In their experiments, they tracked the magnetic islands created by the tearing mode as a trap of energetic particles, and calculated the relevant parameters of these islands. According to the results of the experiments, they duplicated the dynamic spectra of the plasma emission from magnetic islands that were consistent with observations.

In several radio bands, on the other hand, observations indicated the occurrence of very fast pulses, but their synthetic dynamic spectra did not display these features. This is probably because the successive production of magnetic islands of various sizes could occur naturally in reality, but small scale structures may not be able to appear in the numerical experiments due to the numerical diffusion as a result of the finite size of the grid. Another reason that causes the difference could be the fact that the plasma beta is large, say $\beta = 0.1$, in numerical experiments compared to that in the realistic coronal environment, say $\beta \leq 10^{-3}$.

Karlický and Bárta (2011) further investigated the production and the coalescence of plasma blobs inside the CS. They found that the free electrons could be apparently accelerated by the coalescence of blobs to very high energy that is enough to produce HXR emission on the top of flare loops. In addition, they noticed that a small CS was created and then got fragmented as two plasma blobs coalesced into one another in the large scale flare CS (see also Nishizuka and Shibata 2013), and particles accelerated in the small CS might be responsible for narrow band decimetric spikes. Further studies and experiments were also performed for the observational consequences of the above process seen in Hα, and good agreement was obtained (e.g., see Bárta et al. 2011a).

On the basis of their previous works (Bárta et al. 2008), Bárta et al. (2011b) enhanced the resolution of grids in their experiments, and studied in detail how the magnetic field could be quickly dissipated in the reconnection process. Their results support the theory of Shibata and Tanuma (2001), indicating that the energy associated with large scale structures was successively transferred to that associated with small structures until the kinetic process dominates the energy conversion. This process is similar to that occurring in an incompressible fluid, like water, in which the vortices are created so that the large scale kinetic energy of the fluid flow could be quickly dissipated. Bárta et al. (2011b) found that the coalescence of magnetic islands was as important as the tearing mode instability itself. This is because of the further dissipation of magnetic field by reconnection as plasma blobs coalesce in the CS (see also experiments of Zhong et al. 2010 and the consequences). As the two processes balanced each other, they noticed that, a cascading process, which is suggested by the power-law distribution of the energy versus blob sizes, was eventually set up (Fig. 23). This confirmed both qualitatively and quantitatively the turbulent property of the magnetic reconnection process in the CS, in which large scale structures fill with small structures (e.g., see also detailed discussions of Bemporad 2008).

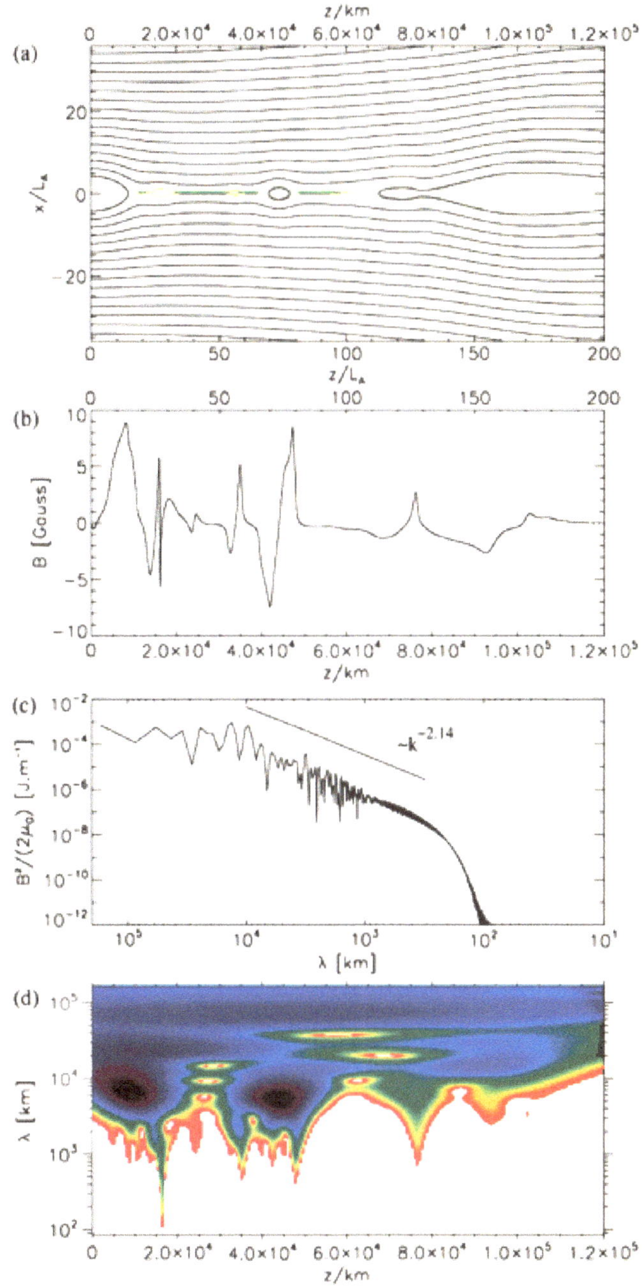

Figure 23: One-dimensional scale analysis of magnetic field structure along the line x = 0. (a) Magnetic field lines and current-density structure (green) in the computational domain at t = 316. The z-axis shows positions both in units of LA (top) and in kilometers according to the scaling adopted in Sect. 2. (b) Profile of the B_x component of magnetic field along the line x = 0. (c) Fourier power spectrum of the B_x profile. (d) Wavelet power spectrum of the B_x profile. From Bárta et al. (2011b)

Statistical descriptions for magnetic islands formed in CS have recently been developed. Using the flux contained in the magnetic island and the area it encloses as two parameters, Fermo et al. (2010, 2012) studied the distribution functions of the islands. Their results indicated that the distribution function was determined by the generation of secondary islands, the growth rate of islands, and their merging (or coalescence). The exponential tail of the distribution function as a function of flux was found in their numerical experiments. Numerical simulations of Loureiro et al. (2012) confirmed the inverse square law found by Uzdensky et al. (2010) between the self-similar distribution function of plasmoid size and fluxes.

Loureiro et al. (2012) specified that the plasmoid flux and half-width distribution functions scale as an inverse square law.

Later, Shen et al. (2013) performed a set of two dimensional resistive MHD simulations to study the statistical properties of the fine structure and the dependence of the spectral energy on these properties in the two-ribbon flare CS on the basis of their previous work (Shen et al.2011). They found that the flux and size distribution functions of plasmoids roughly follow inverse square power laws at large scales (Fig. 24a), that the mass distribution function was steep at large scales and shallow at small scales (Fig. 24b), and that the size distribution also showed that plasmoids were highly asymmetric soon after being formed, while older plasmoids tended to be more circular (Fig. 24c). Their results indicated that spectral profiles of magnetic and kinetic energy inside the current sheet were both consistent with a power law. The corresponding spectral indices were found to vary with the Lundquist number S of the system, but tended to approach a constant for large S(> 10^5).. It may be worth noting that SADs also manifested round and elongated geometric characteristics (e.g., see Hanneman and Reeves2014) although plasmoids and SADs are different from one another.

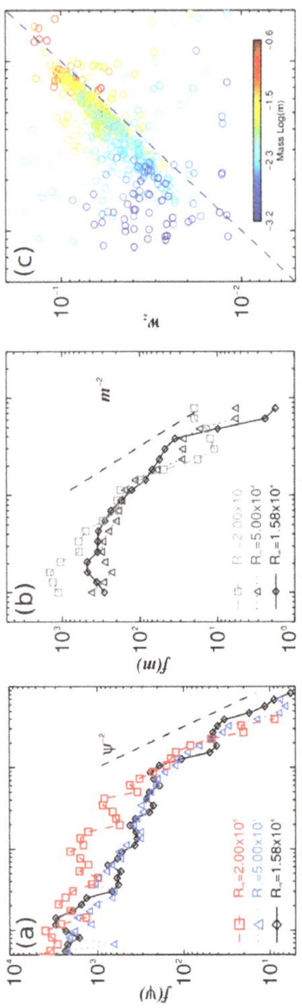

Figure 24: (a) Plasmoid flux distribution function f (Ψ) for different Rm. The dashed straight line represents f (Ψ) ~ Ψ $^{-2}$. (b) Plasmoid mass distribution function f (m) for different R_m. Here, m is the amount mass contained inside each plasmoid. The dashed straight line shows an m^{-2} power law for reference. (c) Plasmoid width distribution in two directions for the case of $R_m = 5 \times 10^4$. Here, w_z and w_x are the widths of each plasmoid along the z- and x-directions, respectively. Color-bar indicates the mass contained in each plasmoid. From Shen et al. (2013)

The motion and growth of blobs changed the spectral index such that the growth of new islands and the coalescence of islands caused the power spectrum to steepen, but it became shallower when old and large plasmoids left the computational domain. A very important result obtained by Shen et al. (2013) is that cascading not only happens to the magnetic energy, but happens to the kinetic energy of the plasma flow in the CS as well, which implies a double-diffusive process in which dissipations in both magnetic field and plasma flow take place simultaneously, and the reconnection process occurring in such an environment would diffuse the magnetic field and convert the associated energy into the other types of energy in a significantly efficient way.

Similar motion patterns and behaviors of plasmoids inside the flare CS have also reported previously by Forbes and Priest (1982), Riley et al. (2007), Bárta et al. (2011a, 2011b), and so on, cascading of the magnetic energy was also found by Bárta et al. (2011a, 2011b). Meanwhile, Ni et al. (2012b, 2013) performed similar analyses for the magnetic reconnection process in the Harris sheet with R_m up to 10^6, and noticed the same cascading feature of the kinetic energy as well. Hence, cascading in the kinetic energy should be a natural and common phenomenon of magnetic reconnection that takes place in the environment of large R_m.

In addition to the complex pattern that consists of plasma blobs and secondary turbulence due to the blob interactions (Fig. 25a), Mei et al. (2012) noticed the formation of the Petschek slow mode shocks that were located near the edge of the CS. Looking into the plasma flow and magnetic structures around the CS, they recognized the characteristics of Petschek-type reconnection, together with the plasma blob and the consequence of the interaction among blobs (Fig. 25b). The plasma flow pattern and the magnetic field in that region manifested clearly the slow-mode shock of the Petschek type. This was, however, different from the result of the standard theory of the Petschek reconnection, which predicted the angle spanning between two slow shocks of about 7°, the angle measured in their experiments ranged from 6. °5 to 14. °1. They ascribed this further expansion of the region between the slow shocks to the turbulence of the plasma in the region. Therefore, broadening of the reconnecting CS could be due to the combined effect of the turbulence and the development of the slow-mode shock (see also discussions of Lin et al. 2008, 2009).

No matter whether the detailed physical properties of the Petschek-type reconnection have been fully understood, the work of Mei et al. (2012) revealed that a CME/flare CS possesses much more complex internal structures than expected. The role of these structures are two-fold. First, they are small compared to the whole CS, which apparently enhances the diffusion in the CS; and second, they provide extra pressure to balance that of the reconnection inflow, allowing the CS to have a finite thickness. Therefore, the rate of magnetic reconnection could remain high even in a thick CS, and a reasonably high rate of energy conversion required for the major eruption could be guaranteed. More important, Mei et al. (2012) found that the reconnection outflow between a pair of Petschek shocks is faster than that near the shock, therefore, the magnetic field inside the CS manifests a W- or

M-shape configuration (Fig. 25c) and the magnetic field near the CS center does not monotonically vanish, instead it shows a small overshoot with its amplitude compared to that in the inflow region being roughly M_A (T.G. Forbes 2012, private communications). The W-shape magnetic field inside a CS was predicted several decades ago by Vasyliunas (1975) in theory, but had never been confirmed by the numerical experiment before Mei et al. (2012).

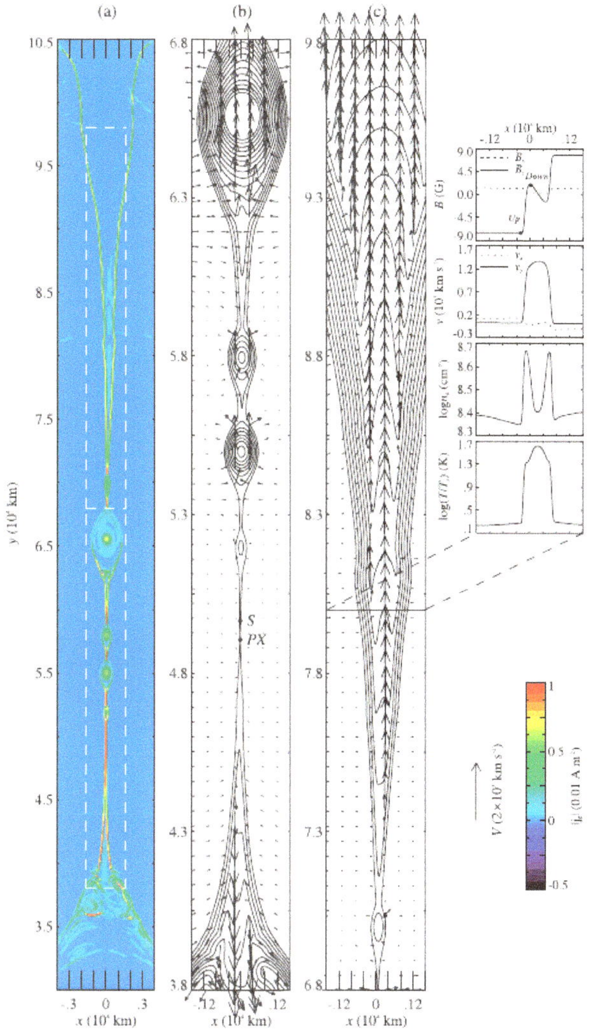

Figure 25: Detailed features of various parameters around the CS seen at t = 29 min in the experiment of Mei et al. (2012). Distribution of the electric current is shown in (**a**), and two smallerrectangular regions surrounded by dashed lines are specified for studying the other parameters in more details. The lower region (**b**) includes several magnetic islands, the post-flare loop system, as well as the S- and PX-points, which manifests a scenario of the turbulent reconnection; the upper region (**c**) displays two slow-mode shocks at either side of the outflow region, and shows typical features of the Petschek-type reconnection. The distributions of parameters along a cut marked on (**c**) are shown by an inset at the upper right-hand corner. From Mei et al. (2012)

All of these facts suggest that the CS is a complex assembly that includes magnetic islands (see also Drake et al. 2006; Lazarian et al. 2012), turbulent flows and termination shocks produced as islands collide with one another and with the closed magnetic structures near two tips of the large CS (e.g., see also Mei et al. 2012; Ni et al. 2012b; and references therein), respectively, and even the slow mode shocks of the Petschek-type in some segments of the CS (e.g., see also theoretical discussions of Vasyliunas 1975; and experimental results of Mei et al. 2012). These macro and micro structures could give rise to important observational consequences manifesting in various aspects of either the CS itself (e.g., see discussions of Bemporad 2008; Lin et al. 2007, 2009; Vršnak et al. 2009) or the output of the reconnection process, namely rapid energy release that results in the intensive heating and acceleration of the plasma as well as production of energetic particles (e.g., see Martens and Young 1990; Litvinenko 1996; Li and Lin 2009; Shibata and Magara 2011; Zharkova et al. 2011; Li et al. 2013b; Nishizuka and Shibata 2013; and references therein).

Before ending this part of work, we need to note here that we are mainly focusing on 2D and 2.5D models dealing with the formation of the plasmoid in the CS, and we did not discuss the full 3D scenario in detail. This is because the three-dimensional nature of the tearing mode and related plasmoid instability is not well understood in the context of the solar atmosphere although several groups have made some progress in developing fully 3D MHD models. Analytically, Baalrud et al. (2012) used reduced MHD to investigate the oblique nature of the 3D plasmoid instability. Beresnyak (2013) presents 3D MHD resistive simulations in which turbulence driven by reconnection inhibits the development of plasmoids and flux ropes. Wyper and Pontin (2014a, 2014b) present simulations of null point reconnection that show the formation of flux ropes which are the 3D analogs of 2D magnetic islands. Fully kinetic 3D simulations by Daughton et al. (2011) show that the current sheet develops into a tangled web of interconnected flux ropes, which is likely to also occur in 3D resistive MHD simulations. Over all, developing 3D model is the inevitable tendency, it is worth investing more effort in this issue in the future.

SUMMARY

We briefly reviewed and looked back on studies of the CS occurring in solar eruptive events in both theory and observations, especially those performed after the catastrophe model for the solar eruption was developed and the CS was identified for the first time in observations. [We may need to point out here as well that observations of Aurass et al. (2013) and Aurass (2014) showed the occurrence of magnetic reconnection both at the top of the disrupting magnetic configuration and above the post-flare loops. This suggests that the large-scale CS could also develop in the fashion of the break-out model. But these observations are very difficult, and only possible if the eruptive process is near but not at the limb.] Further studies through observations of the CS region revealed that the CME/flare CS could be as thick as a few times 10^4 km, even up to 10^5 (Lin et al. 2007)! This result was not expected previously from the traditional theory of magnetic reconnection and the current sheet. We discussed the question what modifications are needed in the theory and in data interpretation to consider the observed broad tail behind CMEs as a signature of the predicted long CS.

Traditionally, it was believed that the thickness of the CS should be as small as the skin depth of the plasma, which is about several meters, not bigger than 1 km, in the coronal environment because the energy conversion or the magnetic field dissipation occurring inside the CS must be efficient enough to support the fast energy release in a typical solar eruption that produces both CME and solar flare. To our surprise, the measurement of the CS thickness, d, by Lin et al. (2007) indicated that the value of d could be up to several tens of thousands km, which is significantly larger than what was previously expected. More interesting and more important, one always ended up with the similar results for d for different events no matter in which way the value of d was deduced, and no matter in which wavelength the CS was observed as long as it could be recognized (e.g., see Ciaravella et al. 2013; Ling et al. 2014; and references therein). We do not know yet whether this "agreement" could reveal physically valuable information, we shall keep an eye on this issue, and continue to look into it in more detail from different viewpoints in the future. Studies of the projection effects indicated that the impact of the projection effects on the direct measurement of d could be limited, and might not be able to result in an apparent value of d that is more than, at least, 4 orders of magnitude larger than the true value. This is suggestive of that an observed thick CS may not be due to projection effects and complex structures of the CS, but due to the intrinsic physical properties of the reconnection process.

Both observational features of a CS and the related theoretical investigations indicated that the reconnection process is governed by the turbulence and the resultant small structures (also known as the fractal CS) inside the CS. So the thickness of the CS and the rate of reconnection could not be determined by dynamic behaviors and properties of individual particles, but should be controlled by the tearing mode, together with the resultant plasmoid instability. The magnetic energy cascades from large scale turbulent eddies (or magnetic islands) to smaller ones until the diffusion becomes significant. In addition, vorticities, S-points, and velocity shears were also seen in the reconnection outflow (McKenzie 2013), and the secondary instabilities of the Rayleigh-Taylor type has also been used to explain observational features of the CS (Innes et al. 2014).

On the other hand, collisions, or coalescences, among magnetic islands and/or between magnetic islands and the other kind of magnetic structures nearby take place as well (see Mei et al. 2012), which constitutes an inverse cascading process, in the CS. This is not a simple merging of two fluid elements, but accompanied with reconnection between the magnetic fields within different islands, which results in the secondary dissipation of the magnetic field inside the CS. A dynamical balance of cascading and coalescence is usually reached in the reconnection process (e.g., see Bárta et al. 2011a, 2011b). Analyses by Shen et al. (2013) also revealed that the similar cascading behavior occurred to the kinetic energy of the plasma in the CS simultaneously, which implies that the dissipation exists in the fluid motion as well. The laser-driven magnetic reconnection experiments could be considered as possible confirmation of plasmoid collisions and consequences (Zhong et al. 2010) that take place in the flare CS. Furthermore, Mei et al. (2012) found that a CME/flare CS is in fact a complex assembly that includes not only small structures of various sizes but also even the slow mode shocks! We therefore reach the conclusion that the occurrence of various diffusive structures inside the CS allows the reconnection process to take place at a reasonably fast rate even if the CS is thick.

Therefore, it is not easy and straightforward, for the time being, to predict what specific type of reconnection and associated structures of the reconnection would occur in a specific event, or even at a specific stage of an event. The reconnection process could be of either Sweet-Parker, or Petschek, or turbulent, or the combination of these fashions. More effort needs to be invested in looking into detailed processes of magnetic reconnection and the physical properties of the CS in the future, and we have identified the region in which the magnetic diffusion is taking place or has taken place with the CS in a more general sense in our previous and the present works.

The above results improved our knowledge about the internal structure and physical properties of the CME/flare CS in an obvious way, and have apparent significance to works in the other fields. First of all, a long and thick CS is favorable for charged particle acceleration. Currently, the scale of the CS is small, of order a few meters thick and tens of meters long, in existing models of the particle acceleration in the CS (e.g., see Litvinenko 1996; Wood and Neukirch 2005; Dauphin et al. 2007). Particles cannot remain long in such a CS, say $\sim 10^{-5}$ s for electrons and $< 10^{-3}$ s for protons, so the efficiency of acceleration is fairly low and it is difficult to fulfill the requirement for the rate of production of energetic electrons ($>$ 10 keV, 10^{35} s^{-1}) in a typical solar flare (Martens 1988). Recent studies by Mann et al. (2009) found that observations of the 2003 October 28 event by RHESSI indicated that up to 10^{36} electrons with energies 20 keV need to be produced every second to provide a power of about 10^{22} W to drive a major flare.

In the case of a thick CS that includes turbulence, on the other hand, both electrons and the other charged particles would have larger space to move and stay in the acceleration region longer than in the case of a thin CS, which could enormously enhance the acceleration efficiency, and the turbulence would modulate the particle motion and result in the spectra that are often observed. Detailed features of the particle acceleration in a CS are still open question, and more effort is needed in further and deeper research (e.g., see Dauphin et al. 2007; Li and Lin 2009, 2010, 2012, 2013a, 2013b). The role of small scale structures due to turbulence of the CS in particle acceleration has recently studied by Nishizuka and Shibata (2013) who found that the interaction of these structures provides an approach to Fermi acceleration for particles. Mann and Warmuth (2011) pointed out that, in a medium class flare, if about 8 % of the all electrons brought into CS by reconnection inflow are accelerated, they are able to bring about 12 % of the total released energy into the flare region; in a major flare, on the other hand, up 60 % total inflow electrons need to be accelerated to the energy $>$ 20 keV to account for 60 % the total released energy observed in the flare region. In both cases, they found that the slow-mode shock plays a role in heating and the termination shock plays a role in particle accelerations.

In addition, our previous understanding of the plasma instabilities and turbulence is on the scale of laboratory or the skin depth of protons, which is usually several meters in the coronal environment (Drake et al. 2006). However, observations of the CME/flare CS indicate that the similar processes or phenomena could take place on the scale from 10^4 to 10^5 km, which implies that some theories and understandings on the basis of the lab scale need to be modified and improved. Bemporad (2008) provided observational evidence for such works that have already been started (e.g., see the recent serial of the works by Bárta and co-workers, Shen et al. 2011; Mei et al. 2012; Shen et al. 2013). Meanwhile, the work by Shibata and Tanuma (2001) deserves more attention and some issues discussed by them are also worth further investigation.

The CME/flare CS is a relatively new topic that started drawing our attention about 15 years ago. Research on many important issues have commenced just recently, and our knowledge on the internal structure and the corresponding properties is still fairly preliminary. Among the issues that draws most debate is probably the value of the CS thickness. Although observations continue to confirm the significantly large thickness of the CS, all the results obtained so far were the FWHM of the brightness profile of the CS, not the FWHM of the electric current in the CS. This is because all we can currently observe is the plasma emission, not the electric current sheet. Therefore, there are effects like the thermal halo that could create a sheath of plasma around the CS, yielding uncertainties in determining d if the measurement is on the basis of the brightness at a specific wavelength that also depends on the plasma temperature. However, we may need to note here as well that the brightness in white-light depends on the electron density only, and hence the measurement of d on the basis of the white-light data is not affected by the thermal halo. Furthermore, Ciaravella and Raymond (2008) measured d based on the WL data and the UVCS data that were taken simultaneously at the same location. Their results show that the values of d measured from WL and UVCS data are the same, which implies that if a thermal halo exists, the gas in this region is less dense than that inside the CS and may not contribute much to the EM. Hence the thermal halo is unlikely to have a big effect on measurement of d.

There is currently no existing technique that allows us to directly measure or detect the distribution of the electric current (density or intensity) inside the CS. So for the time being, studying the brightness distribution is the only approach via which more or less, we are able to look into the geometric structure and physical properties of the CS. In addition to the thermal halo, it is possible that, like the case of the helmet streamer, the CME/flare CS itself is actually very thin and embedded in a thick plasma sheet that could be recognized according to the brightness, and the CS thickness is actually much smaller than that of the plasma sheet (e.g., see also Song et al. 2012).

But in the framework of Lazarian and Vishniac (1999), magnetic reconnection is turbulent, and the reconnection region fills with turbulent eddies, which results in a broad reconnection region, so the plasma sheet and the CS fill the same region in space. Kowal et al. (2009) studied numerically the effects of turbulence on magnetic reconnection using three-dimensional direct numerical simulations, and tested the model of fast magnetic reconnection developed by Lazarian and Vishniac (1999). The numerical simulations duplicated the Sweet-Parker CS in the absence of turbulence. Kowal et al. (2009) noticed that the effect of the Ohmic dissipation and that of the Hall MHD disappeared as turbulence was turned on in the CS. Further investigations of the internal structures of CS were performed by Vishniac et al. (2012), and the fact was noticed that peaks of the electric current were distributed throughout the reconnection region. This confirmed that the CS and the reconnection region (or the plasma sheet) are nearly co-located in space with one another, and are filled with turbulent structures. But their results were not tested for the case of weak turbulence, which could be a weakness of the study (e.g., see also Lazarian et al. 2014).

Overall, the geometric structure and the internal features of the CME/flare CS is still an open question, and the true value of d as well as correct approaches to measuring d constitutes an important topic that needs us to invest more effort in both observational and theoretical studies. Therefore, we note here that the terms "diffusion", "dissipation", and "current sheet" have actually been used in a more general sense in this and our previous works than were used traditionally to refer to any process that causes magnetic diffusion and any region where such diffusion occurs (e.g., see also discussions of Lin and Forbes 2000). In this sense, some parameters, like η_e and d determined by either high temperatures observed by UVCS or high densities observed by coronagraphs, should be considered effective and average.

Different from our previous concepts that the cascading process transfers the energy associated with large scale structures to that associated with small scales until the kinetic process dominates, recent numerical experiments for magnetic reconnection revealed that an inverse process, namely merging or coalescence of magnetic islands, takes place simultaneously (e.g., see Bárta et al. 2011a, 2011b; Shen et al. 2011, 2013; Mei et al. 2012; Ni et al. 2012a, 2012b, 2015). It is probably this inverse process that causes the energy spectra, which is usually a power law, to possess an index steeper than that of the Kolmogorov value of 5/3. The indices of the energy spectra appearing in these experiments usually ranges from 2 to 6. More attention has been theoretically and numerically drawn to this issue recently, and further and deeper understanding about it also requires high quality and high resolution observations. Great progress could be expected should the small scale features of a CS and the associated properties be recognized and understood in the future.

Here the high quality observations first refer to the direct measurement of the coronal magnetic field with which we are able to perform the spectral analyses about the magnetic energy in the CS as we have done for the corresponding numerical results, second to the 3D measurements of various parameters for the CS so that the kinetic energy of plasma blobs can be obtained and the similar spectral analyses about the kinetic energy could be performed, and third to spending more time and effort in analyzing and studying the existing data in order to acquire important information that might have been missed in previous works; the high resolution observations refers to the observations of the CS with spatial resolution, velocity resolution, and time cadence like those obtained with the Interface Region Imaging Spectrograph (IRIS, De Pontieu et al. 2014), or even better. IRIS provides simultaneous spectra and images of the photosphere, chromosphere, transition region, and corona with 0.33–0.4 arcsec spatial resolution, 2 s time cadence, and 1 km s^{-1} velocity resolution over a FOV up to 175 × 175 $arcsec^2$

Another important remaining topic for future work is the nonlinear evolution of the tearing and plasmoid instabilities in three dimensions. Particle-in-cell (PIC) simulations of small-scale reconnection by Daughton et al. (2011) show that plasmoids and magnetic islands in 2D correspond to highly structured flux ropes in 3D. In the future, it will be important to couple these small-scale, fully kinetic simulations to larger scales where the resistive MHD approximation is more appropriate. The nonlinear evolution of the plasmoid instability in 3D has important implications on the interpretation of blobs observed in current sheets (e.g., Ko et al. 2003; Guo et al. 2013).

Last but not least, we note that the CS in the real eruptive process forms and quickly develops in the early stage of the solar flare, which is the most dynamic phase of the eruption, and magnetic reconnection takes place in a region that itself is during a violently changing process. However, most numerical experiments on reconnection so far known were started with a pre-existing CS that starts in mechanical and thermal equilibrium, and the reconnection process commences after a perturbation to the system is introduced. To our knowledge, Riley et al. (2007) and Mei et al. (2012) are probably so far the known numerical works that paid significant attention to the details of the interior of the CS that was created in a disrupting magnetic configuration. In the work of Reeves et al. (2012), formation of plasma blobs in the CS that was developed dynamically were noticed, but no further investigation on internal structures of the CS was performed. Therefore, our knowledge about magnetic reconnection occurring in a dynamically developing CS must be improved and/or renewed. We need to study in the future the occurrence of the plasma instabilities and the resultant turbulence in a fast evolving CS, to look into their properties that might be different from those appearing in a static CS, and to investigate possible implications of these features to particle acceleration in the CME/flare CS as well.

ACKNOWLEDGEMENTS

JL thanks the organizer of the Hinode-7 Science Meeting for invitation to give an overview on the related topic, and the main content of this paper was reported at the meeting. This work was supported by Program 973 grant 2013CBA01503, NSFC grants 11203069, 11273055, 11333007, and 11403100, CAS grant XDB09040202, Yunnan Province grant 2011FB113, as well as NSF SHINE grants AGS-1156076 and AGS-1358342.

REFERENCES

1. T. Amari, J.F. Luciani, J.J. Aly, Astrophys. J. 629, L37 (2005)
2. T. Amari, J.J. Aly, Z. Mikić, J. Linker, Astrophys. J. 717, L26 (2010)
3. J. Ambrosiano, W.H. Matthaeus, M.L. Goldstein, D. Plante, J. Geophys. Res. 93, 14383 (1988)
4. S.K. Antiochos, Astrophys. J. 502, L181 (1998)
5. S.K. Antiochos, C.R. DeVore, J.A. Klimchuk, Astrophys. J. 510, 485 (1999)
6. Asai, T. Yokoyama, M. Shimojo, K. Shibata, Astrophys. J. 605, L77 (2004)
7. M.J. Aschwanden, Astrophys. J. 608, 554 (2004)
8. H. Aurass, Sol. Phys. 289, 4517 (2014)
9. H. Aurass, G. Mann, Astrophys. J. 615, 526 (2004)
10. H. Aurass, B. Vršnak, G. Mann, Astron. Astrophys. 384, 273 (2002)
11. H. Aurass, G. Mann, G. Rausche, A. Warmuth, Astron. Astrophys. 457, 658 (2006)
12. H. Aurass, F. Landini, S. Braune, G. Poletto, Astron. Astrophys. 506, 901 (2009)
13. H. Aurass, G. Mann, P. Zlobec, M. Karlicky, Astrophys. J. 730, 57 (2011)
14. H. Aurass, G. Holman, S. Braune, G. Mann, P. Zlobec, Astron. Astrophys. 555, A40 (2013)
15. S.D. Baalrud, A. Bhattacharjee, Y.-M. Huang, Phys. Plasmas 19, 022101 (2012)
16. H.M. Bain, S. Krucker, L. Glesener, R.P. Lin, Astrophys. J. 750, 44 (2012)
17. M. Bárta, M. Karlický, R. Žemlička, Sol. Phys. 253, 173 (2008)
18. M. Bárta, J. Büchner, M. Karlický, Adv. Space Res. 45, 10 (2010)
19. M. Bárta, J. Büchner, M. Karlický, P. Kotrč, Astrophys. J. 730, 47 (2011a)
20. M. Bárta, J. Büchner, M. Karlický, J. Skála, Astrophys. J. 737, 24 (2011b)
21. H. Baty, E.R. Priest, T.G. Forbes, Phys. Plasmas 16, 060701 (2009)
22. Bemporad, Astrophys. J. 689, 572 (2008)
23. Bemporad, G. Poletto, S.T. Suess et al., Astrophys. J. 638, 1110 (2006)
24. A.R. Beresnyak, arXiv:1301.7424 (2013)
25. Bhattacharjee, Y. Yuan, Astrophys. J. 449, 739 (1995)
26. Bhattacharjee, Y.M. Huang, H. Yang, B. Rogers, Phys. Plasmas 16, 112102 (2009)
27. L. Biermann, Z. Naturforsch. 5a, 65 (1950)
28. D. Biskamp, Phys. Fluids 29, 1520 (1986)
29. D. Biskamp, Nonlinear Magnetohydrodynamics (Cambridge University Press, Cambridge, 1993), p. 22
30. D. Biskamp, Magnetic Reconnection in Plasmas (Cambridge University Press, Cambridge, 2000), p. 256
31. N.A. Bobrova, S.I. Syrovatskii, Sov. J. Plasma Phys. 6, 1 (1980)
32. H. Carmichael, in AAS-NASA Symp. Phys. Solar Flares (1964), p. 451
33. Caspi, R.P. Lin, Astrophys. J. 725, L161 (2010)
34. P.A. Cassak, M.A. Shay, Space Sci. Rev. 182, 283 (2012)
35. P.A. Cassak, J.F. Drake, J.T. Gosling, T.D. Phan, M.A. Shay, L.S. Shepherd, Astrophys. J. 775, L14 (2013)
36. M. Cécere, M. Schneiter, A. Costa, S. Elaskar, S. Maglione, Astrophys. J. 759, 79 (2012)
37. P. Chen, K. Shibata, D.H. Brooks, H. Isobe, Astrophys. J. 602, L61 (2004)
38. X. Cheng, J. Zhang, Y. Liu, M.D. Ding, Astrophys. J. 732, L25 (2011)
39. Ciaravella, J.C. Raymond, Astrophys. J. 686, 1372 (2008)
40. Ciaravella, J.C. Raymond, J. Li, P. Reiaer et al.,

Astrophys. J. 575, 1116 (2002)
41. Ciaravella, D.F. Webb, S. Giordano, J.C. Raymond, Astrophys. J. 766, 65 (2013)
42. W. Daughton, V. Roytershteyn, H. Karimabadi, L. Yin, B.J. Albright, B. Bergen, K.J. Bowers, Nat. Phys. 7, 539 (2011)
43. Dauphin, N. Vilmer, A. Anastasiadis, Astron. Astrophys. 468, 273 (2007)
44. J.A. Davies, R.A. Harrison, A.P. Rouillard, N.R. Sheeley Jr., C.H. Perry, D. Bewsher et al., Geophys. Res. Lett. 36, L02102 (2009)
45. C.J. Davis, J.A. Davies, M. Lockwood, A.P. Rouillard, C.J. Eyles, R.A. Harrison, Geophys. Res. Lett. 36, L08102 (2009)ADS
46. De Pontieu, A.M. Title, J.R. Lemen, G.D. Kushner, D.J. Akin, B. Allard et al., Sol. Phys. 289, 2933 (2014)
47. P. Démoulin, J.C. Henoux, E.R. Priest, C.H. Mandrini, Astron. Astrophys. 308, 648 (1996)
48. G.A. Doschek, D.E. McKenzie, H.P. Warren, Astrophys. J. 788, 26 (2014)
49. J.F. Drake, M. Swisdak, K.M. Schoeffler et al., Geophys. Res. Lett. 33, L13105 (2006)
50. J.W. Dungey, Philos. Mag. 44, 725 (1953)
51. J.P. Eastwood, T.-D. Phan, F.S. Mozer, M.A. Shay, M. Fujimoto, A. Retinó et al., J. Geophys. Res. 112, 6235 (2007)
52. G. Einaudi, S. Chibbaro, R.B. Dahlburg, M. Velli, Astrophys. J. 547, 1167 (2001)
53. D.F. Elmore, J.T. Burkepile, J.A. Darnell, A.R. Lecinski, A.L. Stanger, in Polarimetry in Astronomy, ed. by S. Fineschi. Proceedings of the SPIE, vol. 4843 (2003), p. 66
54. Y. Fan, S.E. Gibson, Astrophys. J. 668, 1232 (2007)
55. R.L. Fermo, J.F. Drake, M. Swisdak, Phys. Plasmas 17, 010702 (2010)
56. R.L. Fermo, J.F. Drake, M. Swisdak, Phys. Rev. Lett. 108, 255005 (2012)
57. T.G. Forbes, Geophys. Astrophys. Fluid Dyn. 62, 15 (1991)
58. T.G. Forbes, L.W. Acton, Astrophys. J. 459, 330 (1996)
59. T.G. Forbes, P.A. Isenberg, Astrophys. J. 373, 294 (1991)
60. T.G. Forbes, J. Lin, J. Atmos. Sol.-Terr. Phys. 62, 1499 (2000)
61. T.G. Forbes, J.M. Malherbe, Astrophys. J. 302, L67 (1986)
62. T.G. Forbes, J.M. Malherbe, Sol. Phys. 135, 361 (1991)
63. T.G. Forbes, E.R. Priest, Sol. Phys. 81, 303 (1982)
64. T.G. Forbes, E.R. Priest, Sol. Phys. 84, 169 (1983)
65. T.G. Forbes, E.R. Priest, Rev. Geophys. 25, 1583 (1987)
66. T.G. Forbes, E.R. Priest, Astrophys. J. 446, 377 (1995)
67. T.G. Forbes, E.R. Priest, P.A. Isenberg, Sol. Phys. 150, 245 (1994)
68. T.G. Forbes, J.A. Linker, J. Chen, C. Cid, J. Kota, M.A. Lee et al., Space Sci. Rev. 123, 251 (2006)
69. H.P. Furth, J. Killeen, M.N. Rosenbluth, Phys. Fluids 6, 459 (1963)
70. P.T. Gallagher, Sol. Phys. 210, 341 (2002)
71. G. Gao, M. Wang, J. Lin, N. Wu, C. Tan, B.H. Kliem, Y. Su, Res. Astron. Astrophys. 14, 843 (2014)
72. L. Glesener, S. Krucker, H.M. Bain, R.P. Lin, Astrophys. J. 779, L29 (2013)
73. N. Gopalswamy, S. Yashiro, G. Michalek, G. Stenborg, A. Vourlidas, S. Freeland, R. Howard, Earth Moon Planets 104, 295 (2009)
74. S.E. Guidoni, D.W. Longcope, Astrophys. J. 730, 90 (2011)
75. L.J. Guo, A. Bhattacharjee, Y.M. Huang, Astrophys. J. 771, L14 (2013)
76. L.-J. Guo, Y.-M. Huang, A. Bhattacharjee, D.E. Innes, Astrophys. J. 796, L29 (2014)
77. W.J. Hanneman, K.K. Reeves, Astrophys. J. 786, 95 (2014)
78. H. Hara, T. Watanabe, L.K. Harra et al., Astrophys. J. 741, 107 (2011)
79. Y.M. Huang, A. Bhattacharjee, Phys. Plasmas 17, 062104 (2010)
80. Y.M. Huang, A. Bhattacharjee, Phys. Rev. Lett. 109, 265002 (2012)
81. G. Huang, J. Lin, Astrophys. J. 639, L99 (2006)
82. Y.M. Huang, A. Bhattacharjee, E.G. Zweibel, Phys. Plasmas 17, 055707 (2010)
83. S.Y. Huang, A. Vaivads, Y.V. Khotyaintsev, M. Zhou, H.S. Fu, A. Retinó et al., Geophys. Res. Lett. 39, L11103 (2012)ADS
84. S. Ichimaru, Astrophys. J. 202, 524 (1975)
85. D.E. Innes, D.E. McKenzie, T.J. Wang, Sol. Phys. 217, 267 (2003a)
86. D.E. Innes, D.E. McKenzie, T.J. Wang, Sol. Phys. 217, 247 (2003b)
87. D.E. Innes, L. Guo, A. Bhattacharjee, Y.M. Huang (2014)
88. Isavnin, A. Vourlidas, E.K.J. Kilpua, Sol. Phys. 284, 203 (2013)
89. P.A. Isenberg, T.G. Forbes, P. Demoulin, Astrophys. J. 417, 368 (1993)
90. M. Karlický, F. Fárník, S. Krucker, Astron. Astrophys. 419, 365 (2004)
91. M. Karlický, P. Zlobec, H. Mészárosová, Sol. Phys. 261, 281 (2010)
92. M. Karlický, M. Bárta, Astrophys. J. 733, 107 (2011)
93. M. Karlický, B. Kliem, Sol. Phys. 266, 71 (2010)
94. J.T. Karpen, S.K. Antiochos, C.R. DeVore, Astrophys. J. 760, 81 (2012)
95. Kliem, T. Török, Phys. Rev. Lett. 96, 255002 (2006)
96. Kliem, M. Karliký, A.O. Benz, Astron. Astrophys. 360, 715 (2000)ADS
97. Kliem, M.G. Linton, T. Török, M. Karliký, Sol. Phys. 266, 39 (2010)
98. Kliem, J. Lin, T.G. Forbes, E.R. Priest, T. Török, Astrophys. J. 789, 46 (2014)

99. Y.-K. Ko, J.C. Raymond, J. Lin, G. Lawrence, J. Li, A. Fludra, Astrophys. J. 594, 1068 (2003)
100. Y.-K. Ko, J.C. Raymond, B. Vršnak, E. Vujić, Astrophys. J. 722, 625 (2010)
101. J.L. Kohl, R. Esser, L.D. Gardner, S. Habbal, P.S. Daigneau, E.F. Dennis et al., Sol. Phys. 162, 313 (1995)
102. J.L. Kohl, N. Giancarlo, S.R. Cranmer, J.C. Raymond, Astron. Astrophys. Rev. 13, 31 (2006)
103. S. Kolomański, M. Karlický, Astron. Astrophys. 475, 685 (2007)
104. R.A. Kopp, G.W. Pneuman, Sol. Phys. 50, 85 (1976)
105. G. Kowal, A. Lazarian, E.T. Vishniac, K. Otmianowska-Mazur, Astrophys. J. 700, 63 (2009)
106. S. Krucker, M. Battagalia, P.J. Cargill, L. Fletcher, H.S. Hudson, A.L. MacKinnon et al., Astron. Astrophys. Rev. 16, 155 (2008)
107. R.M. Kulsrud, Earth Planets Space 53, 417 (2001)
108. Landi, J.C. Raymond, M.P. Miralles, H. Hara, Astrophys. J. 711, 75 (2010)
109. Landi, J.C. Raymond, M.P. Miralles, H. Hara, Astrophys. J. 751, 21 (2012)
110. Lazarian, E.T. Vishniac, Astrophys. J. 517, 700 (1999)
111. Lazarian, L. Vlahos, G. Kowal, H. Yan, A. Beresnyak, E.M. de Gouveia Dal Pino, Space Sci. Rev. 173, 557 (2012)
112. Lazarian, G. Eyink, E. Vishniac, G. Kowal, arXiv:1407.6356 (2014)
113. Y. Li, W. Gan, Astrophys. J. 629, L137 (2005)
114. Y. Li, W. Gan, Astrophys. J. 644, L97 (2006)
115. Y. Li, J. Lin, Acta Astron. Sin. 50, 271 (2009)ADS
116. Y. Li, J. Lin, Chin. Astron. Astrophys. 34, 48 (2010)
117. Y. Li, J. Lin, Sol. Phys. 279, 91 (2012)
118. J. Li, D. Jewitt, B. LaBonte, Astrophys. J. 539, L67 (2000)
119. C.K. Li, F.H. Seguin, J.A. Frenje, J.R. Rygg, R.D. Petrasso, R.P.J. Town et al., Phys. Rev. Lett.99, 055001 (2007)
120. Li, X. Kong, G. Zank, Y. Chen, Astrophys. J. 769, 22 (2013a)
121. Y. Li, H.D. Winter, N.A. Murphy, J. Lin, N. Wu, Publ. Astron. Soc. Jpn. 65, 101 (2013b)
122. J. Lin, Chin. J. Astron. Astrophys. 2, 539 (2002)
123. J. Lin, Sol. Phys. 222, 115 (2004)
124. J. Lin, T.G. Forbes, J. Geophys. Res. 105, 2375 (2000)
125. J. Lin, W. Soon, New Astron. 47, 53 (2004)
126. J. Lin, T.G. Forbes, E.P. Priest, T.N. Bungey, Sol. Phys. 159, 275 (1995)
127. J. Lin, T.G. Forbes, P.A. Isenberg, P. Démoulin, Astrophys. J. 504, 1006 (1998)
128. J. Lin, A.A. van Ballegooijen, T.G. Forbes, J. Geophys. Res. 107, 1438 (2002)
129. J. Lin, W. Soon, S.L. Baliunas, New Astron. Rev. 47(2), 53 (2003)
130. J. Lin, J.C. Raymond, A.A. van Ballegooijen, Astrophys. J. 602, 422 (2004)
131. J. Lin, Y.K. Ko, L. Sui et al., Astrophys. J. 622, 1251 (2005)
132. J. Lin, S. Mancuso, A. Vourlidas, Astrophys. J. 649, 1110 (2006)
133. J. Lin, J. Li, T.G. Forbes, Y.K. Ko, J.C. Raymond, A. Vourlidas, Astrophys. J. 658, L123 (2007)
134. J. Lin, S.R. Cranmer, C.J. Farrugia, J. Geophys. Res. 113(A11), A11107 (2008)
135. J. Lin, J. Li, Y.K. Ko, J.C. Raymond, Astrophys. J. 693, 1666 (2009)
136. A.G. Ling, D.F. Webb, J.T. Burkepile, E.W. Cliver, Astrophys. J. 784, 91 (2014)
137. J.A. Linker, Z. Mikić, R. Lionello, P. Riley, T. Amari, D.J. Odrstrcil, Plasma Phys. 10(5), 1971 (2003)
138. Y. Litvinenko, Astrophys. J. 462, 997 (1996)
139. R. Liu, Mon. Not. R. Astron. Soc. 434, 1309 (2013)
140. R. Liu, J. Lee, T. Wang et al., Astrophys. J. 723, L28 (2010)
141. Y.H. Liu, J.F. Drake, M. Swisdak, Phys. Plasmas 19, 022110 (2012)
142. W. Liu, Q. Chen, V. Petrosian, Astrophys. J. 773, 166 (2013)
143. D.W. Longcope, T.G. Forbes, Sol. Phys. 289, 2091 (2014)
144. N.F. Loureiro, A.A. Schekochihin, S.C. Cowley, Phys. Plasmas 14, 100703 (2007)
145. N.F. Loureiro, R. Samtaney, A.A. Schekochihin, D.A. Uzdensky, Phys. Plasmas 19, 042303 (2012)
146. B.J. Lynch, S.K. Antiochos, P.J. MacNeice, T.H. Zurbuchen, L.A. Fisk, Astrophys. J. 617, 589 (2004)
147. B.J. Lynch, S.K. Antiochos, C.R. DeVore, J.G. Luhmann, T.H. Zurbuchen, Astrophys. J. 683, 1192 (2008)
148. B.J. Lynch, Y. Li, A.F.R. Thernisien, E. Robbrecht, G.H. Fisher, J.G. Luhmann, A. Vourlidas, J. Geophys. Res. 115, A07106 (2010)ADS
149. Mann, A. Warmuth, Astron. Astrophys. 528, A104 (2011)
150. Mann, A. Warmuth, H. Aurass, Astron. Astrophys. 494, 669 (2009) MATH
151. P.C.H. Martens, Astrophys. J. 330, L131 (1988)
152. P.C.H. Martens, N.P.M. Kuin, Sol. Phys. 122, 263 (1989)
153. P.C.H. Martens, A. Young, Astrophys. J. Suppl. Ser. 73, 333 (1990)
154. S. Masuda, T. Kosugi, H. Hara, S. Tsuneta, Y. Ogawara, Nature 371, 495 (1994)
155. D.E. McKenzie, Sol. Phys. 195, 381 (2000)
156. D.E. McKenzie, Astrophys. J. 766, 39 (2013)
157. D.E. McKenzie, H.S. Hudson, Astrophys. J. 519, L93 (1999)
158. Z. Mei, C. Shen, N. Wu et al., Mon. Not. R. Astron. Soc. 425, 2824 (2012)
159. Z. Mikić, J.A. Linker, Astrophys. J. 430, 898 (1994)
160. Z. Mikić, D.C. Barnes, D.D. Schnack, Astrophys. J. 328, 830 (1988)

161. R.O. Milligan, R.T.J. McAteer, B.R. Dennis, C.A. Young, Astrophys. J. 713, 1292 (2010)
162. R.L. Moore, A.C. Sterling, H.S. Hudson, J.R. Lemen, Astrophys. J. 552, 833 (2001)
163. N.A. Murphy, A.K. Young, C.C. Shen, J. Lin, L. Ni, Phys. Plasmas 20, 061211 (2013)
164. N. Narukage, K. Shibata, Astrophys. J. 637, 1122 (2006)
165. L. Ni, K. Germaschewski, Y.M. Huang et al., Phys. Plasmas 17, 052109 (2010)
166. L. Ni, I.I. Roussev, J. Lin, U. Ziegler, Astrophys. J. 758, 20 (2012a)
167. L. Ni, U. Ziegler, Y.M. Huang, J. Lin, Z. Mei, Phys. Plasmas 19, 072902 (2012b)
168. L. Ni, J. Lin, N.A. Murphy, Phys. Plasmas 20, 061206 (2013)
169. L. Ni, B. Kliem, J. Lin, N. Wu, Astrophys. J. 799, 79 (2015)
170. P.N. Nilson, L. Willingale, M.C. Kaluza, C. Kamperidis, S. Minardi, M.S. Wei et al., Phys. Rev. Lett. 97, 255001 (2006)
171. P.N. Nilson, L. Willingale, M.C. Kaluza, C. Kamperidis, S. Minardi, M.S. Wei et al., Phys. Plasmas 15, 092701 (2008)
172. N. Nishizuka, K. Shibata, Phys. Rev. Lett. 110, 051101 (2013)
173. G. Noci, J.L. Kohl, G.L. Withbroe, Astrophys. J. 315, 706 (1987)
174. M. Opher, J.F. Drake, M. Swisdak et al., Astrophys. J. 734, 71 (2011)
175. S. Patsourakos, A. Vourlidas, Astron. Astrophys. 525, 27 (2011)
176. S. Patsourakos, A. Vourlidas, G. Stenborg, Astrophys. J. 764, 125 (2013)
177. H.E. Petschek, in The Physics of Solar Flares, ed. by W.N. Hess (1964), p. 425. NASA Spec. Publ. SP-50
178. G. Poletto, S.T. Suess, A. Bemporad, N.A. Schwadron, H.A. Elliott, T.H. Zurbuchen, Y.K. Ko, Astrophys. J. 613, L173 (2004)
179. E.R. Priest, Solar Magnetohydrodynamics. Geophys. Astrophys. Monogr., vol. 21 (Reidel, Dordrecht, 1982)
180. E.R. Priest, Rep. Prog. Phys. 48, 955 (1985)
181. E.R. Priest, Astrophys. Space Sci. 264, 77 (1988)
182. E.R. Priest, T.G. Forbes, Magnetic Reconnection: MHD Theory and Applications (Cambridge University Press, Cambridge, 2000) MATH
183. E.R. Priest, T.G. Forbes, Astron. Astrophys. Rev. 10, 313 (2002)
184. P.L. Pritchett, Y.C. Lee, J.F. Drake, Phys. Fluids 23, 1368 (1980)
185. J.C. Raymond, A. Ciaravella, D. Dobrzycka et al., Astrophys. J. 597, 1106 (2003)
186. K.K. Reeves, L. Golub, Astrophys. J. 727, L52 (2011)
187. K.K. Reeves, T.B. Guild, W.J. Hughes et al., J. Geophys. Res. 113(A7), A00B02 (2008a)ADS
188. K.K. Reeves, D.B. Seaton, T.G. Forbes, Astrophys. J. 675, 868 (2008b)
189. K.K. Reeves, J.A. Linker, Z. Mikić, T.G. Forbes, Astrophys. J. 721, 1547 (2010)
190. K.K. Reeves, S.E. Gibson, T.A. Kucera, H.S. Hudson, R. Kano, Astrophys. J. 746, 146 (2012)
191. B.A. Remington, R.P. Drake, D.D. Ryutov, Rev. Mod. Phys. 78, 755 (2006)
192. Retinó, D. Sundkvist, A. Vaivads, F.S. Mozer, M. André, C.J. Owen, Nat. Phys. 3, 236 (2007)
193. Retinó, R. Nakamura, A. Vaivads, Y. Khotyaintsev, T. Hayakawa, K. Tanaka et al., J. Geophys. Res. 113, 12215 (2008)
194. P. Riley, R. Lionello, Z. Mikić et al., Astrophys. J. 655, 591 (2007)
195. I.W. Roxburgh, Mon. Not. R. Astron. Soc. 132, 201 (1966)
196. D.D. Ryutov, R.P. Drake, B.A. Remington, Astrophys. J. Suppl. Ser. 127, 456 (2000)
197. P. Saint-Hilaire et al., Astrophys. J. 696, 941 (2009)
198. S.L. Savage, D.E. McKenzie, Astrophys. J. 730, 98 (2011)
199. S.L. Savage, D.E. McKenzie, K.K. Reeves et al., Astrophys. J. 722, 329 (2010)
200. S.L. Savage, D.E. McKenzie, K.K. Reeves, Astrophys. J. 747, L40 (2012a)
201. S.L. Savage, G. Holman, K.K. Reeves, D.B. Seaton, D.E. McKenzie, Y. Su, Astrophys. J. 754, 13 (2012b)
202. Schmieder, T.G. Forbes, J.M. Malherbe, M.E. Machado, Astrophys. J. 317, 956 (1987)
203. Schmieder, P. Demoulin, G. Aulanier, Adv. Space Res. 51, 1967 (2013)
204. R.B. Scott, D.W. Longcope, D.E. McKenzie, Astrophys. J. 776, 54 (2013)
205. D.B. Seaton, T.G. Forbes, Astrophys. J. 701, 348 (2009)
206. V.D. Shafranov, Rev. Plasma Phys. 2, 103 (1966)ADS
207. M.A. Shay, J.F. Drake, J.P. Eastwood, T.D. Phan, Phys. Rev. Lett. 107(6), 065001 (2011)
208. N.R. Sheeley, Y.M. Wang, Astrophys. J. 579, 874 (2002)
209. N.R. Sheeley, Y.M. Wang, Astrophys. J. 655, 1142 (2007)
210. N.R. Sheeley, H.P. Warren, Y.M. Wang, R.A. Howard, J. Geophys. Res. 104, 24739 (1999)
211. N.R. Sheeley, H.P. Warren, Y.M. Wang, Astrophys. J. 616, 1224 (2004)
212. Shen, J. Lin, N.A. Murphy, Astrophys. J. 737, 14 (2011)
213. Shen, J. Lin, N.A. Murphy, J.C. Raymond, Phys. Plasmas 20, 072114 (2013)
214. K. Shibata, T. Magara, Living Rev. Sol. Phys. 8, 6 (2011)
215. K. Shibata, S. Tanuma, Earth Planets Space 53, 473 (2001)
216. Shiota, H. Isobe, P.F. Chen, T. Yamamoto, T. Sakajiri, K. Shibata, Astrophys. J. 634, 663 (2005)
217. G.M. Simnett, Astron. Astrophys. 416, 759 (2004)
218. E.C. Sittler Jr., M. Guhathakurta, Astrophys. J. 523, 812 (1999)

219. M. Skender, G. Lapenta, Phys. Plasmas 17, 2905 (2010)
220. Song, X. Kong, Y. Chen et al., Sol. Phys. 276, 261 (2012)
221. J.A. Stamper, B.H. Ripin, Phys. Rev. Lett. 34, 138 (1975)
222. H.R. Strauss, Phys. Fluids 29, 3668 (1986) MATH
223. H.R. Strauss, Astrophys. J. 326, 412 (1988)
224. Y. Su, A.M. Veronig, G.D. Holman, B.R. Dennis, T. Wang, M.G. Temmer, W. Gan, Nat. Phys. 9, 489 (2013)
225. L. Sui, G.D. Holman, Astrophys. J. 596, L251 (2003)
226. L. Sui, G.D. Holman, B.R. Dennis, Astrophys. J. 612, 546 (2004)
227. L. Sui et al., Astrophys. J. 633, 1175 (2005)
228. L. Sui, G.D. Holman, B.R. Dennis, Astrophys. J. 645, L157 (2006)
229. R. Susino, A. Bemporad, S. Krucker, Astrophys. J. 777, 93 (2013)
230. Z. Švestka, Solar Flares (Springer, Berlin, 1976), p. 168
231. Z. Švestka, Sol. Phys. 169, 403 (1996)
232. Z. Švestka, E.W. Cliver, Eruptive Solar Flares (Springer, New York, 1992)
233. Z. Švestka, J.M. Fontenla, M.E. Machado et al., Sol. Phys. 108, 237 (1987)
234. Z. Švestka, F. Fárník, H.S. Hudson, P. Hick, Sol. Phys. 182, 179 (1998)
235. S. Takasao, A. Asai, H. Isobe, K. Shibata, Astrophys. J. 745, L6 (2012)
236. V.S. Titov, P. Démoulin, Astron. Astrophys. 351, 707 (1999)ADS
237. T. Török, B.H. Kliem, Astrophys. J. 630, L97 (2005)
238. D.A. Uzdensky, N.F. Loureiro, A.A. Schekochihin, Phys. Rev. Lett. 105, 235002 (2010)
239. V.M. Vasyliunas, Rev. Geophys. Space Phys. 13, 303 (1975)
240. A.M. Veronig, M. Karliký, B. Vršnak et al., Astron. Astrophys. 446, 675 (2006)
241. E.T. Vishniac, S. Pillsworth, G.L. Eyink, G. Kowal, A. Lazarian, S. Murray, Nonlinear Process. Geophys. 19, 605 (2012)
242. Vršnak, A. Warmuth, M. Temmer, A. Veronig, J. Magdelenić, A. Hillaris, M. Karlický, Astron. Astrophys. 448, 739 (2006)
243. Vršnak, G. Poletto, E. Vujić et al., Astrophys. J. 499, 905 (2009)
244. H.P. Warren, C.M. O'Brien, N.R. Sheeley, Astrophys. J. 742, 92 (2011)
245. D.F. Webb, E.W. Cliver, J. Geophys. Res. 100, 5853 (1995)
246. D.F. Webb, J. Burkepile, T.G. Forbes, P. Riley, J. Geophys. Res. 108(A12), SSH6-1 (2003)
247. L.M. Widrow, Rev. Mod. Phys. 74, 775 (2002)
248. P. Wood, T. Neukirch, Sol. Phys. 226, 73 (2005)
249. P.F. Wyper, D.I. Pontin, Phys. Plasmas 21, 082114 (2014a)
250. P.F. Wyper, D.I. Pontin, Phys. Plasmas 21, 102102 (2014b)
251. X. Yan, Z. Xue, J. Liu, L. Ma, D. Kong, Z. Qu, Z. Li, Astrophys. J. 782, 67 (2014)
252. Yang, Y. Jiang, Y. Bi, H. Li, J. Hong, D. Yang, R. Zheng, B. Yang, Astrophys. J. 749, 12 (2012)
253. M.A. Yates et al., Phys. Rev. Lett. 49, 1702 (1982)
254. T. Yokoyama, K. Shibata, Astrophys. J. 436, L197 (1994)
255. T. Yokoyama, K. Shibata, Astrophys. J. 474, L61 (1997)
256. T. Yokoyama, K. Shibata, Astrophys. J. 494, L113 (1998)
257. T. Yokoyama, K. Shibata, Astrophys. J. 549, 1160 (2001)
258. T. Yokoyama, K. Akita, T. Morimoto et al., Astrophys. J. 546, L69 (2001)
259. Y. Zhang, J. Wang, Astrophys. J. 663, 592 (2007)
260. Y. Zhang, Y. Hu, J. Wang, Astrophys. J. 626, 1096 (2005)
261. Y. Zhang, J. Wang, Y. Hu, Astrophys. J. 641, 572 (2006)
262. V.V. Zharkova et al., Space Sci. Rev. 159, 357 (2011)
263. J.Y. Zhong, Y.T. Li, X.G. Wang, J.Q. Wang, Q.L. Dong, C.J. Xiao et al., Nat. Phys. 6, 984 (2010)
264. E.G. Zweibel, M. Yamada, Annu. Rev. Astron. Astrophys. 47, 291 (2009)

CHAPTER 11 TABLE-TOP SOLAR FLARES PRODUCED WITH LASER DRIVEN MAGNETIC RECONNECTIONS

J.Y. Zhong, Y.T. Li, X.G. Wang, J.Q. Wang, Q.L. Dong, X. Liu, X.X. Lin, D.W. Yuan, F. Du, S.J. Wang, L. Zhang, L. An, C.J. Xiao, H.G. Wei, K. Zhang, F.L. Wang, S.E. Jiang, Y.K. Ding, Z.R. Cao, Z. Yuan, H.Y. Zhang, Z.W. Yang, J.Q. Zhu, X.T. He, H.B. Cai, G. Zhao and J. Zhang

INTRODUCTION

Traditional astrophysical phenomena are pursued with observations and theoretical modelling. The recent advances in observations and modelling provide a solid ground for modern astrophysics. The growing field of high-energy density laboratory astrophysics (HEDLA) seeks to understand the physics involved in fusion applications and astrophysical phenomena in laboratory astrophysics. Contemporary research in laboratory astrophysics utilizes powerful, megajoule class lasers such as at the National Ignition Facility (NIF) and other laser fusion facilities like Omega lasers and GEKKO lasers in Japan. In China, Shenguang-II laser facility is used for the study of HEDLA phenomena. These high energy density laser facilities provide unique opportunities to study fundamental material properties such as opacity [4], equations of state [5], photoionized plasmas [6, 7], as well as the similarity of physics involved in certain hydrodynamic phenomena of jets [8] and shocks [9] when a scaling law between astrophysical and laboratory plasma systems can be applied. Consequently, the main focus of laboratory astrophysics is to simulate fundamental nature of astrophysical plasma physics processes in a laboratory environment so that certain astrophysical phenomenon can be studied in a well-controlled manner [1, 2, 10].

Citation: J.Y. Zhong, Y.T. Li, X.G. Wang, J.Q. Wang, Q.L. Dong, X. Liu, X.X. Lin, D.W. Yuan, F. Du, S.J. Wang, L. Zhang, L. An2, C.J. Xiao, H.G. Wei,K. Zhang, F.L. Wang, S.E. Jiang, Y.K. Ding, Z.R. Cao, Z. Yuan, H.Y. Zhang, Z.W. Yang, J.Q. Zhu, X.T. He, H.B. Cai, G. Zhao and J. Zhang, Table-Top Solar Flares Produced With Laser Driven Magnetic Reconnections, http://dx.doi.org/10.1051/epjconf/20135901010

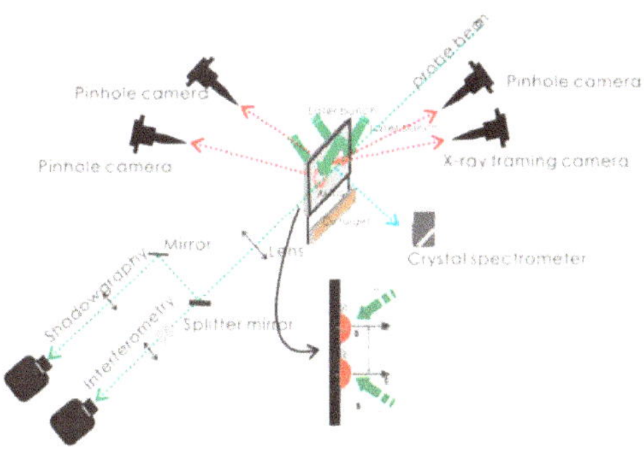

Figure 1: Schematic view of the experimental setup for laser-driven magnetic reconnection. Two laser beams irradiate the target which results in two plasma bubbles. The magnetic fields are anti-parallel between the plasma bubbles. Magnetic reconnection occurs when the two plasma bubbles encounter.

Magnetic reconnection (MR) is a fundamental physical process in a plasma which facilitates the conversion of magnetic energy into kinetic and thermal energies through acceleration and/or heating of particles [11, 12], and is believed to dominate among different astrophysical phenomena from solar flares [13, 14], star formations [15], other astrophysical events [16], laser-produced plasma jets [17–19], tokamak sawteeth [20] to disruptions in laboratories. Considerably large scale differences between astrophysical and laboratory plasmas hinder the direct measurements of magnetic reconnection events in laboratory. However, with proper scaling law [21] between different magnetohydrodynamic (MHD) systems, certain astrophysical phenomena can possibly be envisioned in laboratory such as a solar wind driven magnetosphere [22]. In this article we present a well-scaled laboratory experiment producing a table-top solar flare in which the MR topology is constructed with spontaneously generated magnetic field in high power laser-plasma interactions. This non-equilibrium laser driven magnetic reconnection is quite similar with the real astrophysical MR phenomena, which might open a new direction in the field of laboratory astrophysics.

EXPERIMENTAL SETUP

The experiment is performed on the Shenguang (SG) II laser facility in China, which can deliver a total energy of 2.0 kJ in a nanosecond square pulse. The eight SG-II laser beams (λ_L = 0.351 m) are divided into four bunches and each bunch contains two laser beams. The geometric configuration as shown in Fig. 1 is designed to setup one standard laser driven magnetic reconnection topology. The two synchronized laser bunches separated by 200–800 m are then focused on one side of the Al foil, while the other two bunches shine the other side of the target symmetrically and simultaneously. Each bunch is focused to a focal spot diameter of 50–100 m in full width at half maximum (FWHM) generating an incident laser intensity of ~ 5×10^{15} W/cm^2. A copper target is placed at 250 m away from the 01

Table 1: Typical parameters in solar flare and laser plasmas

Parameters	Solar flare [30, 31]	Laser plasmas	Scaled flare plasmas
Length (cm)	10^9–10^{10}	10^{-1}	10^{-2}–10^{-1}
hline Time (s)	100–1,000	10^{-9}	10^{-10}–10^{-9}
Pressure (Pa)	0.01–10	10^7	10^7–10^{11}
Density (cm-3)	~10^{10}–10^{11}	~10^{19}–10^{20}	10^{19}–10^{21}
Velocity (km s-1)	~10–100	~100	100–1,000
Magnetic field (G)	~10–100	~10^6	10^6–10^7

Al foil edge. The Al foil size is 1, 600 m × 500 m and in thickness of 10–50 m, while the Cu target is 1, 600 m × 250 m with the thickness of 150 m. The X-ray emission is measured by three X-ray pinhole cameras in the forward, side and reverse directions to investigate the reconnection jets as well as their impact on the Cu target. The image is taken with 10-m pinhole filtered with 50 m of beryllium allowing all X-rays with energy above ~1 keV. Most of the signal from high-energy continuum is time-integrated recorded on an X-ray film which is sensitive to X-rays in the 1–10 keV range. A flat crystal spectrometer is set in front of the targets to record the X-ray spectra from the heating plasmas. Shadowgraph and interferometry with a 120 ps green (λ_L= 0.53 m) laser beam are also employed to investigate plasma evolution. A single-frame X-ray framing camera has been set up for imaging of x-ray emissions from laser magnetic reconnection sources. It consists of two parts: an x-ray pinhole camera using an open-ended microchannel plate (MCP) detector coupled to a CCD camera, and a high voltage short duration gate pulse of 70 ps for the MCP. The camera uses a 20-m pinhole aperture for recording image on the MCP detector with a magnification of 3X. The evolutions of 2D X-ray images are recorded with very high-speed time resolution (about 54 ps). 3. RESUL

RESULTS AND DISCUSSIONS

Magnetic reconnection in solar flares and laser driven magnetic reconnection

A solar flare is a fierce explosion in the sun's atmosphere that emits a burst of radiation. One big solar flare can affect space environment and human beings life on earth very seriously. There are many observed evidences that support the magnetic reconnection model of solar flare. In 1992, Masuda et al [23] observed the loop-top X-ray source in solar flares by using the YOHKOH satellite and proposed that two antiparallel magnetic fields were merged above an arcade of closed loops as outflows from the reconnection point which collided with high density plasmas on the loop to produce a hot X-ray region. The observations of Ultraviolet [24] and X-ray [25, 26] plasma jets ejected from the regions above the solar surface also confirm the theoretical models of MR. Table 1 list the typical solar flares parameters.

On the other side, ultra-strong magnetic fields up to mega-gauss scale can be generated via the interactions of ultra-intense lasers with plasmas. The mechanism of self-generated magnetic field in laser plasmas is different due to the laser conditions. Thermoelectric mechanism is most important with the long pulse lasers (nanosecond), where temperature and density gradients on the target surface are nonparallel. Toroidal magnetic fields have been measured successfully with different methods. The main features of self-generated magnetic fields include: (a) strong magnetic field of mega-gauss scale (b) concentrated on a hemispherical plasma bubble (c) surrounding and expanding laterally. Since the magnetic reconnection is the breaking and re-organizing of magnetic field lines which need to satisfy two requirements: firstly magnetized plasmas encounter each other and secondly the anti-parallel pointing B-fields. By considering the features of self-generated magnetic field in laser plasmas, we can construct the MR topology with two laser beams as shown in figure 1. Yates et al [27] observed the hot emission at reconnection point by focussing two laser beams on the gold foil. Nilson et al [17, 18] performed a similar experiment, in which two distinctive plasma jets were observed. Li et al. [28] and Willingale et al. [29] diagnosed the laser-driven magnetic reconnection process with novel proton radiography. The typical laser plasmas parameters used in our experiment are listed in Table 1 along with solar flare parameters in column 2. When compared with the solar flare parameters, one can see that the laser plasma is short lived and is of small size as well.

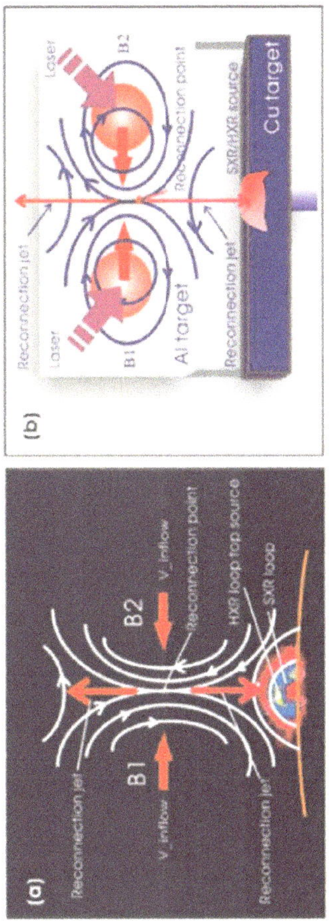

Figure 2: (a) Cartoon shows the loop-top x-ray source in solar flares: white lines show the magnetic reconnection topology and red arrows show the inflows (thick) and outflows (thin). (b) The target is setup to simulate setup (a) where the MR topology can be constructed with intense lasers on the Al target, while the Cu target is presented to simulate the solar limb.

MHD scaling laws

As mentioned in section 3.1, we demonstrate two plasma systems, solar flare and laser plasmas. Magnetic reconnection can occur in both of these systems. The MR process can reasonably be described by MHD, since the magnetic Reynolds number is very high. Ryutov et al [21] reported the scaling relations of two ideal MHD systems (ReM 1). The variables of the systems remain invariant under these transformations, as $r = ar_1, \rho = \rho_1, p = cp_1, t = a\sqrt{\frac{b}{c}}t_1, v = a\sqrt{\frac{c}{b}}v_1, B = \sqrt{c}B_1$, where r is the characteristic length, is the mass density, p is the pressure, v is the velocity, B is the magnetic field of the systems, while a, b, c are transformation coefficients. Bra

Brady et al [22] simulated magnetosphere-solar wind interactions in laboratory with the help of MHD scaling law. By selecting the laser parameters and target materials properly, the magnetic Reynolds number $R_{eM} \approx 0.8\sqrt{\frac{Z+1}{Z-1}}L(cm)[T(eV)]^2$ is around 4,000 (for Z = 13, A = 27, L = 0.1 cm, T = 1 keV), which makes the MHD processes in laser plasmas and solar flares ($R_{eM} \sim 5\times10^8$) comparable. The MHD comparison of solar flares and laser-produced plasmas is shown in Table 1 with the transformation coefficients a ~ 10^{-11}, b ~ 10^8, and c = 10^{10}. The scaled parameters of solar coronal plasmas in the third column are very similar to those of laser-produced plasmas listed in the second column.

Table-top solar flare produced with intense lasers

In astrophysical simulations the geometry of the experimental setup should be similar to that of the astrophysical case. In our experiment, we set up the target which has similar magnetic topology with the solar flare as shown in figure 2(a). The loop-top x-ray source and reconnection jets had been reported in our previous work [3]. Here, we show the dynamic x-ray images observed with framing camera. In solar flare observations, soft x-ray images which are considered to represent magnetic field-line configuration visualize examples of global topology change of plasma configurations. In laser plasmas, we also plot the magnetic field-line over the x-ray emissions. Figure 3 shows x-ray framing images when the two laser spots are separated by 800 m at different time. The most distinctive features of laser driven magnetic reconnection are x-ray emissions at encounter plane of two plasma bubbles and also outflows as shown in figure 2 (b), which will interact with Cu target and re-emit x-rays at the Cu target. At the time of 487 ps when the laser is still on, two weak (see scale bar) x-ray emissions come from two focal spots. In the second row, the laser is just turned off and x-ray emission reaches its maximum peak value at focal spots. Some x-ray emissions (jet-like) are also observed at the corner of the Cu target, while reconnection does not occur because at that time there is no x-ray emissions at middle of two plasma bubbles of Al target. At later time of 1.479 ns (third row), in addition to x-ray emission in diffusion region from the Al target, three bright spots on the Cu target are also found. The side spots are evolution of jet-like x-ray emissions as shown in second row, while the central spot is the result of reconnection outflows interacted with Cu target. Several features are found similar in both systems (the present experimental plasmas and solar flare plasmas) such as reconnection jets and loop top xray source. By considering the scaling relations between these two systems we can conclude that one table-top solar flare could be produced with intense lasers.

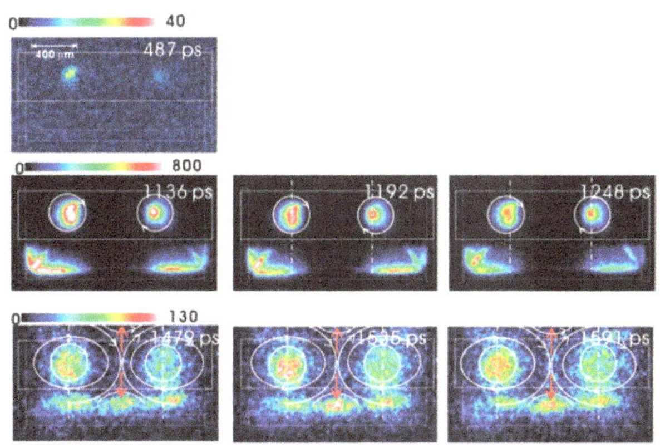

Figure 3: X-ray framing images where the two laser spots are 800 m apart. The large rectangle is Al target and the small rectangle is Cu target, refer to figure 2(b).

CONCLUSIONS

Laser driven magnetic reconnection is an area of exploring a parametric space that was not previously accessible in laboratory experiment. With the application of scaling laws, we can simulate a number of astrophysical phenomena pertaining to MR, such as solar flare, accretion disk etc. In future experiments, we will continue to investigate the MR related particles acceleration. The magnetic field produced by short laser pulse is much higher than that produced with ns lasers. Using these mechanisms for magnetic field generation, we can also study the relativistic magnetic reconnection.

REFERENCES

1. Remington, B. A., Arnett, D, Drake, P & Takabe H., Science 284,1488 (1999)
2. Remington, B. A., Drake, P & Ryutov, Rev. Mod. Phys. 78, 755–807 (2006)
3. Zhong, J. Y. et al., Nature Phys. 6, 984–987 (2010)
4. Takabe, H. Progress of Theoretical Physics Supplement 143, 202–265 (2001)
5. Rogers, Forest J., & Carlos A. Iglesias., Science 263, 50–55 (1994)
6. Celliers, P. M. et al., Phys. Plasmas 11, L41–L44 (2004)
7. Foord, M. E. et al., Phys. Rev. Lett. 93, 055002 (2004)
8. Fujioka, S. et al. Nature Phys. 5, 821–825 (2009)
9. Lebedev, S. V. et al., Astrophys. J. 564, 113–119 (2002)
10. Foster, J. M. et al., Phys. Plasmas 9, 2251–2263 (2002)
11. Yamada, M. Phys. Plasmas 14, 058102 (2007)
12. Zweibel, E & Yamada, M. Annu. Rev. Astron. Astrophys. 47, 291–332 (2009)
13. Sweet, P. A. IAU symp. 6, ed. B. Lehnert, Cambridge Univ. Press, London, p. 123 (1958)
14. Parker, E. N. J. Geophys. Res. 62, 509 (1957)
15. Kulsrud, R. M., Phys. Plasmas 5, 1599 (1998)
16. Goodman, J. & Uzdensky, D., Astrophys. J., 688, 555–558 (2008)
17. Nilson, P. M. et al. Phys. Rev. Lett. 97, 255001 (2006)
18. Nilson, P. M. et al. Phys. Plasmas 15, 092701 (2008)
19. Li, C. K. et al. Phys. Rev. Lett. 102, 205001 (2009)
20. Taylor, J. B. 58, 741–763 (1986)
21. Ryutov, D. D., Drake, R. P., & Remington, B. A., Astrophys. J. 127 (Suppl.), 465–468 (2000)
22. Brady, P., Ditmire, T., Horton, W., Mays, M. L., & Zakharov, Y., Phys. Plasmas 16, 043112 (2009)
23. Masuda, S. et al., Nature 371, 495–497 (1994)
24. Innes, D. E., Inhester, B., Axford, W. I. & Wilhelm, K., Nature 386, 811–813 (1997)
25. Cirtain, J. W., et al., Science, 318, 1580–1582 (2007)
26. Shibata, K., et al., Science, 318, 1591–1594 (2007)
27. Yates, M. A., van Hulsteyn, D. B., Rutkowski, H., Kyrala, G. & Brackbill, J. U., Phys. Rev. Lett.
28. 49, 1702–1704 (1982)
29. Li, C. K. et al., Phys. Rev. Lett. 99, 055001 (2007)
30. L. Willingale et al.,Phys. Plasmas 17, 043104 (2010)
31. Bray, R. J., Cram, L. E., Durrant, C. & Loughhead, R. E., (Cambridge University Press, 1991)
32. Lin, J., et al., Astrophys. J, 658, L123-L126 (2007)

CHAPTER 12 MODERN MICRO AND NANOPARTICLE-BASED IMAGING TECHNIQUES

Marketa Ryvolova, Jana Chomoucka, Jana Drbohlavova, Pavel Kopel, Petr Babula, David Hynek, Vojtech Adam, Tomas Eckschlager, Jaromir Hubalek, Marie Stiborova, Jozef Kaiser and Rene Kizek

Keywords: Imaging techniques, nanoparticles, fluorescence, magnetic resonance imaging

INTRODUCTION

Imaging techniques play an important role in the medical care of all organ systems and better and increased research in medical imaging may be beneficial for the complete health and disease management process. Medical imaging covers many different imaging modalities: X-ray-based methods such as radiography and Computed Tomography (CT), Magnetic Resonance Imaging (MRI), Ultrasound (US), Nuclear medicine with Positron Emission Tomography (PET) and Single Photon Emission Computed Tomography (SPECT), and several optical imaging methods (Figure 1).

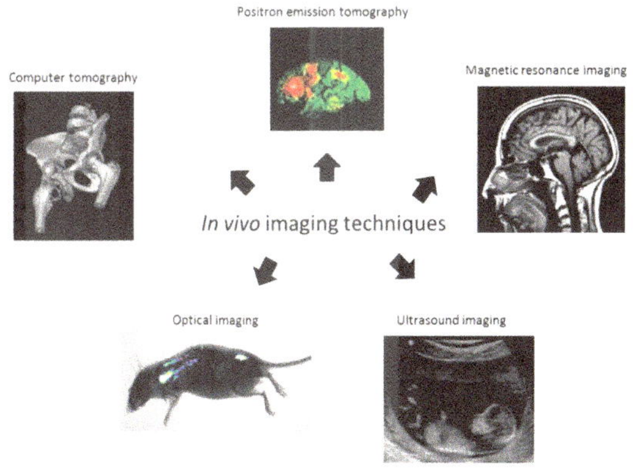

Figure 1: Main *in vivo* imaging techniques.

Citation: Marketa Ryvolova, Jana Chomoucka, Jana Drbohlavova, Pavel Kopel, Petr Babula , David Hynek , Vojtech Adam , Tomas Eckschlager, Jaromir Hubalek, Marie Stiborova, Jozef Kaiser and Rene Kizek, Modern Micro and Nanoparticle-Based Imaging Techniques, doi:10.3390/s121114792.

The important and exciting progress in biotechnology, nanomedicine and new innovative therapies is in many cases highly dependent on the integration of the medical imaging into standard clinical practice. The development of new materials of nanometer dimensions for biomedical applications has been in the focus in the last few years. Their applications became important in medicine, targeted therapies and diagnostics. Modern materials like nanowires [1], quantum dots [2] carbon nanotubes [3], nanoparticles [4,5] or nanomaterials [6] (Figure 2) are in the centre of attention due to the fact that the mechanical, chemical, electrical, optical, magnetic, electro-optical and magneto-optical properties of these particles are different from their bulk properties and depend on the particle size. Nanoparticles (NPs) have been developed as an important approach for delivery of conventional drugs, recombinant proteins, vaccines and more recently, nucleotides. NPs and other colloidal drug-delivery systems modify the kinetics, body distribution and drug release of an associated drug.

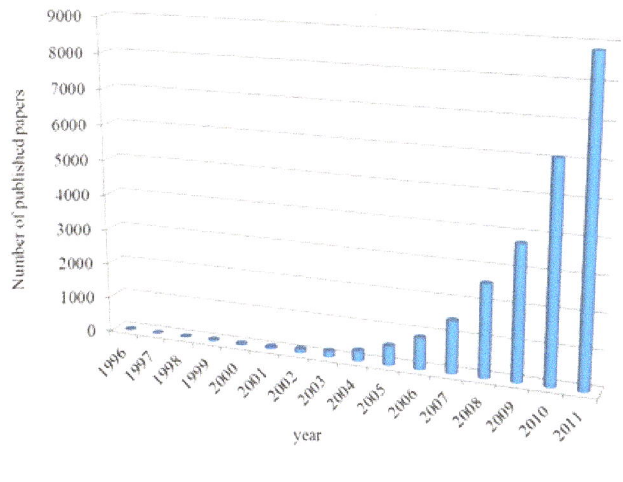

Figure 2: Published items each year containing "imaging and nanoparticle*" in title.

MAGNETIC RESONANCE IMAGING (MRI)

Magnetic Resonance Imaging by Nanoparticles

Magnetic nanoparticles (MNPs) for medical and biological applications are being of great interest due to their unique properties. MNPs can be divided into paramagnetic, ferromagnetic and superparamagnetic particles. Superpara- and ferromagnetic particles are composed of a magnetic core and a surface coating. Paramagnetic particles are mainly based on chelates of paramagnetic ions with no explicit core and surface coating. Thus, their influence on magnetic resonance imaging (MRI) contrast is rather different from that of superpara- and ferromagnetic particles [7, 8].

The switchable magnetic properties of superparamagnetic nanoparticles (SPIONs) make these materials useful for magnetic drug targeting [9], cell tracking [10], hyperthermia [11] and medical imaging [12, 13]. In addition, MNPs are efficiently used for DNA, protein or cell labeling and their subsequent purification in a magnetic field [14]. MRI on its own already offers outstanding soft tissue contrast. The ongoing development of MNPs as contrast agents (CAs) further enhances image contrast. New magnetic contrast agents can be size-tailored to accumulate in specific organs or their surface can be specifically functionalized to target cells (tumor, transplanted cells). Hence, the contrast of MRI is significantly enhanced and diseases can be potentially detected at an earlier stage. Most commonly, a paramagnetic CA, usually a gadolinium-based compound, is used [15]. Gadolinium-doped tissues and fluids appear extremely bright in MR images, and for this reason paramagnetic CAs are called positive CAs.

Biological Application of Superparamagnetic Particles

More recently, superparamagnetic CAs based on iron oxide MNPs have become commercially available. The regions where such agents are delivered appear darker, and therefore, they are called negative CAs. The big advantage of this type of CAs is their higher sensitivity that is expected to reach single cell level [16]. Iron oxides with core/shell structures are the most widely used as sources of magnetic materials [17]. Iron oxides have several crystalline polymorphs known as α-Fe_2O_3 (hematite), β-Fe_2O_3, γ-Fe_2O_3 (maghemite), ε-Fe_2O_3, Fe_3O_4 (magnetite) and some others (amorphous and high pressure forms) [18]. Nevertheless, only maghemite and magnetite have found the greatest interest of bioapplications [19]. Readily, carbonyl iron, which is well-known material with a unique form of elemental iron because of its small particle size, was also used as magnetic core [20].

The surface of MNPs may be coated with number of different functionalities, depending upon the coating material and the reactive groups presented on the targeting ligand. It is desirable that MNPs retain sufficient hydrophilicity and, with coating not exceed 100 nm in size, to avoid rapid clearance by the reticuloendothelial system (RES) [21]. It was found that the surface functionalization also plays a key role in nanoparticle toxicity [17]. Jain *et al.* investigated MNPs consisting of an iron-oxide magnetic core coated with oleic acid and stabilized with a Pluronic® or Tetronic® block copolymer as an effective cancer theranostic (therapy and diagnostic) agent, *i.e.*, an agent with combined drug delivery and imaging properties [22]. The authors revealed that these block copolymers, mainly their molecular structures and weight, significantly influence the physical properties of MNPs, their interactions with biological environment following intravenous administration, and ability to localize to tumors. Schweiger *et al.* investigated γ-Fe_2O_3 MNPs for application as CAs for MRI and as switchable drug delivery vehicles [23]. Namely, the authors studied colloidal stability and cytotoxic effects of these MNPs in adenocarcinomic epithelial cells. They prepared poly (ethylene imine)-g-poly (ethylene glycol) (PEGPEI) as stabilizer for iron oxide MNPs in comparison to branched PEI. Amiri *et al.* presented in their work a series of maghemite/polymer composite ferrofluids with variable magnetic core size, which showed a good efficiency as MRI CA [16]. These ferrofluids were biocompatible and can be proposed as possible platforms for multifunctional biomedical applications. They contain anchoring groups for biofunctionalization, can incorporate fluorescent dyes, and have shown low cellular toxicity. Kumagai*et al.* developed block copolymer-coated magnetite nanoparticles for pancreatic cancer imaging, by means of a chelation between the carboxylic acid groups in PEG-poly (aspartic acid) block copolymer and Fe on the surface of the iron oxide nanoparticles [24]. Masoudi *et al.* synthesized *ex situ* PEG coated iron oxide NPs to be used as potential MRI contrast agents for early detection of lymph nodes and their metastases [25]. According to the important role of nanoparticle size in biodistribution, magnetite nanoparticles coated with different PEG concentrations up to 10/1 PEG/iron oxide weight ratio in an *ex situ*manner incomplete sentence—has no conclusion. To predict the PEG-coated nanoparticle behavior in biological media, such as the blood stream or tissues, colloidal stability was evaluated to estimate the coating endurance under different conditions. Martelli and coworkers synthesized a simple and stable nanoprobe exhibiting magnetic and fluorescent properties for detection of lymph node homing of tumor-specific dendritic cells in a murine breast cancer model [26]. The mechanism of synthesis relies on the covalent coupling of modified organic fluorophores with chitosan, which strongly interact with the surface of the ferric oxide NPs. Jiang *et al.* investigated targeted MR CA transferrin-conjugated SPIONs (Tf-SPIONs) for brain glioma detection [27]. They observed the obvious contrast change of brain glioma before and after administration of Tf-SPIONs in C6 glioma rat model *in vivo* on T2 weighted imaging.

Among the wide variety of iron oxide nanoparticles which are routinely used as MRI CAs, magnetoliposomes (MLs) may be included. The flexibility of the lipid coating allows an efficient functionalization, enabling bimodal imaging or the use of MLs as theranostic agents. Especially the large MLs containing several magnetite cores encapsulated within a single large liposome were found to be highly efficient. A special type of MLs, biogenic MLs, can also be efficiently used in this regard, with potential applications in cancer treatment and imaging. In small MLs the lipid bilayer is immediately attached onto a solid magnetite core. The flexibility of the lipid bilayer allows the incorporation of poly (ethylene glycol)–lipid conjugates to increase blood circulation times and be used as bone marrow contrast agents. Cationic lipids can also be incorporated, leading to high cell uptake and associated strong contrast generation in MRI of implanted cells [28].

FLUORESCENCE-BASED IMAGING TECHNIQUES

In the group of fluorescence-based techniques fluorescence microscopy, flow cytometry, cell sorting fluorescence correlation spectroscopy as well as particle tracking velocimetry may be included. Historically, biological research is based on optical imaging, in particular microscopy, but also several chromophoric assays and photographic approaches. With the recent emergence of methods appropriate for biomarker *in vivo* staining, such as bioluminescence, fluorescent molecular probes and proteins, as well as nanoparticle-based targeted agents, significant attention has been shifted toward *in vivo* interrogations of different dynamic biological processes at the molecular level. This progress has been largely supported by the development of advanced tomographic imaging technologies suitable for obtaining volumetric visualization of bio-marker distributions in small animals at a whole-body or whole-organ scale, an imaging frontier that is not accessible by the existing tissue-sectioning microscopic techniques due to intensive light scattering beyond the depth of a few hundred microns [29].

In Vivo Imaging by Fluorescent Nanoparticles

Semiconductor quantum dots (QDs) are of special interest as tiny light-emitting particles on the nanometer scale that belong to new class of fluorescent probes for *in vivo* biomolecular and cellular imaging [30–33]. In comparison with organic dyes and fluorescent proteins, QDs have unique optical and electronic properties. QDs have molar extinction coefficients that are 10–50 times larger than that of organic dyes, which make them much brighter in photon-limited *in vivo* conditions. The long lifetime in the order of 10–40 ns increases the probability of absorption at shorter wavelengths, and produces a broad absorption spectrum. Further, QD emission wavelengths are size tunable. For example, CdSe/ZnS QDs of approximately 2 nm in diameter produce a blue emission, whereas QDs approximately 7 nm in diameter emit red light [34, 35]. The emission wavelength was extended into the near infrared region (650 nm to 950 nm), to take advantage of the improved tissue penetration depth and reduced background fluorescence at these wavelengths. A key property for *in vivo*imaging is the unusual QD Stokes shift, which can be as large as 300–400 nm, depending on the wavelength of the excitation light [32]. In conjunction with broadband absorption and narrow emission peaks of QDs, this property allows multiplexed imaging applications in which one light source is used to simultaneously excite multicolor QDs without the need for complicated instrumentation. Another important feature is the long-term photostability of QD imaging probes, which opens the possibility of investigating the dynamics of cellular processes over time, such as continuously tracking cell migration, differentiation, and metastasis. These properties have made QDs a topic of intensive interest in cancer biology, molecular imaging, and molecular profiling [36–44]. As expected, *in vitro* investigation of QDs is of a great interest [34, 45–51]. Some examples of application of QDs in imaging are given hereinafter.

A specific advantage of quantum dots for *in vivo* applications is their photostability; QDs allow images to be recorded over a longer period of time than possible with the use of fluorescent dyes or proteins due to their resistance to photobleaching. Maysinger *et al.* visualized CdSe and CdTe QDs detectable at one hour-, one day-, three days-, and seven days post intracortical injection using *in vivo* imaging techniques [52]. Mice were injected subcutaneously and scanned for fluorescence, in particular in the brain, where peak fluorescence was observable at three days post injection and persisted for seven days. Sub-cellular resolution was achieved and allowed the identification of the location of the conjugates. The researchers demonstrated internalization of their QD conjugates, with varied rate and efficacy of internalization for different cell-types. Neuronal internalization is especially challenging, and *in vivo* imaging in neural cells has important relevance to quantum dot toxicity, as this approach would allow the investigation of the neuronal immune response in real time. This approach was novel not only in the *in vivo* imaging aspects but the animal model was transgenic, and was monitored for an astrocyte-responsive luciferase reporter in addition to the injection of quantum dot conjugates. The activation of the astrocyte response was observable as an increase in bioluminescence in response to foreign nanoparticles in mice.

Zimmer *et al.* have employed NIR emitting QDs due to their ability to act as a reporter at a wavelength minimally absorbed by biological species [53]. They prepared a series of InAs/ZnSe core/shell QDs. The small core size, along with variation to the shell thickness and composition offered a range of size tunable emission wavelengths, between 750 and 920 nm. The conjugation of dihydrolipoic acid to the quantum dots allowed for the observations within the interstitial fluid in rats, where the quantum dot conjugates exited the blood vessels. This visualization of the extravasations sites is important, and has potential to interrogate the delivery mechanism of QDs to tumor cells.

Tumor growth requires a supply of nutrients from the blood stream. Angiogenesis is the process by which new vasculature establishes a blood supply to a growing tumor. Receptors, such as integrins, are highly expressed in tumor cells during angiogenesis, and diagnostics targeting such receptors can lend insight into the type and extent of diseases, including cancer. The work by Cai *et al.* detailed the use of a tri-peptide QD conjugate which specifically binds to integrin [54]. The conjugation of arginine-glycine-aspartic acid to a quantum dot with emission at 705 nm yielded a diagnostic tool specific for the integrin receptor that emits in the NIR region. *In vitro* analysis was performed on human glioblastoma and human breast cancer cells, showing specificity for integrin-positive cells only. In addition, binding was inhibited by the presence of an integrin antagonist. Injections of both the conjugates and non-specific controls through the tail vein of mice also demonstrated the specificity, with a maximum fluorescence at six hours post-injection.

Smith *et al.* showed how QDs allow for the non-invasive visualization of blood vessel development over time [55]. The conjugation of biotinylated fibrinogen to QDs showed specificity towards the membrane of blood vessels during angiogenesis. The conjugates had robust biocompatibility, staying in circulation for days without noticeable toxicity. The authors employed QDs with longer emission wavelengths to avoid native auto fluorescence [56]. Jaiswal *et al.* investigated slime mold behavior, a characteristic feature of single-celled organisms in response to starvation. Four distinct cultures of

AX2 amoebae cells were starved for different periods of time, each labeled with a different color QDs. The cells were then mixed with a 10-fold excess of unlabeled, non-starved cells and the QD fluorescence from the labeled cells was tracked for extended observation times. Five melanoma tumor cell populations were labeled with distinct color QDs, using cationic (Lipofectamine 2000) encapsulation, injected them into the tail vein of mice, and tracked the QD emission by multi-photon fluorescence microscopy as the cells extravasated into lung tissues [57].

An alternative to two-photon fluorescence imaging, which requires high excitation powers and a pulsed laser source, deep tissue imaging could rely on NIR fluorescence using nanocrystals that emit at 650–900 nm, a region of the optical spectrum that overlaps with tissue transparency window. NIR fluorescence could allow deep tissue imaging in animals with increased depth and with reduced background [58, 59]. Krishnan *et al.* reported *in vivo* NIR imaging using epidermal growth factor (EGF) conjugated NIR emitting QDs to image the over-expression of epidermal growth factor receptor (EGFR) in tumors compared to normal expression levels in the surrounding normal tissues [60]. They found that following initial incubation, both unconjugated and EGF-coupled QDs exhibited comparable nonspecific uptake into tumor. However, the concentration of free QDs in the tumors gradually decreased, while EGF-QDs progressively accumulated in tumors up to 6 h after injection.

Recently, Kim *et al.* explored the targeting and imaging of lymphatic vessels using QD-conjugated to hyaluronic acid (QD-HA conjugates). They prepared conjugates using electrostatic attractions between anionic HA and amine functionalized DHLA-QDs, and utilized the specific properties of hyaluronic acid to target and visualize lymphatic vessels [61]. Hyaluronic acid (HA) is found in the extracellular matrix and contributes significantly to an array of processes, including cell proliferation, wound healing, metastasis, and prognosis of some malignant tumors. Authors injected HA-QDs into ears of nude mice, and showed that endothelial cells (in lymphatic vessels) could be clearly visualized for up to several days.

Choi *et al.* explored renal clearance of an intravenously administered CdSe-ZnS QDs in mice as a model system [62]. They used several size QDs capped with various hydrophilic small ligands via cap exchange and measured their half-life circulation before eventually being collected in the bladder or other organs (liver, spleen). More recently they followed up on those findings and reported on the correlation between the stabilizing ligands and the biodistribution and clearance of the hydrophilic QDs. They stabilized InAs-ZnS QDs (using InAs instead of a CdSe core) with DHLA and DHLA-PEG ligands with varying PEG chain-length. These results suggested that the biodistribution and clearance of QDs were dependent not only on the core size of the QDs but the state of the stabilizing ligands. Ultra short PEG with two units of ethylene glycol produced QDs with small hydrodynamic diameter (HD) but retained a hydrophobic character and these QDs were rapidly captured by the liver. Conversely, increasing PEG chain length from two to 22 units produced larger HD and uptake into the liver, kidney, pancreas, and lymph nodes, though with varying degrees of uptake. On the other hand, long PEG chains with larger than 22 units are highly hydrophilic and QDs remain in the vasculature for long periods of time. These findings showed that PEG, usually considered as a bio-inert polymer, affected the physical properties of QDs and their biological distribution [63,64]. QD-conjugates with overall small size could also be excreted by the kidney. These studies confirmed that proper surface functionalization and size control of QDs are essential to controlling targeting, biodistribution and excretion from the body. Growing interest in study on QDs can be further demonstrated by preparation and application of different composition of QDs such as CdPbS [65], CdSe, CdSe/CdS [66], CdSeS/ZnS [67], CdTe [68], CdSe [69], Ag_2S [70,71], Ag_2Se [72] or GdS/Eu [73].

Due to the materials used for QDs synthesis, concerns regarding their toxicity have been raised and numerous studies have been carried out [74, 75]. At the same time, methods of synthesis of non- or low toxic QDs have been developed [76].

The excitation of traditional fluorescent labels such as organic dyes, fluorescent proteins and QDs usually requires the use of UV or short-wavelength radiation, which results in serious obstacles including low light-penetration depth inherent to the short-wavelength excitation light and/or low signal-to-noise ratio due to significant tissue autofluorescence. Therefore, it is highly essential to develop more efficient biolabels to overcome these limitations. One of such tags is represented by the "upconversion" paricles. In particular lanthanide doped rare-earth nanocrystals, are characterized by the conversion of long-wavelength radiation, for instance infrared or near infrared radiation, to short-wavelength radiation, usually in the visible range [77–79].

Multimodal Imaging

As mentioned above various imaging modalities have been exploited and are successfully employed to investigate the anatomic or functional dissemination of tissues in the body. However, no single imaging modality allows overall structural, functional, and molecular information as each imaging modality has its own unique strengths and weaknesses. The combination of two popular paging modalities that investigates the strengths of different methods might offer the prospect of improved diagnostic and therapeutic monitoring abilities.

Multimodal molecular imaging can offer a synergistic improvement of diagnostic ability over a single imaging modality. Inorganic nanoparticles such as magnetic iron oxide nanoparticles are studied as MRI contrast agents, QDs or upconversion nanoparticles serve as optical probes and/or metallic gold nanoparticles exhibit unique light scattering patterns for surface-enhanced Raman spectroscopy. Most popular nanostructured multimodal imaging probes are combinations of MRI and optical imaging modalities [76, 80], however also trimodal imaging probes have also been designed by adding another imaging modality. In particular, positron emission tomography isotopes such as ^{111}In or ^{64}Cu, which emit gamma rays from their decay, have been most frequently used as the third probes [81].

IMAGING BY MICROSCOPIC TECHNIQUES

Light and Fluorescence Microscopy

Despite the fact that this technique is the well-known and the well-used in biology, its application in the detection of NPs is relatively limited. The first field is based on the histochemical detection of components of NPs, usually oxides of metal ions. Newer and newer procedures are being proposed and developed [82]. The second possibility is based on the immunohistochemistry, the process based on the detection of antigens in cells and tissues on the principle of monoclonal or polyclonal antibodies binding specifically to antigens [83,84]. However, this technique is usable only for functionalized NPs carrying antigenic structure. Chromogenic reporters are used for the detection of target, in our case NPs. The most common chromogenic reporters are enzymes, such as alkaline phosphatase (AP) and horseradish peroxidase (HRP). In addition, they can be detected not only by chromogenic substrates, such as 3,3′-diaminobenzidine (DAP) or 5-bromo-4-chloro-3-indolyl phosphate/nitro blue tetrazolium (BCIP/NBT), but also by fluorogenic and chemiluminiscent substrates usable in fluorescence microscopy. Light microscopy is used for the visualization of iron oxide-based NPs, which is based mostly on the processing of the semi thin section and Prussian blue/Berlin blue reaction method staining. This technique was used in the visualization of SPIONs in the human cochleae and mouse inner ear cell culture [85]. Incorporation of peroxidase into or onto NPs, which is widely used in the construction of hydrogen peroxide biosensors, may represent useful tool in the visualization of nanoparticles [86, 87]. Like this modified NPs may catalyze the oxidation of peroxidase substrates to produce a color reaction at the site of NPs. Based on this NPs-peroxidase approach, Zhuang *et al.* determined the bio distribution and organ clearance of magnetic NPs in mice by visualizing and quantifying the localization of MNPs within the main organs [88]. This approach has significantly increased the sensitivity of NPs detection compared to convenient Prussian blue staining. Light microscopy may be suitably combined with next imaging techniques, such as magnetic resonance. For example, Neuwelt *et al.* compared histochemical and magnetic resonance detection of Ferumoxtran-10, a dextran-coated iron oxide nanoparticles, in patients suffering from primary metastatic malignant brain tumors [5]. Histochemistry revealed presence of Ferumoxtran-10 NPs in the reactive cells, predominantly in reactive astrocytes and macrophages, and not in tumor cells. This work demonstrates possible benefit of combination of magnetic resonance with histochemistry, which may improve our knowledge about distribution of NPs suffering from malignant disease. Compared to convenient light microscopy, fluorescence microscopy brings some advantages, especially significantly higher quality of images. Samples must be fluorescently stained despite the fact that many compounds are responsible for the intrinsic auto fluorescence of sample. Auto fluorescence may be used for detection of nanoparticles, as NPs are able to affect auto fluorescence of cells. These effects are probably caused by the high surface area of the NPs and ability to catalyze different processes leading to the chemical modification of auto fluorescent biomolecules. El-Sayed *et al.* treated two oral squamous carcinoma cell lines and one immortalized benign epithelial cell line with gold nanoparticles of different shapes, both spheres and rods and monitored changes in auto fluorescence caused by soluble NADH and collagen. They significantly quenched the auto fluorescence of both compounds based on the oxidation of NADH to NAD^+ catalyzed by NPs surface and photonic absorption of light by collagen [89]. This knowledge suggests new potential of NPs in cell and tissue imaging. Fluorescence microscopy enables combination of differently labeled secondary antibodies and multifunctional cell/tissue labeling.

Confocal Microscopy

Conventional immunofluorescence methods may be used for imaging of NPs by confocal (and fluorescence) microscopy. Confocal microscopy is ideally suited to the investigation of biological tissue because it is able to provide high quality images from within a localized region of the cell/tissue. Immunofluorescence methods are based on the specificity of antibodies to antigen to target fluorescence probe to NPs. Nevertheless, NPs must be functionalized by biomolecules which represent a suitable antigen. Secondary, indirect, immunofluorescence is the most commonly used technique in NPs visualization and detection. On the other hand, the complexity of NPs' structure often enables binding of a fluorescent probe or more different fluorescence probes ("doped" NPs) and the NPs can subsequently be visualized directly using confocal fluorescence microscopy techniques. Luminescent nanoparticles represent one of the most interesting opportunities for confocal microscopy. Especially silica-based nanoparticles doped with fluorescence molecules [90, 91], gold nanoparticles [92], and quantum dots and rods and luminescent nanocrystals called nanophosphors are discussed [93, 94]. Organically modified silica (ORMOSIL) and nanoparticles doped with two-photon absorbing fluorophores, which exhibit aggregation-enhanced fluorescence, and nanophosphors (ceramic nanoparticles containing luminescent lanthanoid ions) are two most promising luminescent NPs [95–98]. The most important confocal microscopy techniques are briefly summarized in Table 1.

Table 1: The most important confocal microscopy techniques.

Technique	Application	Nanoparticle (s)	Ref.
Confocal Laser Scanning Microscopy	Cellular uptake/transport mechanisms study on Madin-Darby Canine Kidney (MDCK) cells	Coumarin-6 labeled nanoparticles	[100]
Confocal fluorescence microscopy	Cellular uptake/transport mechanisms study on red blood cells	Zwitterionic quantum dots	[101]
Confocal reflection microscopy	Improving of image contrast of tissue structures	60 nm gold nanospheres	[102]
Confocal Raman Microscopy	Cellular uptake and localization study on human hepatocarcinoma HepG2 cells	poly(lactide-co-glycolide) (PLGA) nanoparticles (NPs) and carbon nanotubes (CNTs)	[103]
Confocal interference scattering microscopy	Detection the orientation of individual silver nanorods and measure their relative distances	Silver nanorods	[104]

Studies of cell processes as well as NPs transport and localization are multiparameter problem, so it is highly necessary to examine many parameters within the cells. Due to these needs, microscope systems for multispectral imaging have been developed. Multispectral confocal microscopy enables elegant solving of these needs and enables not only monitoring the cellular uptake, transport and localization of NPs, but also connection of these processes with drug delivery systems [99].

Electron Microscopy

Electron microscopy is an irreplaceable tool in the determination of the size and structure of NPs. The basic application of electron microscopy methods in NPs detection/visualization including the most common scanning (SEM) and transmission electron microscopy (TEM) is summarized in Table 2.

Table 2: The basic application of electron microscopy methods in NPs detection/visualization including the most common scanning (SEM) and transmission electron microscopy (TEM).

Technique	Application	Nanoparticle (s)	Ref.
Transmission electron microscopy (TEM)	Phase characterization of NPs	TiO_2	[140]
Scanning electron microscopy (SEM)	General characterization	Silver nanoparticles	[117]
Scanning transmission electron microscopy (STEM)	General characterization	Platinum/ruthenium nanoparticles	[141]
Reflection electron microscopy (REM)	Nucleation control	Germanium nanoislands on ^{111}Si surface	[142]
Low-voltage electron microscopy (LVEM)	General characterization and investigation of contrast mechanisms	Single-walled carbon nanotubes	[143]

Transmission electron microscopy is the most used technique in the characterization of NPs. TEM was used for the characterization and visualization of different types of nanostructures, such as noble metal- and metal oxide-based NPs [105–107], carbon nanomaterials [108,109], quantum dots [110–112], or polymeric nanoparticles of a dendrimer type [113]. TEM is preferentially used in the study of nanoparticle uptake, cellular compartmentation, transport and accumulation in different tissues and organs. Recently, TEM was used to investigate the presence of metal nanoparticles inside the carbon nanotubes [114]. Scanning electron microscopy is a necessary tool in the determination and structural characterization of NPs, such as size and shape, area, presence of pores, or eccentricity; in addition, SEM is highly necessary in NP quality control [115–117]. Analytical Electron Microscopy (AEM) has become an essential tool for microstructural investigations of nanostructures. It has become one of the most powerful tools to obtain structural, morphological, and compositional information at the nanometer scale by combining imaging, diffraction and spectroscopy, e.g., EDS (energy dispersive X-ray spectrometry) and Electron Energy Loss Spectrometry (EELS). Field-emission gun (FEG) electron source instruments enable highly effective study of nanostructures using various methods, such as high resolution transmission electron microscopy (HRTEM) [118,119], high-resolution Z-contrast (STEM/HAADF) imaging [120], different techniques of electron diffraction (SAED, CBED, nanodiffraction) [121], X-ray energy dispersive spectroscopy (XEDS) [122,123] and electron energy loss spectroscopy (EELS) [65,124,125].

In addition to all above-mentioned methods, there are other techniques that have found application in the imaging and visualization of NPs and nanomaterials. The most important techniques are scanning acoustic microscopy (SAM, a technique that uses focused sound to investigate, measure, or image an object and can provide data about the cytoskeleton and movement of the cells in biological and medical applications) [126,127], X-ray microscopy (XM, which uses electromagnetic radiation in the soft X-ray area and is applicable on biological samples in natural state; [128,129], scanning tunneling microscopy (STM) for imaging of surfaces at the atomic level [130–132], scanning probe microscopy (SPM, surface images under the use of physical probes) [133–136] and especially its type atomic force microscopy (AFM, an application used especially in biophysics in the monitoring of protein-ligand interactions) [133,137–139].

Electrochemical Microscopy

Another imaging technology, which is in the interest of scientists, is Scanning Electrochemical Microscopy (SECM). SECM can provide information about the topography and electrochemical properties of a surface. Measured samples do not have to be special prepared. Using this method it is possible to measure the cellular activity and the effects of cellular inhibitors by observing the oxygen reduction current [144], to investigate the anticancer drug sensitivity [145], to detect genetic mutations [146] or select metastatic/nonmetastatic cells base on their physiologic effects [147]. The topography and intracellular redox behavior of non-metastatic and metastatic cell lines was investigated too [148]. Liebetrau *et al.* [149] showed that SECM could record real-time morphological changes evoked by modulating the ionic strength of surrounding medium. This was done via using hydrophobic mediators (e.g., to obtain quantitative information about morphology of living cells PC12 by studying changes in the tip current arising from topography variations).

IMAGING BY LASER-INDUCED BREAKDOWN SPECTROSCOPY (LIBS)

Although similar to microscopy, LIBS cannot be applied to higher living organisms such as animals, but it can be potentially applied to analyze cells and therefore we believe that it should be mentioned in this review.

Laser-Induced Breakdown Spectroscopy (LIBS) is a spectrochemical analytical technique that utilizes high power densities employing focused radiation, usually from a pulsed, fixed-frequency laser, in order to generate luminous plasma from a sample (solid, liquid, and gaseous samples) [150]. During the LIBS analysis the stoichiometric sample ablation is assumed, so that the plasma composition reflects the elemental composition of the ablated target. Comparing LIBS to another, more traditional laser-ablation based techniques, *i.e.*, to LA-ICP-OES/MS highlights that LIBS setup is in general simpler and this approach is more cost reliable. Considering the capabilities for spatially-resolved analysis, LIBS gives an instantaneous signal directly related to the location at which a single ablation event occurred, while mapping using LA-ICP-OES/MS usually represents more experimental difficulties, because this technique involves sample transport, so the signal produced in the mass (or optical) spectrometer without considerable care in the analysis is not directly attributable to a specific location on the sample [151]. LIBS is uniquely capable for light-element detection, useful for material identification, depth profiling, and elemental surface mapping, and so it can solve numerous industrial and scientific problems in real-time [152–154]. On the other hand, the limits of detection (LOD) of LIBS, which in its basic configuration are typically of the order of $mg \cdot kg^{-1}$ or tens of $mg \cdot kg^{-1}$ for most of the elements, are poorer with respect to other more traditional analytical techniques (e.g., ICP-MS or ICP-OES) [146]. One of the suggested approaches to improve LIBS LODs is the double-pulse (DP) configuration. The DP-LIBS [155,156] approach was introduced by Niemax and his co-workers from ISAS Dortmund in 1991 [157], and nowadays it is a common technique used in many LIBS research laboratories. DP-LIBS allows better coupling of the laser energy to the target and to the ablated material, leading so to a more efficient production of analyte atoms in the excited state. The necessary LIBS instrumentation can be summarized as short duration pulsed laser (or lasers in the case of DP-LIBS technique), focusing optics, collecting optics for the radiation emitted by the laser-induced plasma, a wavelength analyzer and a gated detector (capable temporarily resolved analysis), all computer controlled [152].

LIBS in Nanoparticles Analysis

The LIBS capabilities for rapid analysis of multi-component pharmaceutical tablets were demonstrated using several examples earlier [158,159]. LIBS was evaluated as a rapid and direct method for measuring the drug content in pharmaceutical tablets. Although the tablet components are all organic, in a large fraction of cases the drug molecule contains an element which distinguishes it from the other tablet components. Measuring the drug content can therefore be carried out by quantifying this element [159].

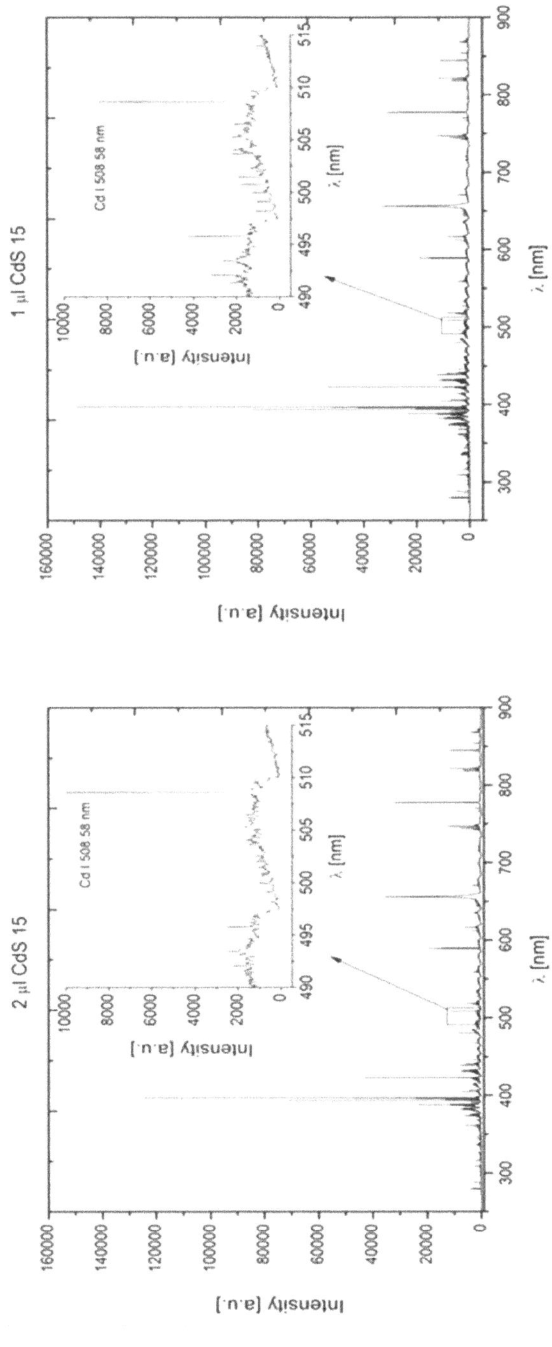

Figure 3: Comparison of DP-LIBS spectrum for the CdS QDs sample of 1 μL and 2 μL volume. First laser—532 nm, pulse duration ~10 ns, perpendicular direction to the sample surface, second laser—1,064 nm, pulse duration ~6 ns, parallel direction to the sample. Spectral lines 508.58 nm for cadmium and 276.67 nm for tellurium.

Based on our experiences the extension of the LIBS capabilities for determination of Quantum Dots (QDs) in liquid solution is feasible. LIBS has not been used for this purpose, in spite of the fact that it could cast the light on qualitative and quantitative properties of produced nanoparticles. In our work, we prepared QDs as particles floating in liquid medium, so it has to be ensured that they are deployed equally during the time of LIBS experiment in the whole volume. Basically, there are two ways how to achieve this: circulate the medium or deposit the liquid medium on some kind of carrier medium. Circulating the medium has a disadvantage in a relatively big volume of necessary liquid (~20 mL). On the contrary, application of liquids on suitable carrier medium needs small amount of sample (~1 μL) and also it can be more easily manipulated. Materials suitable for carrier medium have to comply with following parameters - chemical stability, high absorption, well defined elemental composition (most important is that any spectral line of included elements does not interfere with spectral lines of elements of interests), low costs and easy manipulation.

The ~10 ns long, 532 nm laser pulse of ablation laser was directed perpendicular to the sample surface by mirrors and focused by a 30 mm focal length glass triplet. The beam of second 1,064 nm laser pulse, with a pulse width ~6 ns was directed parallel to the sample surface to the plasma. Spectral lines 508.58 nm for cadmium and 276.67 nm for tellurium were studied (Figure 3).

IN VIVO IMAGING OF TARGETED DRUG DELIVERY

Drug nanodelivery belongs to one of the most important part of nanomedicine that is already implemented. Difficulties with the drug delivery, such as low solubility and/or bioavailability, short half-life or inconvenient distribution to the target location may cause drug inefficiency. Nanoparticles designed for tumor targeted therapies consist of various components, in most cases a nanocarrier (nanovector) and a drug [160].

Nanoparticle carriers are mostly composed of iron oxides, gold, biodegradable polymers, dendrimers, lipid based carriers such as liposomes or micelles, viruses (viral nanoparticles) or organometallic compounds [161–164]. The majority of anticancer therapeutics is water insoluble and need to be dissolved in an organic solvent that is usually toxic [165]. The low molecular weight of anticancer drugs causes their rapid excretion, requiring the administration of escalating doses [166]. The drug encapsulation in nanocarriers provides better biocompatibility. Several such engineered drugs are already used in clinical practice as registered drugs e.g., liposomal doxorubicin [167]. The main advantage of these particles in the clinic is minimum toxicity of the carrier itself and decreased toxicity of the anticancer drug [168]. Concerning the nanoparticles shapes used are nanoshells, nanorods, nanocages, nanocubes or nanotubes (Figure 4). Nanoparticle-mediated targeted delivery of drugs might significantly reduce the dosage of the drugs with keeping of the effect that reduce side effects and increase its specificity and bioavailability [12,164,169].

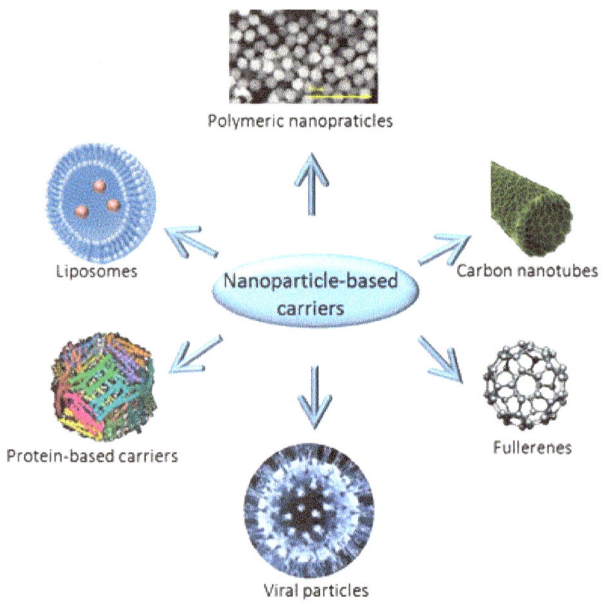

Figure 4: Nanoparticle-based carriers used for targeted drug/gene delivery.

Therapeutic nanosystems have four unique properties: (i) they themselves may have therapeutic properties; (ii) nanosystems can be attached to targeting ligands, which yield high affinity and specificity for target cells; (iii) nanosystems can contain combination of drugs; (iv) nanosystems can bypass some of drug resistance mechanisms. By using both passive and active targeting strategies, the nanocarriers can achieve increased intracellular concentration of drugs in cancer cells while minimizing toxicity in normal cells, simultaneously enhancing anticancer effects and reducing systemic toxicity [170,171].

Targeted delivery in anticancer therapy is one of the more intensively studied areas. The above mentioned improvements are based on the enhancement of permeability and retention of the vasculature surrounding tumors caused by poor quality of newly developing tumor vessels which carry of nanoparticles to target molecules expressed by tumor cells [172]. Antibodies, peptides, nucleic acid aptamers, carbohydrates and small molecules can be used as ligands. The key features of anticancer nanoparticles are mainly their size, surface properties and in some types of them also targeting ligands. The size of nanoparticles varies between 10 to 200 nm. Nanoparticle property requirements also depend on tumor characteristics such as cancer type, clinical stage, or location [173]. The precise and fast diagnostic of therapeutic targets is one of the basic conditions for implementation of personalized medicine. For targets detection nanoparticles can be used [174,175]. Photodynamic therapy (PDT) is based on utilization of light-sensitive drugs or photosensitizers that can be preferentially localized in tumors and their therapeutic effect is activated by excitation of the photosensitizer to generate a cytotoxic species, usually singlet oxygen, free radicals, or peroxides. This caused relatively selective damage of cancer cells but spare adjacent healthy ones. There are several technical difficulties in the application of PDT, but some of them could be overcome by using of nanoparticles. Nanoparticles in PDT can be divided into passive carriers and active participants in photosensitizer excitation. Active nanoparticles are classified into photosensitizer nanoparticles, self-illuminating nanoparticles and upconverting nanoparticles [176]. QDs can transfer energy to surrounding O_2, generate singlet oxygen with consequent toxicity to cells [177]. The quality of a photosensitising drug for PDT depends on its ability to transfer energy and/or electron to molecular oxygen and the subsequent production of reactive oxygen intermediates (ROI). Photostability, broad absorption band and large two-photon absorption cross-section of QDs are advantages for PDT. Although targeted delivery of QDs in cancer cells and tumor milieu is possible, compared with conventional photosensitising drugs, the efficiency of QDs to produce ROI induced by direct photoactivation is lower [177]. Therefore conjugates between QDs and conventional photosensitising drugs are subjects of intensive research [178].

Other application of nanoparticules is hyperthermia. Thermal ablation involves heating cells at temperatures above 55 °C, which causes coagulative necrosis and protein denaturation that kill targeted cells. The temperature of carbon nanotubes can rise to over 60 °C within 2 min when they are exposed to near infrared wavelengths of 700–1,100 nm [179]. Research has been focused on development of nanostructures that produce heat on activation by near infrared optical excitation. Such nanoparticules target specifically the tumor and selectively ablate it, leaving surrounding healthy tissue undamaged. This therapeutic modality has been shown to be efficacious for the treatment of several tumors, e.g., lung, liver, and prostate carcinoma [180]. Magnetic nanoparticles can also be used for hyperthermia. Magnetic nanoparticles can be localized in deep tissue and fixed by external magnetic field and alteration of the AC field increase the temperature of the magnetic nanoparticles [179]. Radiosensitizers are drugs that enhance the effects of radiotherapy, and some of them are metallic nanoparticles. Such nanoparticles should provide enhanced interaction cross-section with gamma photons. They increase effect of radiotherapy by scattering and/or absorbing X and γ rays causing localized damage of cancer cells and thus decreasing total radiation dose to minimize side effects [181]). This approach of using nanoparticules as radiosensitizers is termed Nanoparticle Enhanced X-ray Therapy (NEXT). It is planned to test CaF, LaF, ZnS or ZnO QD, carbon dots, liposomes with some radiosensitizers or antioxidants, nanoparticle-photosensitizer conjugates and silicon nanoparticles [182]. Radiosensitization by 1–12 nm gold nanoparticles irradiated by high energy electrons generated low energy electrons that caused increased DNA damage [183]. In another study, gold microspheres of 1.5–3.0 µm size have been tested both *in vitro* and *in vivo* studied for enhancement of X-rays that resulted in radiosensitization by gold particles [184]. Furthermore, the application of gold nanorods for therapeutic purposes has been investigated [185–187]. Contrast nanoparticules may also increase efficiency and/or safety of radiotherapy by improvement of planning by MRI [188].

Nanotechnology may also be applied in diagnostics for surgical oncology. Nanoparticles can be accumulated in solid tumors that enable one to localize tumors, detect margins, identify important adjacent structures, map sentinel lymph nodes, and detect residual tumor cells or micrometastases. Intraoperative imaging overcomes the tissue penetration problem of optical methods [189].

OUTLOOK

Imaging techniques are progressing rapidly due to the development of new types of contrast agents as well as progress in the development of instrumentation, electronics and software. This is a multi/transdisciplinary approach combining the knowledge of chemists, biologists, physicists, and mathematicians as well as engineers and IT specialists. It is highly improbable that QDs will fully replace the well-established technologies using fluorophores or fluorescent protein-fusion, but will rather complement them for applications needing better photostability, sensitivity and wide variability. Magnetic nanoparticles will also broaden the area of already employed contrast agents and find more and more biomedical applications for their unique properties. The advantages of these types of nanoparticles are their low cost synthesis, huge modification possibilities and easy application. Such symbiosis of deep-rooted and modern methods will lead to improvement in early diagnostics as well as personalized therapy. However several difficulties have to be overcome before nanoparticles will be used as a routine tool, including the toxicity of metal and semiconductor nanoparticles and/or tissue autofluorescence in the visible spectral range. Nevertheless, the potential of the *in vivo* imaging is great and opens new possibilities for faster and better diagnostics on the one hand, and improved treatment strategies on the other hand.

ACKNOWLEDGMENTS

Financial support from the following projects CEITEC CZ.1.05/1.1.00/02.0068, GACR 102/10/P618, NanoBioTECell GACR P102/11/1068 project for conceptual development of research organization 00064203 is highly acknowledged.

REFERENCES

1. Llinas, R.R.; Walton, K.D.; Nakao, M.; Hunter, I.; Anquetil, P.A. Neuro-vascular central nervous recording/stimulating system: Using nanotechnology probes. *J. Nanopart. Res.* **2005**, *7*, 111–127.
2. Santra, S.; Yang, H.; Stanley, J.T.; Holloway, P.H.; Moudgil, B.M.; Walter, G.; Mericle, R.A. Rapid and effective labeling of brain tissue using TAT-conjugated CdS: Mn/ZnS quantum dots. *Chem. Commun.* **2005**, *2005*, 3144–3146.
3. Keefer, E.W.; Botterman, B.R.; Romero, M.I.; Rossi, A.F.; Gross, G.W. Carbon nanotube coating improves neuronal recordings. *Nat. Nanotechnol.* **2008**, *3*, 434–439.
4. Neuwelt, E.A.; Varallyay, C.G.; Manninger, S.; Solymosi, D.; Haluska, M.; Hunt, M.A.; Nesbit, G.; Stevens, A.; Jerosch-Herold, M.; Jacobs, P.M.; *et al.* The potential of ferumoxytol nanoparticle magnetic resonance imaging, perfusion, and angiograpgy in central nervous system malignancy: A pilot study. *Neurosurgery* **2007**, *60*, 601–611.
5. Neuwelt, E.A.; Varallyay, P.; Bago, A.G.; Muldoon, L.L.; Nesbit, G.; Nixon, R. Imaging of iron oxide nanoparticles by MR and light microscopy in patients with malignant brain tumours. *Neuropathol. Appl. Neurobiol.* **2004**, *30*, 456–471.
6. Clavijo-Jordan, V.; Kodibagkar, V.D.; Beeman, S.C.; Hann, B.D.; Bennett, K.M. Principles and emerging applications of nanomagnetic materials in medicine. *Wiley Interdiscip. Rev. Nanomed. Nanobiotechnol.* **2012**, *4*, 345–365.
7. Rumenapp, C.; Gleich, B.; Haase, A. Magnetic nanoparticles in magnetic resonance imaging and diagnostics. *Pharm. Res.* **2012**, *29*, 1165–1179.
8. Thanh, N.T.K. *Magnetic Nanoparticles from Fabrication to Clinical Applications*; Taylor and Francis: Oxon, UK, 2012; p. 583.
9. Arias, J.L.; Reddy, L.H.; Couvreur, P. Fe_3O_4/chitosan nanocomposite for magnetic drug targeting to cancer. *J. Mater. Chem.* **2012**, *22*, 7622–7632.
10. Banerjee, R.; Katsenovich, Y.; Lagos, L.; McIintosh, M.; Zhang, X.; Li, C.Z. Nanomedicine: Magnetic nanoparticles and their biomedical applications. *Curr. Med. Chem.* **2010**, *17*, 3120–3141.
11. Khandhar, A.P.; Ferguson, R.M.; Simon, J.A.; Krishnan, K.M. Tailored magnetic nanoparticles for optimizing magnetic fluid hyperthermia. *J. Biomed. Mater. Res. A* **2012**, *100A*, 728–737.
12. Chomoucka, J.; Drbohlavova, J.; Huska, D.; Adam, V.; Kizek, R.; Hubalek, J. Magnetic nanoparticles and targeted drug delivering. *Pharmacol. Res.* **2010**, *62*, 144–149.
13. Silva, A.K.A.; Wilhelm, C.; Kolosnjaj-Tabi, J.; Luciani, N.; Gazeau, F. Cellular transfer of magnetic nanoparticles via cell microvesicles: Impact on cell tracking by magnetic resonance imaging. *Pharm. Res.* **2012**, *29*, 1392–1403.
14. Wenzel, D.; Rieck, S.; Vosen, S.; Mykhaylyk, O.; Trueck, C.; Eberbeck, D.; Trahms, L.; Zimmermann, K.; Pfeifer, A.; Fleischmann, B.K. Identification of magnetic nanoparticles for combined positioning and lentiviral transduction ofendothelial cells. *Pharm. Res.* **2012**, *29*, 1242–1254.
15. Abou-Montet, K.; Montet, X.; Weissleder, R.; Josephson, L. Transfection agent induced nanoparticle cell loading. *Mol. Imaging* **2005**, *4*, 165–171.
16. Amiri, H.; Bustamante, R.; Millan, A.; Silva, N.J.O.; Pinol, R.; Gabilondo, L.; Palacio, F.; Arosio, P.; Corti, M.; Lascialfari, A. Magnetic and relaxation properties of pultifunctional polymer-based nanostructured bioferrofluids as MRI contrast agents. *Magn. Reson. Med.* **2011**, *66*, 1715–1721.
17. Drbohlavova, J.; Hrdy, R.; Adam, V.; Kizek, R.; Schneeweiss, O.; Hubalek, J. Preparation and properties of various magnetic nanoparticles. *Sensors* **2009**, *9*, 2352–2362.
18. Zboril, R.; Mashlan, M.; Petridis, D. Iron(III) oxides from thermal processes-synthesis, structural and magnetic properties, mossbauer spectroscopy characterization, and applications. *Chem. Mat.* **2002**, *14*, 969–982.
19. Tucek, J.; Zboril, R.; Petridis, D. Maghemite nanoparticles by view of mossbauer spectroscopy. *J. Nanosci. Nanotechnol.***2006**, *6*, 926–947.
20. Reshmi, G.; Kumar, P.M.; Malathi, M. Preparation, characterization and dielectric studies on carbonyl

iron/cellulose acetate hydrogen phthalate core/shell nanoparticles for drug delivery applications. *Int. J. Pharm.* **2009**, *365*, 131–135.

21. Shubayev, V.I.; Pisanic, T.R.; Jin, S.H. Magnetic nanoparticles for theragnostics. *Adv. Drug Deliv. Rev.* **2009**, *61*, 467–477.

22. Jain, T.K.; Foy, S.P.; Erokwu, B.; Dimitrijevic, S.; Flask, C.A.; Labhasetwar, V. Magnetic resonance imaging of multifunctional pluronic stabilized iron-oxide nanoparticles in tumor-bearing mice. *Biomaterials* **2009**, *30*, 6748–6756.

23. Schweiger, C.; Pietzonka, C.; Heverhagen, J.; Kissel, T. Novel magnetic iron oxide nanoparticles coated with poly(ethylene imine)-g-poly(ethylene glycol) for potential biomedical application: Synthesis, stability, cytotoxicity and MR imaging. *Int. J. Pharm.* **2011**, *408*, 130–137.

24. Kumagai, M.; Kano, M.R.; Morishita, Y.; Ota, M.; Imai, Y.; Nishiyama, N.; Sekino, M.; Ueno, S.; Miyazono, K.; Kataoka, K. Enhanced magnetic resonance imaging of experimental pancreatic tumor *in vivo* by block copolymer-coated magnetite nanoparticles with TGF-beta inhibitor. *J. Control. Release* **2009**, *140*, 306–311.

25. Masoudi, A.; Hosseini, H.R.M.; Shokrgozar, M.A.; Ahmadi, R.; Oghabian, M.A. The effect of poly(ethylene glycol) coating on colloidal stability of superparamagnetic iron oxide nanoparticles as potential MRI contrast agent. *Int. J. Pharm.* **2012**, *433*, 129–141.

26. Martelli, C.; Borelli, M.; Ottobrini, L.; Rainone, V.; Degrassi, A.; Russo, M.; Gianelli, U.; Bosari, S.; Fiorini, C.; Trabattoni, D.; *et al.* In vivo imaging of lymph node migration of MNP- and In-111-labeled dendritic cells in a transgenic mouse model of breast cancer (MMTV-Ras). *Mol. Imaging Biol.* **2012**, *14*, 183–196.

27. Jiang, W.L.; Xie, H.; Ghoorah, D.; Shang, Y.L.; Shi, H.J.; Liu, F.; Yang, X.L.; Xu, H.B. Conjugation of functionalized SPIONs with transferrin for targeting and imaging brain glial tumors in rat model. *PLoS One* **2012**, *7*, e37376.

28. Soenen, S.J.; Vande-Velde, G.; Ketkar-Atre, A.; Himmelreich, U.; De Cuyper, M. Magnetoliposomes as magnetic resonance imaging contrast agents. *Wiley Interdiscip. Rev. Nanomed. Nanobiotechnol.* **2011**, *3*, 197–211.

29. Razansky, D.; Deliolanis, N.C.; Vinegoni, C.; Ntziachristos, V. Deep tissue optical and optoacoustic molecular imaging technologies for pre-clinical research and drug discovery. *Curr. Pharm. Biotechnol.* **2012**, *13*, 504–522.

30. Alivisatos, A.P.; Gu, W.W.; Larabell, C. Quantum dots as cellular probes. In *Annual Review of Biomedical Engineering*; Annual Reviews: Palo Alto, CA, USA, 2005; Volume 7, pp. 55–76.

31. Alivisatos, P. The use of nanocrystals in biological detection. *Nat. Biotechnol.* **2004**, *22*, 47–52.

32. Gao, X.H.; Yang, L.L.; Petros, J.A.; Marshal, F.F.; Simons, J.W.; Nie, S.M. *In vivo* molecular and cellular imaging with quantum dots. *Curr. Opin. Biotechnol.* **2005**, *16*, 63–72.

33. Michalet, X.; Pinaud, F.F.; Bentolila, L.A.; Tsay, J.M.; Doose, S.; Li, J.J.; Sundaresan, G.; Wu, A.M.; Gambhir, S.S.; Weiss, S. Quantum dots for live cells, *in vivo* imaging, and diagnostics. *Science* **2005**, *307*, 538–544.

34. Drbohlavova, J.; Adam, V.; Kizek, R.; Hubalek, J. Quantum dots—Characterization, preparation and usage in biological systems. *Int. J. Mol. Sci.* **2009**, *10*, 656–673.

35. Yu, W.W.; Qu, L.H.; Guo, W.Z.; Peng, X.G. Experimental determination of the extinction coefficient of CdTe, CdSe, and CdS nanocrystals. *Chem. Mat.* **2003**, *15*, 2854–2860.

36. Baruah, S.; Ortinero, C.; Shipin, O.V.; Dutta, J. Manganese doped zinc sulfide quantum dots for detection of Escherichia coli. *J. Fluoresc.* **2012**, *22*, 403–408.

37. Drummen, G. Quantum dots-from synthesis to applications in biomedicine and life sciences. *Int. J. Mol. Sci.* **2010**, *11*, 154–163.

38. Jin, Z.W.; Hildebrandt, N. Semiconductor quantum dots for *in vitro* diagnostics and cellular imaging. *Trends Biotechnol.***2012**, *30*, 394–403.

39. Liu, Q.H.; Deng, R.P.; Ji, X.L.; Pan, D.C. Alloyed Mn-Cu-In-S nanocrystals: A new type of diluted magnetic semiconductor quantum dots. *Nanotechnology* **2012**, *23*, 2–6.

40. Mattoussi, H.; Palui, G.; Na, H.B. Luminescent quantum dots as platforms for probing *in vitro* and *in vivo* biological processes. *Adv. Drug Deliv. Rev.* **2012**, *64*, 138–166.

41. Mukerjee, A.; Ranjan, A.P.; Vishwanatha, J.K. Combinatorial nanoparticles for cancer diagnosis and therapy. *Curr. Med. Chem.* **2012**, *19*, 3714–3721.

42. Nie, S.M.; Xing, Y.; Kim, G.J.; Simons, J.W. Nanotechnology applications in cancer. In *Annual Review of Biomedical Engineering*; Annual Reviews: Palo Alto, CA, USA, 2007; Volume 9, pp. 257–288.

43. Pericleous, P.; Gazouli, M.; Lyberopoulou, A.; Rizos, S.; Nikiteas, N.; Efstathopoulos, E.P. Quantum dots hold promise for early cancer imaging and detection. *Int. J. Cancer* **2012**, *131*, 519–528.

44. Singhal, M.; Sharma, J.K.; Kumar, S. Effect of biocompatible glutathione capping on core-shell ZnS quantum dots. *J. Mater. Sci. Mater. Electron.* **2012**, *23*, 1387–1392.

45. Chan, W.C.W.; Nie, S.M. Quantum dot bioconjugates for ultrasensitive nonisotopic detection. *Science* **1998**, *281*, 2016–2018.

46. Jaiswal, J.K.; Mattoussi, H.; Mauro, J.M.; Simon, S.M. Long-term multiple color imaging of live cells using quantum dot bioconjugates. *Nat. Biotechnol.* **2003**, *21*, 47–51.

47. Larson, D.R.; Zipfel, W.R.; Williams, R.M.; Clark, S.W.; Bruchez, M.P.; Wise, F.W.; Webb, W.W. Water-soluble quantum dots for multiphoton fluorescence imaging *in vivo*. *Science* **2003**, *300*, 1434–1436.

48. Medintz, I.L.; Clapp, A.R.; Mattoussi, H.; Goldman, E.R.; Fisher, B.; Mauro, J.M. Self-assembled nanoscale biosensors based on quantum dot FRET donors. *Nat. Mater.* **2003**, *2*, 630–638.

49. Medintz, I.L.; Uyeda, H.T.; Goldman, E.R.; Mattoussi, H. Quantum dot bioconjugates for imaging, labelling and sensing. *Nat. Mater.* **2005**, *4*, 435–446.
50. Ryvolova, M.; Chomoucka, J.; Janu, L.; Drbohlavova, J.; Adam, V.; Hubalek, J.; Kizek, R. Biotin-modified glutathione as a functionalized coating for bioconjugation of CdTe based quantum dots. *Electrophoresis* **2011**, *32*, 1619–1622.
51. Wu, X.Y.; Liu, H.J.; Liu, J.Q.; Haley, K.N.; Treadway, J.A.; Larson, J.P.; Ge, N.F.; Peale, F.; Bruchez, M.P. Immunofluorescent labeling of cancer marker Her2 and other cellular targets with semiconductor quantum dots. *Nat. Biotechnol.* **2003**, *21*, 41–46.
52. Maysinger, D.; Behrendt, M.; Lalancette-Herbert, M.; Kriz, J. Real-time imaging of astrocyte response to quantum dots: *In vivo* screening model system for biocompatibility of nanoparticles. *Nano Lett.* **2007**, *7*, 2513–2520.
53. Zimmer, J.P.; Kim, S.W.; Ohnishi, S.; Tanaka, E.; Frangioni, J.V.; Bawendi, M.G. Size series of small indium arsenide-zinc selenide core-shell nanocrystals and their application to *in vivo* imaging. *J. Am. Chem. Soc.* **2006**, *128*, 2526–2527.
54. Cai, W.B.; Shin, D.W.; Chen, K.; Gheysens, O.; Cao, Q.Z.; Wang, S.X.; Gambhir, S.S.; Chen, X.Y. Peptide-labeled near-infrared quantum dots for imaging tumor vasculature in living subjects. *Nano Lett.* **2006**, *6*, 669–676.
55. Smith, J.D.; Fisher, G.W.; Waggoner, A.S.; Campbell, P.G. The use of quantum dots for analysis of chick CAM vasculature. *Microvasc. Res.* **2007**, *73*, 75–83.
56. Walling, M.; Novak, J.; Shepard, J.R.E. Quantum dots for live cell and *in vivo* imaging. *Int. J. Mol. Sci.* **2009**, *10*, 441–491.
57. Voura, E.B.; Jaiswal, J.K.; Mattoussi, H.; Simon, S.M. Tracking metastatic tumor cell extravasation with quantum dot nanocrystals and fluorescence emission-scanning microscopy. *Nat. Med.* **2004**, *10*, 993–998.
58. Pic, E.; Pons, T.; Bezdetnaya, L.; Leroux, A.; Guillemin, F.; Dubertret, B.; Marchall, F. Fluorescence imaging and whole-body biodistribution of near-infrared-emitting quantum dots after subcutaneous injection for regional lymph node mapping in mice. *Mol. Imaging. Biol.* **2010**, *12*, 394–405.
59. Pons, T.; Pic, E.; Lequeux, N.; Cassette, E.; Bezdetnaya, L.; Guillemin, F.; Marchal, F.; Dubertret, B. Cadmium-free CuLnS2/ZnS quantum dots for sentinel lymph mode imaging with reduced toxicity. *ACS Nano* **2010**, *4*, 2531–2538.
60. Diagaradjane, P.; Orenstein-Cardona, J.M.; Colon-Casasnovas, N.E.; Deorukhkar, A.; Shentu, S.; Kuno, N.; Schwartz, D.L.; Gelovani, J.G.; Krishnan, S. Imaging epidermal growth factor receptor expression *in vivo*: Pharmacokinetic and biodistribution characterization of a bioconjugated quantum dot nanoprobe. *Clin. Cancer Res.* **2008**, *14*, 731–741.
61. Bhang, S.H.; Won, N.; Lee, T.J.; Jin, H.; Nam, J.; Park, J.; Chung, H.; Park, H.S.; Sung, Y.E.; Hahn, S.K.; *et al.* Hyaluronic acid-quantum dot conjugates for *in vivo* lymphatic vessel imaging. *ACS Nano* **2009**, *3*, 1389–1398.
62. Choi, H.S.; Liu, W.; Misra, P.; Tanaka, E.; Zimmer, J.P.; Ipe, B.I.; Bawendi, M.G.; Frangioni, J.V. Renal clearance of quantum dots. *Nat. Biotechnol.* **2007**, *25*, 1165–1170.
63. Choi, H.S.; Ipe, B.I.; Misra, P.; Lee, J.H.; Bawendi, M.G.; Frangioni, J.V. Tissue- and organ-selective biodistribution of NIR fluorescent quantum dots. *Nano Lett.* **2009**, *9*, 2354–2359.
64. Choi, H.S.; Liu, W.H.; Liu, F.B.; Nasr, K.; Misra, P.; Bawendi, M.G.; Frangioni, J.V. Design considerations for tumour-targeted nanoparticles. *Nat. Nanotechnol.* **2010**, *5*, 42–47.
65. Moorthy, S.K.E.; Rousseau, O.; Viret, M.; Kociak, M. Nanoscale chemical and structural characterization of transient metallic nanowires using aberration-corrected STEM-EELS. *Nano Lett.* **2012**, *12*, 2732–2739.
66. Bao, Y.J.; Li, J.J.; Wang, Y.T.; Yu, L.; Wang, J.; Du, W.J.; Lou, L.; Zhu, Z.Q.; Peng, H.; Zhu, J.Z. Preparation of water soluble CdSe and CdSe/CdS quantum dots and their uses in imaging of cell and blood capillary. *Opt. Mater.* **2012**, *34*, 1588–1592.
67. Zhan, H.J.; Zhou, P.J.; He, Z.Y.; Tian, Y. Microwave-assisted aqueous synthesis of small-sized, highly luminescent CdSeS/ZnS core/shell quantum dots for live cell imaging. *Eur. J. Inorg. Chem.* **2012**, *2012*, 2487–2493.
68. Dong, C.Q.; Ren, J.C. Water-soluble mercaptoundecanoic acid (MUA)-coated CdTe quantum dots: One-step microwave synthesis, characterization and cancer cell imaging. *Luminescence* **2012**, *27*, 199–203.
69. Fang, X.L.; Han, M.; Lu, G.F.; Tu, W.W.; Dai, Z.H. Electrochemiluminescence of CdSe quantum dots for highly sensitive competitive immunosensing. *Sens. Actuator B Chem.* **2012**, *168*, 271–276.
70. Jiang, P.; Zhu, C.N.; Zhang, Z.L.; Tian, Z.Q.; Pang, D.W. Water-soluble Ag_2S quantum dots for near-infrared fluorescence imaging *in vivo*. *Biomaterials* **2012**, *33*, 5130–5135.
71. Zhang, Y.; Hong, G.S.; Zhang, Y.J.; Chen, G.C.; Li, F.; Dai, H.J.; Wang, Q.B. Ag_2S quantum dot: A bright and biocompatible fluorescent nanoprobe in the second near-infrared window. *ACS Nano* **2012**, *6*, 3695–3702.
72. Gu, Y.P.; Cui, R.; Zhang, Z.L.; Xie, Z.X.; Pang, D.W. Ultrasmall near-infrared Ag_2Se quantum dots with tunable fluorescence for *in vivo* imaging. *J. Am. Chem. Soc.* **2012**, *134*, 79–82.
73. Jung, J.; Kim, M.A.; Cho, J.H.; Lee, S.J.; Yang, I.; Cho, J.; Kim, S.K.; Lee, C.; Park, J.K. Europium-doped gadolinium sulfide nanoparticles as a dual-mode imaging agent for T-1-weighted MR and photoluminescence imaging.*Biomaterials* **2012**, *33*, 5865–5874.
74. Ghaderi, S.; Ramesh, B.; Seifalian, A.M. Fluorescence nanoparticles "quantum dots" as drug delivery system and their toxicity: A review. *J. Drug Target.* **2011**, *19*, 475–486.
75. Hoshino, A.; Hanada, S.; Yamamoto, K. Toxicity of nanocrystal quantum dots: The relevance of surface

modifications.*Arch. Toxicol.* **2011**, *85*, 707–720.

76. Ruan, J.; Song, H.; Qian, Q.R.; Li, C.; Wang, K.; Bao, C.C.; Cui, D.X. Her2 monoclonal antibody conjugated RNase-A-associated CdTe quantum dots for targeted imaging and therapy of gastric cancer. *Biomaterials* **2012**, *33*, 7093–7102.

77. He, M.; Huang, P.; Zhang, C.L.; Hu, H.Y.; Bao, C.C.; Gao, G.; He, R.; Cui, D.X. Dual phase-controlled synthesis of uniform lanthanide-doped $NaGdF_4$ upconversion nanocrystals via an OA/ionic liquid two-phase system for *in vivo*dual-modality imaging. *Adv. Funct. Mater.* **2011**, *21*, 4470–4477.

78. Vetrone, F.; Capobianco, J.A. Lanthanide-doped fluoride nanoparticles: Luminescence, upconversion, and biological applications. *Int. J. Nanotechnol.* **2008**, *5*, 1306–1339.

79. Wang, F.; Banerjee, D.; Liu, Y.S.; Chen, X.Y.; Liu, X.G. Upconversion nanoparticles in biological labeling, imaging, and therapy. *Analyst* **2010**, *135*, 1839–1854.

80. Koole, R.; Mulder, W.J.M.; van Schooneveld, M.M.; Strijkers, G.J.; Meijerink, A.; Nicolay, K. Magnetic quantum dots for multimodal imaging. *Wiley Interdiscip. Rev. Nanomed. Nanobiotechnol.* **2009**, *1*, 475–491.

81. Kim, J.; Piao, Y.; Hyeon, T. Multifunctional nanostructured materials for multimodal imaging, and simultaneous imaging and therapy. *Chem. Soc. Rev.* **2009**, *38*, 372–390.

82. Smith, A.A. Hematein chelates of unusual metal ions for tinctorial histochemistry. *Biotech. Histochem.* **2010**, *85*, 43–54.

83. Aoki, V.; Fukumori, L.M.I.; Freitas, E.L.; Sousa, J.X.; Perigo, A.M.; Oliveira, Z.N.P. Direct and indirect immunofluorescence. *An. Brasil. Dermatol.* **2010**, *85*, 490–499.

84. Shakes, D.C.; Miller, D.M.; Nonet, M.L. Immunofluorescence microscopy. In *Caenorhabditis Elegans: Cell Biology and Physiology*, 2nd ed.; Elsevier: Oxford, UK, 2012; Volume 107, pp. 35–66.

85. Thaler, M.; Roy, S.; Fornara, A.; Bitsche, M.; Qin, J.; Muhammed, M.; Salvenmoser, W.; Rieger, G.; Fischer, A.S.; Glueckert, R. Visualization and analysis of superparamagnetic iron oxide nanoparticles in the inner ear by light microscopy and energy filtered TEM. *Nanomed. Nanotechnol. Biol. Med.* **2011**, *7*, 360–369.

86. Hua, M.Y.; Lin, Y.C.; Tsai, R.Y.; Chen, H.C.; Liu, Y.C. A hydrogen peroxide sensor based on a horseradish peroxidase/polyaniline/carboxy-functionalized multiwalled carbon nanotube modified gold electrode. *Electrochim. Acta* **2011**, *56*, 9488–9495.

87. Zheng, L.; Jiang, F.H.; Ma, G.R.; Zhuang, Q.F.; Li, F. Hydrogen peroxide sensor based on horseradish peroxidase combined with $CaCO_3$ microspheres and gold nanoparticles. *Chem. Res. Chin. Univ.* **2011**, *27*, 875–879.

88. Zhuang, J.; Fan, K.L.; Gao, L.Z.; Lu, D.; Feng, J.; Yang, D.L.; Gu, N.; Zhang, Y.; Liang, M.M.; Yan, X.Y. *Ex vivo* detection of iron oxide magnetic nanoparticles in mice using their intrinsic peroxidase-mimicking activity. *Mol. Pharm.* **2012**, *9*, 1983–1989.

89. El-Sayed, I.; Huang, X.; Macheret, F.; Humstoe, J.O.; Kramer, R.; El-Sayed, M. Effect of plasmonic gold nanoparticles on benign and malignant cellular autofluorescence: A novel probe for fluorescence based detection of cancer. *Technol. Cancer Res. Treat.* **2007**, *6*, 403–412.

90. Intartaglia, R.; Bagga, K.; Scotto, M.; Diaspro, A.; Brandi, F. Luminescent silicon nanoparticles prepared by ultra short pulsed laser ablation in liquid for imaging applications. *Opt. Mater. Express* **2012**, *2*, 510–518.

91. Zhao, Y.Y.; Li, Y.; Li, W.; Wu, Y.Q.; Wu, L.X. Preparation, structure, and imaging of luminescent SiO_2 nanoparticles by covalently grafting surfactant-encapsulated europium-substituted polyoxometalates. *Langmuir* **2010**, *26*, 18430–18436.

92. Zhang, J.; Fu, Y.; Lakowicz, J.R. Luminescent images of single gold nanoparticles and their labeling on silica beads.*Opt. Express* **2007**, *15*, 13415–13420.

93. Ohulchanskyy, T.Y.; Roy, I.; Yong, K.T.; Pudavar, H.E.; Prasad, P.N. High-resolution light microscopy using luminescent nanoparticles. *Wiley Interdiscip. Rev. Nanomed. Nanobiotechnol.* **2010**, *2*, 162–175.

94. Santra, S.; Xu, J.S.; Wang, K.M.; Tan, W.H. Luminescent nanoparticle probes for bioimaging. *J. Nanosci. Nanotechnol.***2004**, *4*, 590–599.

95. Jung, C.Y.; Kim, J.S.; Kim, H.S.; Ha, J.M.; Kim, S.T.; Lim, H.J.; Koo, S.M. Selective surface reactions for janus ORMOSIL particles with multiple functional groups using an ordered monolayer film at liquid-liquid interface. *J. Colloid Interface Sci.* **2012**, *367*, 257–263.

96. Paschoal, A.R.; Ayala, A.P.; Pinto, R.C.F.; Paschoal, C.W.A.; Tanaka, A.A.; Boaventura, J.S.; Jose, N.M. About the SDS inclusion in PDMS/TEOS ORMOSIL: A vibrational spectroscopy and confocal raman scattering study. *J. Raman Spectrosc.* **2011**, *42*, 1601–1605.

97. Xiong, L.Q.; Chen, Z.G.; Tian, Q.W.; Cao, T.Y.; Xu, C.J.; Li, F.Y. High contrast upconversion luminescence targeted imaging *in vivo* using peptide-labeled nanophosphors. *Anal. Chem.* **2009**, *81*, 8687–8694.

98. Yu, M.X.; Li, F.Y.; Chen, Z.G.; Hu, H.; Zhan, C.; Yang, H.; Huang, C.H. Laser scanning up-conversion luminescence microscopy for imaging cells labeled with rare-earth nanophosphors. *Anal. Chem.* **2009**, *81*, 930–935.

99. Prow, T.W.; Kotov, N.A.; Lvov, Y.M.; Rijnbrand, R.; Leary, J.F. Nanoparticles, molecular biosensors, and multispectral confocal microscopy. *J. Mol. Histol.* **2004**, *35*, 555–564.

100. Zhao, S.S.; Yuan, L.; Wang, J.C.; Zhang, X.; He, Z.G.; Zhang, Q. A novel and facile approach to imaging nanoparticles transport across transwell filter grown cell mono layer in real-time and *in situ* under confocal laser scanning microscopy. *Biol. Pharm. Bull.* **2012**, *35*, 335–345.

101. Wang, T.T.; Bai, J.; Jiang, X.; Nienhaus, G.U. Cellular uptake of nanoparticles by membrane penetration: A study combining confocal microscopy with FTIR

spectroelectrochemistry. *ACS Nano* **2012**, *6*, 1251–1259.

102. Veksler, B.A.; Lemelle, A.; Kozhevnikov, I.S.; Akchurin, G.G.; Meglinski, I.V. Improving image quality in reflection confocal microscopy involving gold nanoparticles and osmotically active immersion liquids. *Opt. Spectrosc.* **2011**, *110*, 483–488.

103. Romero, G.; Rojas, E.; Estrela-Lopis, I.; Donath, E.; Moya, S.E. Spontaneous confocal raman microscopy—A tool to study the uptake of nanoparticles and carbon nanotubes into cells. *Nanoscale Res. Lett.* **2011**, *6*, 1–4.

104. Failla, A.V.; Jager, S.; Zuchner, T.; Steiner, M.; Meixner, A.J. Topology measurements of metal nanoparticles with 1 nm accuracy by confocal interference scattering microscopy. *Opt. Express* **2007**, *15*, 8532–8542.

105. Murawska, M.; Skrzypczak, A.; Kozak, M. Structure and morphology of gold nanoparticles in solution studied by TEM, SAXS and UV-Vis. *Acta Phys. Pol. A* **2012**, *121*, 888–892.

106. Sanada, T.; Okumura, K.; Murakami, C.; Oyama, T.; Isoda, A.; Katada, N. Spontaneous dispersion of gold nanoparticles loaded on USY zeolites as analyzed by XAFS, XRD, and TEM. *Chem. Lett.* **2012**, *41*, 337–339.

107. Tran, D.T.; Jones, I.P.; Preece, J.A.; Johnston, R.L.; van den Brom, C.R. TEM characterization of chemically synthesized copper-gold nanoparticles. *J. Nanopart. Res.* **2011**, *13*, 4229–4237.

108. Hayashi, T.; Muramatsu, H.; Kim, Y.A.; Kajitani, H.; Imai, S.; Kawakami, H.; Kobayashi, M.; Matoba, T.; Endo, M.; Dresselhaus, M.S. TEM image simulation study of small carbon nanotubes and carbon nanowire. *Carbon* **2006**, *44*, 1130–1136.

109. Qin, C.; Peng, L.M. Measurement accuracy of the diameter of a carbon nanotube from TEM images. *Phys. Rev. B* **2002**,*65*, 1–7.

110. Kashtiban, R.J.; Bangert, U.; Sherliker, B.; Halsall, M.P.; Harvey, A.J. Study of InGaN/GaN Quantum Dot Systems by TEM techniques and Photoluminescence Spectroscopy. Proceedings of the 16th International Conference on Microscopy of Semiconducting Materials, Oxford, UK, 17– 20 March 2010. Volume 209.

111. Kirmse, H.; Hausler, I.; Neumann, W.; Strittmatter, A.; Reissmann, L.; Bimberg, D. TEM characterization of self-organized (In,Ga)N quantum dots. In *Microscopy of Semiconducting Materials*; Springer: Berlin, Germany, 2008; Volume 120, pp. 255–258.

112. Sychugov, I.; Lu, J.; Elfstrom, N.; Linnros, J. Structural imaging of a Si quantum dot: Towards combined PL and TEM characterization. *J. Lumines.* **2006**, *121*, 353–355.

113. Jackson, C.L.; Chanzy, H.D.; Booy, F.P.; Drake, B.J.; Tomalia, D.A.; Bauer, B.J.; Amis, E.J. Visualization of dendrimer molecules by transmission electron microscopy (TEM): Staining methods and cryo-TEM of vitrified solutions.*Macromolecules* **1998**, *31*, 6259–6265.

114. Pumera, M. Carbon nanotubes contain residual metal catalyst nanoparticles even after washing with nitric acid at elevated temperature because these metal nanoparticles are sheathed by several graphene sheets. *Langmuir* **2007**, *23*, 6453–6458.

115. Alexander, S.K.; Azencott, R.; Bodmann, B.C.; Bouamrani, A.; Chiappini, C.; Ferrari, M.; Liu, X.; Tasciotti, E. SEM Image Analysis for Quality Control Of Nanoparticles. Proceedings of the 13th International Conference on Computer Analysis of Images and Patterns, Berlin, Germany, 2–4 September 2009; Volume 5702. pp. 590–597.

116. Lemine, O.M. Microstructural characterisation of alpha-Fe_2O_3 nanoparticles using, XRD line profiles analysis, FE-SEM and FT-IR. *Superlattices Microstruct.* **2009**, *45*, 576–582.

117. Strohmeier, B.R.; Bunker, K.L.; Lopano, C.L.; Marquis, J.P.; Piasecki, J.D.; Bennethum, K.E.; White, R.G.; Nunney, T.; Lee, R.J. XPS and SEM/STEM characterization of silver nanoparticles formed from the X-ray-induced and thermal reduction of silver behenate. *Microsc. Microanal.* **2009**, *15*, 1298–1299.

118. Gomez-Villalba, L.S.; Delgado, M.L.; Ruiz-Navas, E.M. High resolution transmission electron microscopy study on the development of nanostructured precipitates in Al-Cu obtained by mechanical alloying. *Mater. Chem. Phys.* **2012**, *132*, 125–130.

119. Nguyen, V.L.; Ohtaki, M.; Matsubara, T.; Cao, M.T.; Nogami, M. New experimental evidences of Pt-Pd bimetallic nanoparticles with core-shell configuration and highly fine-ordered nanostructures by high-resolution electron transmission microscopy. *J. Phys. Chem. C* **2012**, *116*, 12265–12274.

120. Rouviere, J.L.; Bougerol, C.; Amstatt, B.; Bellet-Almaric, E.; Daudin, B. Measuring local lattice polarity in AlN and GaN by high resolution Z-contrast imaging: The case of (0001) and (1(1)over-bar00) GaN quantum dots. *Appl. Phys. Lett.* **2008**, *92*, 1–5.

121. Witkowski, S.; Ruszak, M.; Sayag, C.; Pielaszek, J.; Djega-Mariadassou, G. Nanocrystalline NbC formation from mesostructured niobium oxide studied by HRTEM, SAED and *in situ* XRD. *Appl. Catal. A Gen.* **2006**, *307*, 205–211.

122. Falke, M.; Mogilatenko, A.; Neumann, W.; Brombacher, C.; Rohrmann, H.; Kratzer, M.; Albrecht, M.; Bleloch, A.; Terborg, R.; Kroemer, R.; *et al.* Element distribution in novel hedgehog-like magnetic nanostructures studied by Cs-corrected STEM-EELS and uncorrected STEM-XEDS using SDD-technology. *Microsc. Microanal.* **2009**, *15*, 1214–1215.

123. Zhao, W.W.; Graca, S. Evidence of Ti-related inclusions in an al alloy interconnecting layer for nanometer 256 MBit DRAM semiconductor devices characterized by TEM, STEM, EELS elemental mapping, and XEDS linescan. In*Quantum Confined Semiconductor Nanostructures*; Materials Research Society: Warrendale, PA, USA, 2003; Volume 737, pp. 649–654.

124. Estrade, S.; Yedra, L.; Lopez-Ortega, A.; Estrader, M.; Salazar-Alvarez, G.; Baro, M.D.; Nogues, J.; Peiro, F. Distinguishing the core from the shell in MnOx/MnOy and FeOx/MnOx core/shell nanoparticles through quantitative electron energy loss spectroscopy (EELS) analysis. *Micron* **2012**, *43*, 30–36.

125. Stender, A.S. Plasmonic behavior of quantum-size metallic nanoparticles as investigated with STEM-EELS. *MRS Bull.* **2012**, *37*, 543–544.

126. Brandt, J.; Klemenz, A.; Raum, K.; Seidler, S. Scanning acoustic microscopy for micromeasurement of elastomechanical bone properties—Comparison with nanoindentation results. In *Acoustical Imaging*; Springer: Berlin, Germany, 2002; Volume 26, pp. 53–59.

127. Rupin, F.; Saied, A.; Dalmas, D.; Peyrin, F.; Haupert, S.; Raum, K.; Barthel, E.; Boivin, G.; Laugier, P. Assessment of microelastic properties of bone using scanning acoustic microscopy: A face-to-face comparison with nanoindentation. *Jpn. J. Appl. Phys.* **2009**, *48*, 1–5.

128. Burke, K.B.; Stapleton, A.J.; Vaughan, B.; Zhou, X.J.; Kilcoyne, A.L.D.; Belcher, W.J.; Dastoor, P.C. Scanning transmission X-ray microscopy of polymer nanoparticles: Probing morphology on sub-10 nm length scales. *Nanotechnology* **2011**, *22*, 1–6.

129. Burke, K.B.; Vaughn, B.; Stapleton, A.J.; Belcher, W.J.; Zhou, X.J.; Kilcoyne, A.L.D.; Dastoor, P.C. Scanning transmission X-ray microscopy of polymer nanoparticles: Probing morphology on sub-10 nm length scales. *Abstr. Paper. Am. Chem. Soc.* **2011**, *242*, 1.

130. Bartling, S.; Barke, I.; Sell, K.; Polei, S.; von Oeynhausen, V.; Meiwes-Broer, K.H. Structure of AuSi nanoparticles on Si(111) from reflection high-energy electron diffraction and scanning tunneling microscopy. *Eur. Phys. J. D* **2011**, *63*, 225–230.

131. Rim, K.T.; Eom, D.; Liu, L.; Stolyarova, E.; Raitano, J.M.; Chan, S.W.; Flytzani-Stephanopoulos, M.; Flynn, G.W. Charging and chemical reactivity of gold nanoparticles and adatoms on the (111) surface of single-crystal magnetite: A scanning tunneling microscopy/spectroscopy study. *J. Phys. Chem. C* **2009**, *113*, 10198–10205.

132. Zhou, Y.H.; Zhou, J. Growth and sintering of Au-Pt nanoparticles on oxidized and reduced CeOx(111) thin films by scanning tunneling microscopy. *J. Phys. Chem. Lett.* **2010**, *1*, 609–615.

133. Park, K.W.; Dasika, V.D.; Nair, H.P.; Crook, A.M.; Bank, S.R.; Yu, E.T. Conductivity and structure of ErAs nanoparticles embedded in GaAs pn junctions analyzed via conductive atomic force microscopy. *Appl. Phys. Lett.* **2012**, *100*, 1–6.

134. Romanchenko, A.S.; Mikhlin, Y.L.; Makhova, L.V. Investigation of gold nanoparticles immobilized on the surface of pyrite by scanning probe microscopy, scanning tunneling spectroscopy, and X-ray photoelectron spectroscopy. *Glass Phys. Chem.* **2007**, *33*, 417–421.

135. Silva-Pinto, E.; Gomes, A.P.; Pinheiro, C.B.; Ladeira, L.O.; Pimenta, M.A.; Neves, B.R.A. Controlled growth and positioning of metal nanoparticles via scanning probe microscopy. *Langmuir* **2009**, *25*, 3356–3358.

136. Vakarelski, I.U.; Brown, S.C.; Moudgil, B.M.; Higashitani, K. Nanoparticle-terminated scanning probe microscopy tips and surface samples. *Adv. Powder Technol.* **2007**, *18*, 605–614.

137. Hubert, C.; Amrani, H.; Khan, M.A.; Vocanson, F.; Destouches, N. Electrical growth of metallic nanoparticles in mesoporous silica films using atomic force microscopy. *Appl. Phys. Lett.* **2012**, *100*, 1–6.

138. Kent, R.D.; Vikesland, P.J. Controlled evaluation of silver nanoparticle dissolution using atomic force microscopy. *Environ. Sci. Technol.* **2012**, *46*, 6977–6984.

139. Onochi, Y.; Nakamura, M.; Hoshi, N. Atomic force microscopy of the dissolution of cubic and tetrahedral Pt nanoparticles in electrochemical environments. *J. Phys. Chem. C* **2012**, *116*, 15134–15140.

140. Roshanak, G.Z.; Rahele, J.; Zahra, S.; Sarmad, N.M.; Masoud, S.M.; Mehdi, B.M.M.; Mohammad, M.A. Phase characterization of TiO_2 nanoparticles by transmission electron microscopy (TEM). *Clin. Biochem.* **2012**, *44*, S215–S216.

141. Chang, L.Y.; Lazar, S.; Baranova, E.A.; Bock, C.; Botton, G.A. Quantitative characterisation of surface defects and composition on PtRu nanoparticles using aberration-corrected TEM/STEM. *Microsc. Microanal.* **2009**, *15*, 1416–1417.

142. Shibata, M.; Shklyaev, A.A.; Ichikawa, M. Observation and nucleation control of Ge nanoislands on Si(111) surfaces using scanning reflection electron microscopy. *J. Electron Microsc.* **2000**, *49*, 217–223.

143. Zhang, R.Y.; Wei, Y.; Nagahara, L.A.; Amlani, I.; Tsui, R.K. The contrast mechanism in low voltage scanning electron microscopy of single-walled carbon nanotubes. *Nanotechnology* **2006**, *17*, 272–276.

144. Roberts, W.S.; Lonsdale, D.J.; Griffiths, J.; Higson, S.P.J. Advances in the application of scanning electrochemical microscopy to bioanalytical systems. *Biosens. Bioelectron.* **2007**, *23*, 301–318.

145. Torisawa, Y.S.; Kaya, T.; Takii, Y.; Oyamatsu, D.; Nishizawa, M.; Matsue, T. Scanning electrochemical microscopy-based drug sensitivity test for a cell culture integrated in silicon microstructures. *Anal. Chem.* **2003**, *75*, 2154–2158.

146. Lacina, K.; Skladal, P.; Nagy, G. The scanning electrochemical microscopy. *Chem. Listy* **2012**, *106*, 253–263.

147. Edwards, M.A.; Martin, S.; Whitworth, A.L.; Macpherson, J.V.; Unwin, P.R. Scanning electrochemical microscopy: Principles and applications to biophysical systems. *Physiol. Meas.* **2006**, *27*, R63–R108.

148. Liu, B.; Cheng, W.; Rotenberg, S.A.; Mirkin, M.V. Scanning electrochemical microscopy of living cells—Part 2. Imaging redox and acid/basic reactivities. *J. Electroanal. Chem.* **2001**, *500*, 590–597.

149. Liebetrau, J.M.; Miller, H.M.; Baur, J.E. Scanning electrochemical microscopy of model neurons: Imaging and real-time detection of morphological changes. *Anal. Chem.* **2003**, *75*, 563–571.

150. Hahn, D.W.; Omenetto, N. Laser-induced breakdown spectroscopy (LIBS), part I: Review of basic diagnostics and plasma-particle interactions: Still-challenging issues within the analytical plasma community. *Appl. Spectrosc.* **2010**, *64*, 335A–366A.

151. Galiova, M.; Kaiser, J.; Novotny, K.; Novotny, J.; Vaculovic, T.; Liska, M.; Malina, R.; Stejskal, K.; Adam,

V.; Kizek, R. Investigation of heavy-metal accumulation in selected plant samples using laser induced breakdown spectroscopy and laser ablation inductively coupled plasma mass spectrometry. *Appl. Phys. A Mater. Sci. Process.* **2008**, *93*, 917–922.

152. Hahn, D.W.; Omenetto, N. Laser-induced breakdown spectroscopy (LIBS), part II: Review of instrumental and methodological approaches to material analysis and applications to different fields. *Appl. Spectrosc.* **2012**, *66*, 347–419.

153. Porizka, P.; Prochazka, D.; Pilat, z.; Krajcarova, L.; Kaiser, J.; Malina, R.; Novotny, J.; Zemanek, P.; Jezek, J.; Sery, M.; *et al.* Application of laser-induced breakdown spectroscopy to the analysis of algal biomass for industrial biotechnology.*Spectroc. Acta Pt. B Atom. Spectr.* **2012**, *74–75*, 169–176.

154. Vitkova, G.; Novotny, K.; Prokes, L.; Hrdlicka, A.; Kaiser, J.; Novotny, J.; Malina, R.; Prochazka, D. Fast identification of biominerals by means of stand-off laser-induced breakdown spectroscopy using linear discriminant analysis and artificial neural networks. *Spectroc. Acta Pt. B Atom. Spectr.* **2012**, *73*, 1–6.

155. Babushok, V.I.; DeLucia, F.C.; Gottfried, J.L.; Munson, C.A.; Miziolek, A.W. Double pulse laser ablation and plasma: Laser induced breakdown spectroscopy signal enhancement. *Spectroc. Acta Pt. B Atom. Spectr.* **2006**, *61*, 999–1014.

156. De Giacomo, A.; Dell'Aglio, M.; Bruno, D.; Gaudiuso, R.; De Pascale, O. Experimental and theoretical comparison of single-pulse and double-pulse laser induced breakdown spectroscopy on metallic samples. *Spectroc. Acta Pt. B Atom. Spectr.* **2008**, *63*, 805–816.

157. Uebbing, J.; Brust, J.; Sdorra, W.; Leis, F.; Niemax, K. Reheating of a laser-produced plasma by a 2nd pulse laser. *Appl. Spectrosc.* **1991**, *45*, 1419–1423.

158. Mowery, M.D.; Sing, R.; Kirsch, J.; Razaghi, A.; Bechard, S.; Reed, R.A. Rapid at-line analysis of coating thickness and uniformity on tablets using laser induced breakdown spectroscopy. *J. Pharm. Biomed. Anal.* **2002**, *28*, 935–943.

159. St-Onge, L.; Kwong, E.; Sabsabi, M.; Vadas, E.B. Quantitative analysis of pharmaceutical products by laser-induced breakdown spectroscopy. *Spectroc. Acta Pt. B-Atom. Spectr.* **2002**, *57*, 1131–1140.

160. Ferrari, M. Cancer nanotechnology: Opportunities and challenges. *Nat. Rev. Cancer* **2005**, *5*, 161–171.

161. Cho, K.J.; Wang, X.; Nie, S.M.; Chen, Z.; Shin, D.M. Therapeutic nanoparticles for drug delivery in cancer. *Clin. Cancer Res.* **2008**, *14*, 1310–1316.

162. Mishra, B.; Patel, B.B.; Tiwari, S. Colloidal nanocarriers: A review on formulation technology, types and applications toward targeted drug delivery. *Nanomed.-Nanotechnol. Biol. Med.* **2010**, *6*, 9–24.

163. Peer, D.; Karp, J.M.; Hong, S.; FaroKhzad, O.C.; Margalit, R.; Langer, R. Nanocarriers as an emerging platform for cancer therapy. *Nat. Nanotechnol.* **2007**, *2*, 751–760.

164. Zhang, L.F.; Chan, J.M.; Gu, F.X.; Rhee, J.W.; Wang, A.Z.; Radovic-Moreno, A.F.; Alexis, F.; Langer, R.; Farokhzad, O.C. Self-assembled lipid-polymer hybrid nanoparticles: A robust drug delivery platform. *ACS Nano* **2008**, *2*, 1696–1702.

165. Kwon, G.S. Polymeric micelles for delivery of poorly water-soluble compounds. *Crit. Rev. Ther. Drug Carr. Syst.* **2003**,*20*, 357–403.

166. Walko, C.M.; McLeod, H. Pharmacogenomic progress in individualized dosing of key drugs for cancer patients. *Nat. Clin. Pract. Oncol.* **2009**, *6*, 153–162.

167. Slingerland, M.; Guchelaar, H.J.; Gelderblom, H. Liposomal drug formulations in cancer therapy: 15 years along the road. *Drug Discov. Today* **2012**, *17*, 160–166.

168. Puri, A.; Loomis, K.; Smith, B.; Lee, J.H.; Yavlovich, A.; Heldman, E.; Blumenthal, R. Lipid-based nanoparticles as pharmaceutical drug carriers: From concepts to clinic. *Crit. Rev. Ther. Drug Carr. Syst.* **2009**, *26*, 523–580.

169. Patra, C.R.; Bhattacharya, R.; Mukhopadhyay, D.; Mukherjee, P. Fabrication of gold nanoparticles for targeted therapy in pancreatic cancer. *Adv. Drug Deliv. Rev.* **2010**, *62*, 346–361.

170. Acharya, S.; Dilnawaz, F.; Sahoo, S.K. Targeted epidermal growth factor receptor nanoparticle bioconjugates for breast cancer therapy. *Biomaterials* **2009**, *30*, 5737–5750.

171. Misra, R.; Acharya, S.; Sahoo, S.K. Cancer nanotechnology: Application of nanotechnology in cancer therapy. *Drug Discov. Today* **2010**, *15*, 842–850.

172. Wang, J.Q.; Sui, M.H.; Fan, W.M. Nanoparticles for tumor targeted therapies and their pharmacokinetics. *Curr. Drug Metab.* **2010**, *11*, 129–141.

173. Adiseshaiah, P.P.; Hall, J.B.; McNeil, S.E. Nanomaterial standards for efficacy and toxicity assessment. *Wiley Interdiscip. Rev. Nanomed. Nanobiotechnol.* **2010**, *2*, 99–112.

174. Jain, K.K. Nanobiotechnology-based cancer diagnosis preface. *Clin. Lab. Med.* **2012**, *32*, IX–X.

175. Youns, M.; Hoheisel, J.D.; Efferth, T. Therapeutic and diagnostic applications of nanoparticles. *Curr. Drug Targets* **2011**,*12*, 357–365.

176. Chatterjee, D.K.; Fong, L.S.; Zhang, Y. Nanoparticles in photodynamic therapy: An emerging paradigm. *Adv. Drug Deliv. Rev.* **2008**, *60*, 1627–1637.

177. Samia, A.C.S.; Chen, X.B.; Burda, C. Semiconductor quantum dots for photodynamic therapy. *J. Am. Chem. Soc.* **2003**,*125*, 15736–15737.

178. Morosini, V.; Bastogne, T.; Frochot, C.; Schneider, R.; Francois, A.; Guillemin, F.; Barberi-Heyo, M. Quantum dot-folic acid conjugates as potential photosensitizers in photodynamic therapy of cancer. *Photochem. Photobiol. Sci.* **2011**, *10*, 842–851.

179. Kam, N.W.S.; O'Connell, M.; Wisdom, J.A.; Dai, H.J. Carbon nanotubes as multifunctional biological transporters and near-infrared agents for selective cancer cell destruction. *Proc. Natl. Acad. Sci. USA* **2005**, *102*, 11600–11605.

180. Madani, S.Y.; Naderi, N.; Dissanayake, O.; Tan, A.; Seifalian, A.M. A new era of cancer treatment: Carbon

nanotubes as drug delivery tools. *Int. J. Nanomed.* **2011**, *6*, 2963–2979.

181. Carter, J.D.; Cheng, N.N.; Qu, Y.Q.; Suarez, G.D.; Guo, T. Nanoscale energy deposition by X-ray absorbing nanostructures. *J. Phys. Chem. B* **2007**, *111*, 11622–11625.

182. Juzenas, P.; Chen, W.; Sun, Y.P.; Coelho, M.A.N.; Generalov, R.; Generalova, N.; Christensen, I.L. Quantum dots and nanoparticles for photodynamic and radiation therapies of cancer. *Adv. Drug Deliv. Rev.* **2008**, *60*, 1600–1614.

183. Zheng, Y.; Hunting, D.J.; Ayotte, P.; Sanche, L. Radiosensitization of DNA by gold nanoparticles irradiated with high-energy electrons. *Radiat. Res.* **2008**, *169*, 19–27.

184. Herold, D.M.; Das, I.J.; Stobbe, C.C.; Iyer, R.V.; Chapman, J.D. Gold microspheres: A selective technique for producing biologically effective dose enhancement. *Int. J. Radiat. Biol.* **2000**, *76*, 1357–1364.

185. Huang, P.; Bao, L.; Zhang, C.L.; Lin, J.; Luo, T.; Yang, D.P.; He, M.; Li, Z.M.; Gao, G.; Gao, B.; *et al.* Folic acid-conjugated silica-modified gold nanorods for X-ray/CT imaging-guided dual-mode radiation and photo-thermal therapy. *Biomaterials* **2011**, *32*, 9796–9809.

186. Chen, F.; Huang, P.; Zhu, Y.J.; Wu, J.; Zhang, C.L.; Cui, D.X. The photoluminescence, drug delivery and imaging properties of multifunctional $Eu3^{+}/Gd3^{+}$ dual-doped hydroxyapatite nanorods. *Biomaterials* **2011**, *32*, 9031–9039.

187. Pan, B.F.; Cui, D.X.; Sheng, Y.; Ozkan, C.G.; Gao, F.; He, R.; Li, Q.; Xu, P.; Huang, T. Dendrimer-modified magnetic nanoparticles enhance efficiency of gene delivery system. *Cancer Res.* **2007**, *67*, 8156–8163.

188. Vilarino-Varela, M.J.; Taylor, A.; Rockall, A.G.; Reznek, R.H.; Powell, M.E.B. A verification study of proposed pelvic lymph node localisation guidelines using nanoparticle-enhanced magnetic resonance imaging. *Radiother. Oncol.* **2008**,*89*, 192–196.

189. Singhal, S.; Nie, S.M.; Wang, M.D. Nanotechnology applications in surgical oncology. *Ann. Rev. Med.* **2010**, *61*, 359–373.

CHAPTER 13

COUPLING OF TRANSIENT NEAR INFRARED PHOTONIC WITH MAGNETIC NANOPARTICLE

Vidya Sagar, V. S. R.Atluri, A.Tomitaka, P. Shah, A. Nagasetti, S. Pilakka-Kanthikeel, N. El-Hage, A. McGoron, Y.Takemura & M. Nair

INTRODUCTION

Magnetic nanoparticles (MNPs) have been intensively investigated for various biomedical applications which includes therapeutic drugs targeting, gene delivery, bio-separation of biological entities, hyperthermia induced destruction of cells and tumors, magnetic resonance imaging (MRI), stem cell tracking, tissue repair, bio-sensing, etc.[1,2,3,4,5,6,7,8,9,10,11,12,13]. MNPs possess a distinct advantage over other nanocarriers because of their inherent superparamagnetism which allows control over its magnetization and therefore its movement/speed can be regulated. By applying remote, non-invasive magnetic forces of required intensity at the desired site it is possible to achieve tissue/cell-specific targeting with MNPs. Other characteristics of MNPs which make them popular are feasibility in production[14] that they can be used as a contrast agent for MRI[4,14], and their amphoterism in aqueous medium[15,16]. In aqueous solution, MNPs develop a positive or negative charge at the surface-water interface in a pH-dependent manner which allows ionic bonding of varieties of molecules at their surface[17]. Higher immobilization of molecules on MNPs can be achieved by coating or functionalization of MNPs with various surfactants[4]. Thus, the well-defined and rigid structures of MNPs serve as a solid binding platform for various ligands of diagnostic or therapeutical importance. MNPs can also be encapsulated in liposomes to create magnetoliposomes[18]. This can prevent MNPs bound drugs from direct exposure to phagocytic cells of reticuloendothelial system and other detrimental enzymatic activity in blood circulation and, in turn, physiological bioavailability of therapeutics can be significantly increased. Importantly, external control over the movement of MNPs exponentially improves the ability of the nanocarrier to reach the target site by reducing its peripheral circulation time compared to other nanocarriers[3]. Moreover, the iron content in MNPs–in particular the magnetite and maghemite- can be readily metabolized by cellular regulation using the transferrin pathway. This makes MNPs easily degradable and able to pass in and out of cells across the plasma membrane[19]. Thus, MNPs within the permissible dose limit should have non-significant safety concerns and can be extremely suitable for *in vivo* applications[20].

In the past decade, several studies have been carried out on the development of stimuli responsive materials or techniques to design stimuli-responsive nano-devices for biomedical applications. These devices can be sensitive to a range of stimuli, which include change in pH, glutathione concentration or enzyme concentration, changes associated with the pathological situation, and extracorporeal physical stimuli via photo-, thermo- or ultrasound-targeting. These stimuli cause specific protonation, hydrolytic cleavage, molecular or supramolecular conformational changes in the material to exert the desired effect[21,22,23]. Laser-initiated photo-targeting has shown tremendous potential for cancer therapy, gene delivery, imaging, and on-demand drug delivery[24,25,26,27]. In most cases phototargeting is achieved by hybridizing a light source with other existing techniques. As such light sensitive hydrogels[28,29,30] and liposomes[31] have been discovered in recent years. Some studies used light in the UV and visible spectral range for optoporation of macromolecules in cells[32,33,34]. However, light in the UV-visible range potentiates damage to the cellular organelles, DNA and proteins. Moreover, deeper penetration of light in the UV-visible wavelength into *in vivo*

Citation: Vidya Sagar, V. S. R.Atluri, A.Tomitaka, P. Shah, A. Nagasetti, S. Pilakka-Kanthikeel, N. El-Hage, A. McGoron, Y.Takemura & M. Nair, Coupling of transient near infrared photonic with magnetic nanoparticle, doi:10.1038/srep29792.

tissues or organs is not possible due to higher scattering and absorption. Recently, near infrared (NIR) region light in the wavelength range of 700–1000 nm has been experimented for several biological applications. This wavelength range is referred to as transparency "therapeutic window" because of deeper *in vivo* penetration and minimum absorption and scattering in compare to UV-VIS light[35,36,37,38,39]. Nonetheless, second (1100–1350 nm) and third (1600–1870 nm) NIR spectral window may be more superior[40]. Different energy levels of NIR light beam are applied from femtoseconds to several minutes as per the necessity of application[19,26,30,37,40,41,42].

NIR phototargeting, in conjugation with MNPs, has largely been restricted for peripheral cancer therapy by photothermal effects where targeted irradiation is applied for more than 15 minutes[19,42,43,44]. Considering the sophistication and interdependence of brain cells networks in driving nuances of body physiology a damaging thermal effect should be minimized or avoided while targeting brain. As such, transient or intermittent NIR exposure to brain cells can be more accommodating for their physiological ambience. A recent study suggests MNPs-NIR assisted improved gene delivery with no cytotoxicity[26]. Similarly, a magnetic/NIR-responsive on-demand, targeted drug delivery and multicolor imaging system have been invented[27]. Again, applications of this unique (combined) approach have been limited to peripheral body regions. Almost all neurological disorders remain untreated, primarily due to lack of a technique that can deliver therapeutic devices for disease diagnosis and/or treatment across the impenetrable blood-brain barrier (BBB) as and when required. Several ongoing studies showing safe use of MNPs for imaging diagnosis, drug delivery, etc. in the brain region remain at the pre-clinical stage. An improvement by combining magnetic and NIR-responsive techniques may be beneficial in this regard. The application can range from brain cell specific gene delivery, imaging and on-demand drug targeting to magnetized photobiomodulation for treating various neuro-disorders. Nonetheless, physiological implications of combined MNP/NIR phototargeting on different brain cells need to be examined. As such, we studied the effect of NIR exposure on different brain cells with or without MNP treatment. Herein, for the first time, we report that short exposure of NIR light with a wavelength of 808 nm does not affect the viability and growth behavior of three major brain cells, namely, human primary astrocytes, the SKNMC neuronal cell line, and CHME microglia cell lines. Also, combined MNP/NIR phototargeting did not affect the spinal plasticity of SKNMC neuroepithelioma cells. Thus, we believe that this combined approach can be of safe for their potential in varieties of CNS related biomedical application.

MATERIALS AND METHODS

Synthesis of magnetic nanoparticles

The co-precipitation method was used for synthesis of magnetic nanoparticles[18]. Briefly, 3 ml $FeCl_3$ (0.487 g dissolved in 2 mol l^{-1} HCl) was thoroughly mixed in 10.33 ml H_2O and subsequent drop-by-drop addition of 2 ml Na_2SO_3 (0.126 g in 2 ml of water) to this solution was stir-mixed within a minute. Gradually the reaction solution turns from yellow to red-light yellow. Now 80 ml of ammonium hydroxide solution (0.80 mol^{-1}) is added with vigorous stirring which lead to black precipitation. The solution is kept under continuous stirring for additional 30 minutes. The resultant MNPs crystals are washed and suspended in H_2O which measures a pH of 7.5. The stability of MNPs can be achieved by adjusting the pH to 3.0 and subsequent heating at 90 °C and 100 °C for 5 and 60 min, respectively. All process was performed at room temperature.

Characterization of magnetic nanoparticles

Structural conformation of synthesized MNPs was verified using Bruker GADDS/D8 X-ray diffraction system with Apex Smart CCD Detector and Mo direct-drive rotating anode (50 kV; 20 mA). Diffraction patterns were analyzed and indexed using ICDD PDF 2015 database and Match software. Further, to confirm the elemental composition of MNPs, energy dispersive spectroscopy (EDS) was conducted in scanning electron microscopy (JEOL JSM 5900LV) at 15 kV and working distance of 10 mm.

The hydrodynamic radius and size distribution of MNPs were analyzed using dynamic laser scattering (DLS) (90 Plus particle size analyzer, Brookhaven Instruments, USA) at room temperature. Further, to examine the original crystal size, transmission electron microscopy (TEM) analysis was performed with the JEOL 1010 Transmission Electron microscope operated at 100 kV. The magnetization curve of MNPs was measured using vibrating sample magnetometer (VSM-3, Toei Kogyo, Tokyo, Japan) equipped with an electromagnet (TEM-WFR7, Toei Kogyo, Tokyo, Japan) and a gaussmeter (Model 421, Lake Shore Cryotronics, Inc.). The measurement was conducted at room temperature with a maximum field of 780 kA/m.

The Agilent 8453 UV-Visible Spectrometer with Quartz-1 cm path length was used for evaluating absorbance of MNPs from 200 to 1000 nm wavelength.

Cell Culture

SK-N-MCs, a neuroepithelioma cell line derived from a metastatic supra-orbital human brain tumor, were cultured in minimum essential medium (MEM). MEM was supplemented with 10% fetal bovine serum (FBS), 100 U/ml penicillin, and 100 mg/ml streptomycin (Gibco-BRL, Gaithersburg, MD). Cells were incubated at 37 °C in a 5%CO_2 incubator. Similarly, human primary astrocytes (HA) and CHME-5 human microglia cells were cultivated as per provider's recommendations.

NIR Exposure

NIR exposure was performed with collimated NIR Laser Module source (RLDH808–1200-5, Roithner Laserthchnik Gmbh, Vienna, Austria) as described by Tang and McGoron[45]. 808 nm NIR with ~1.5 W/cm^2 power density was focused for 2 minutes on brain cells (human primary astrocytes, SKNMC neuronal cells, and CHME-5 glia cells) cultured in 96 well plates in the presence or absence of MNPs (50 μg MNPs/ml). This fixed laser source has spot size of 5 mm which approximately covers central 80% cells of a well in 96 well culture plates. Cells in the periphery of a well are exposed due to potential beam-spread upon surface hitting of NIR during irradiation and as such whole well is illuminated (Supplementary Fig. 3). Cells were cultured in alternate wells so that potential crosstalk of NIR to a specific well was minimized for an adjacent well. Cells were pre-treated with MNPs 12 hrs before NIR targeting. Temperature of a specific well was measured using a thermocouple (0.22 mm diameter) for the entire 0–2 min of NIR exposure.

Cell viability assay

The MTT (Thiazolyl blue tetrazolium bromide) cell proliferation assay was performed as described previously[17,18,46]. Briefly, cells after NIR treatment were re-incubated at 37 °C for 3–6 hours (to imitate a real-time situation where cells will be under the natural condition post NIR treatment). Cells from different experimental groups were given a 200 µl media change with 20 µl MTT solution added and gently rocked in the dark at room temperature for 2–3 hrs. One volume of STOP solution containing 20% SDS in 50% dimethyl formamide was added to the rocking cell suspension in MTT solution and further gently rocked in the dark at room temperature for 1–2 hrs. The cell suspension was centrifuged at 2000 rpm for 10 minutes and the supernatant was collected for the optical density determination of the solubilized formazan at 550 nm using Spectronic Genesys Bio10 spectrophotometer. The optical density of formazan in each treatment groups is directly proportional to the cell viability.

Cell growth resistance/impedance (Ω) measurement

Astrocytes growth resistance/impedance was measured with the help of the Electric cell-substrate impedance sensing instrument (model 1600RE, Applied Biophysics, USA) using 8W10E PET chips (Applied Biophysics), containing 8 cell culture wells[47,48]. Each well of the chip contains 10 working electrodes (250 µm diameter) embedded in parallel on a gold connection pad and all wells share a common reference electrode. Astrocytes cultured with or without MNPs treatment were photo-targeted with NIR light and seeded in chip wells (5×10^4 cells/well). Both, the working and the reference electrodes were connected to a phase-sensitive lock-in amplifier through a 1 MΩ resistor before applying the AC signal. An electric potential of 1 V at 4 KHz was used for cell growth resistance measurements at 37 °C in a humidified incubator for 0–10 hrs.

Confocal microscopy and Characterization of neuro-spine density

Membrane staining of neuronal cells for confocal microscopy and measurement of spine density was performed according to the method adopted from Atluri *et al.*[49]. Cells were imaged using TCS SP2 Confocal Laser Scanning Microscope (Leica Microsystems, Germany) at 488 nm using 60X oil immersion objectives and 2.5X confocal electronic zoom.

Biostatistical analysis

Data in different figures are presented as mean ± standard error of three experiments (n = 3). Student's t-test was performed to compare means of two groups using GraphPad prism6 (San Diego, Ca) and *P*values ≤0.05 were considered as significant.

RESULTS AND DISCUSSION

Short-term MNPs-NIR exposure does not affect the temperature of cell culture ambience

We herein investigated the combined effect of MNPs and NIR treatment on growth dynamics of three major brain cells *i.e.* astrocytes, microglia and neuronal cells. MNPs were synthesized using the co-precipitation method which is regarded as one of the most efficient ways to prepare MNPs. In the co-precipitation method, either Na_2SO_3 or $FeSO_4$ is used to reduce ferrous ion from $FeCl_3$. While $FeSO_4$ based reduction results in rod-shaped nanoparticles, the relatively gentle reduction ability of Na_2SO_3 in an aqueous solution produces round MNPs. The primary product of this reduction reaction is magnetite, which can be further acid-oxidized at 100 °C for maghemite as the more chemically stable end product. Nonetheless, both magnetite and maghemite has similar magnetic properties[50]. The crystalline structure and phase purity of synthesized nanoparticles were evaluated using x-ray diffraction spectroscopy which shows magnetite/maghemite specific diffraction peaks (220, 311, 400, 511, and 440 planes; JCPDS 00-089-0691) (Fig. 1A). Further, Energy Dispersive X-Ray Spectorscopy (EDS) analysis confirmed FeO specific elemental composition. Observation of both FeL and FeK peaks for Fe_3O_4 in EDS is an expected outcome because many elements can be observed with more than one shell in a specific energy range (Supplementary Fig. 1). The polydispersity index of 0.19 in DLS suggests a very narrow size distribution of these particles. Nonetheless, average hydrodynamic size was estimated as 127 nm (Fig. 1B) which is higher than the TEM size of <15 nm (Insert image in Fig. 1A). This size difference between TEM and DLS can be attributed to dried and aqueous solution of particles used during respective analysis. The hydrodynamic size of nanoparticles in colloidal suspension is always greater than TEM due to adsorbed aqueous molecules. Water molecules (H-O-H) influence the surface charge ($Fe\text{-}OH + H^+ = Fe\text{-}OH_2^+/Fe\text{-}OH = Fe\text{-}O^- + H^+$) which allows binding of molecules (e.g. drugs, nucleic acid, etc.) to MNPs via ionic interaction for various biomedical applications. The magnetic hysteresis loops of these particles display high saturation magnetism at room temperature when measured between +780 and −780 Oe (Oersted). The magnetization curves displayed typical superparamagnetic behavior with no hysteresis and zero coercivity (Fig. 1C). The superparamagnetism can be utilized to manipulate MNPs movement for target-specific delivery by an external, non-invasive magnetic force and MNPs of this size possess several advantages for neurotargeting. Such as, it is possible to manipulate and target at the subcellular organelles levels[51]. Their higher surface to volume ratio increases loading efficiency of therapeutical agents on MNPs surfaces. More importantly, MNPs between 10–70 nm can readily penetrate capillary vessels17 and therefore it serves as a compatible therapeutical carrier that can transmigrate across the tightly junctioned brain microvascular endothelial cells (BMECs) of BBB along the capillary lining throughout the cerebral microvasculature. Even with coatings of NIR responsive materials, such as hydrogels, liposomes, etc. on MNPs, the BBB transmigration ability of the nanoformulations is expected to be maintained. Magnetoliposomes up to 150 nm size have been reported to cross *in vitro* BBB by the application of 0.3 Tesla magnetic fields[52]. It is expected that the application of a magnetic field dissipate energy into magnetic particles resulting in a thermal impact on the targeted area. Nonetheless, the heating power of MNPs depends on the intensity and frequency of alternating magnetic field and MNPs size and geometry[53,54,55]. Particles within 30 nm size exert zero to minimum hyperthermia due to the magnetic field[48]. Also, many investigations have shown that MNP transmigration across the BBB using an external magnetic force has no adverse effect on the integrity of tight junctions[9,10,11,12,13,17,18,46,56,57].

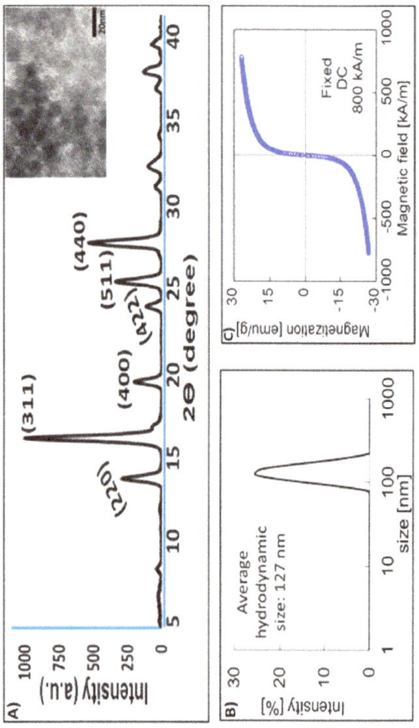

(**A**) XRD spectrum of <20 nm MNPs (in TEM insert image at top right hand side) showing magnetite-specific characteristics plane. (**B**) Dynamic laser scattering (DLS) measurement of hydrodynamic size distribution of MNPs shows an average colloidal size of nanoparticles is 127 nm. (**C**) Magnetic hysteresis loop of MNPs showing superpara-magnetism i.e. zero coercivity at room temperature.

Figure 1: Characterization of magnetite nanoparticles.

NIR-based irradiation in the presence of MNPs was achieved using the collimated NIR laser delivery system as shown in Fig. 2A. The system has been previously used to investigate the *in vitro* effect of the combined therapeutic modalities of chemotherapy and hyperthermia to cancer cell lines[45]. The NIR spectrum in the range of 800–1000 nm wavelength, itself, does not induce DNA damage and is non-toxic to tissues. However, heat may be released following exposure when a material that absorbs energy in the NIR region is present[58,59]. As such, NIR phototargeting in the presence of MNPs may lead to a photothermal effect which primarily depends on the duration and dose of NIR exposure and nanoparticle cluster density[26,44,42,53]. Studies suggest that *in vivo* cytotoxic effects due to hyperthermia are exerted only when NIR with higher power (>3 Wm^{-2}) is exposed for several minutes on a daily basis for several weeks in the presence of MNPs53. In fact, a study by Chu *et al.*[42] shows that NIR exposure for less than 3 minutes on MNP treated cells does not induce any damage or adverse effect[42]. Short-term NIR exposure in conjugation with MNPs can be successfully applied for the gene delivery, imaging, and controlled drug delivery[26,27]. As a first step towards assessing the efficacy of this approach for the brain region, we evaluated the effect on temperature change of an *in vitro* cell culture system upon short-term NIR light exposure (up to 2 minutes). The 808 nm NIR laser has been demonstrated to have superior CNS tissue penetration compared to light of other wavelengths such as 606 nm and 940 nm60. As such we selected this wavelength for our experiments. In clinical NIR based transcranial photobiomodulation ≤2 minute exposure of 808 nm NIR exposure has been shown efficacious in treating major depressive disorder[61,62]. Schematic in Fig. 2A shows the cell culture plate in the presence or absence of MNPs, which were homogenously exposed to 808 nm NIR with power ~1.5 Wcm^{-2}. Typical power intensities reported for photothermal treatment range from 1 to about 100 Wcm^{-2}[63]. Since no previous reports suggest combine use of MNPs and NIR for brain cells, a minimum recommended power density was selected considering higher physiological sensitivity of brain cells in compare to peripheral cells. To simulate the original cell culture condition, the temperature was maintained at 37 °C using the heated platform upon which the culture plates were placed during the NIR laser exposure and increase or decrease in temperature was measured using a thermocouple. As shown in Fig. 2B, NIR laser exposure on the cell culture system without MNPs showed a temperature rise of 0.46 ± 0.152 °C and that with MNPs showed an increase of 0.83 ± 0.057 °C. MNPs absorption at 808 nm is near to bottom line i.e. extremely low (Supplementary Fig. 3). Nonetheless, laser beam energy absorbed by the components of cell-culture media may also add to the 0.46 °C temperature rise in the absence of MNPs. Thus, the net temperature rise upon NIR laser exposure in the presence of MNPs is only 0.37 ± 0.115 °C which is insignificant in terms of hyperthermia generation and can be fairly safe to use in an *in vivo* situation.

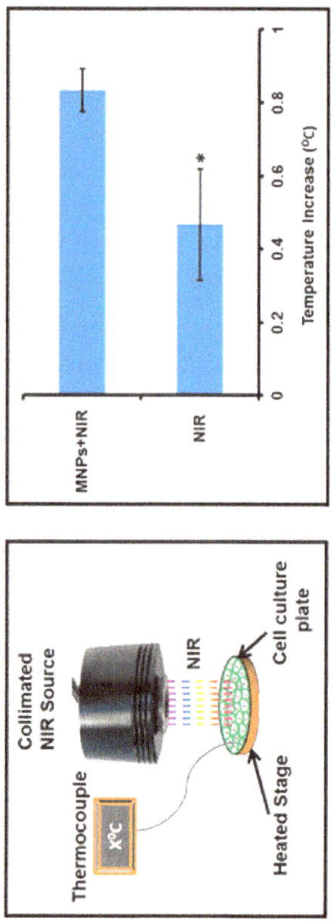

(**A**) Schematic of 808 nm NIR laser module: NIR source is fixed to a holder and connected to an external power supply. Beneath the

NIR source is a heated stage insert to keep the cell-culture system at original culture of 37 °C. Temperature of cell culture system exposed to NIR with or without MNPs is measured with the help of a thermocouple wire. (**B**) Temperature profile of different experimental group: Cell culture with or without MNPs were exposed with ~1.5 Wcm^{-2} NIR for 2 minutes and temperature were recorded via placing thermocouple at the bottom of well throughout the exposure period. Effect of NIR or MNPs-NIR on temperature of cell culture ambience of individual well was obtained by temperature difference at the beginning (0 second) and end (120 seconds) of exposure. NIR exposure on cell culture system without MNPs showed a temperature rise of 0.46 ± 0.152 °C and that of with MNPs showed an increase of 0.83 ± 0.057 °C (*P < 0.0177).

Figure 2: Temperature profiling during NIR exposure.

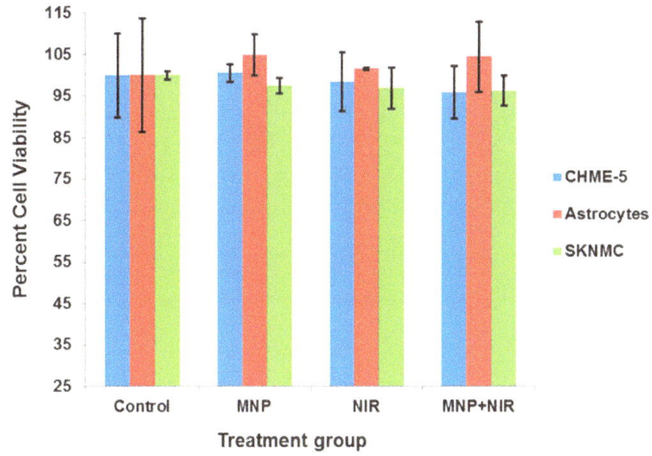

None of the treatment affected the viability of either of cell lines. Percent cell viability in the case of NIR-alone experiment was 98.45 ± 7.05, 101.58 ± 0.36, and 96.90 ± 4.85 for CHME-5 human microglia cell lines, Human primary astrocytes and SKNMC human *neuroepitheliomacells* and that for MNPs-NIR treatment was 96.00 ± 6.25, 104.58 ± 8.40, and 96.35 ± 3.64.

Figure 3: Percent cell viability of three major brain cells, namely, Human primary astrocytes, SKNMC human *neuroepithelioma* cells, and CHME-5 human microglia cell lines were obtained using MTT cytotoxicity assay.

Short-term MNPs-NIR Exposure Does Not Affect the Brain Cell Viability, Growth Behavior and Plasticity

In order to assess the potential use for the brain region, we examined the nonspecific cytotoxicity of this novel MNPs-NIR during short-time (808 nm for 2 minute) treatment to three major brain cells, namely, Human primary astrocytes, SKNMC human neuroepithelioma cells, and CHME-5 human microglia cells. The results of quantitative cell cytotoxicity determined by the MTT cell proliferation assay showed that MNPs-NIR exposure was not toxic to any of the brain cell types used in this study (Fig. 3). Similar to our previous report[17], MNPs alone did not deter the cell viability of Human primary astrocytes, SKNMC human *neuroepithelioma* cells or CHME-5 human microglia cells. NIR light with 800–100 nm wavelength can be transmitted across tissue without inducing any damage to cellular macromolecules or organelles. As such, similar to MNPs-alone treatment, cell viability is expected to remains unaltered upon exposure of NIR-alone. Percent cell viability in the case of NIR-alone experiment was 98.45 ± 7.05, 101.58 ± 0.36, and 96.90 ± 4.85 for CHME-5 human microglia cells, Human primary astrocytes and SKNMC human *neuroepithelioma* cells, respectively. Similar to this result, it has been shown previously that 808 nm irradiation on Eca-109 cells for 20 minutes did not change their viability[48]. Shen *et al.*[19] also showed the same for the A549 cells which were exposed to a 5 Wcm^{-2}808 nm NIR laser in the presence or absence of individually dispersed MNPs for 3 minutes[19]. As shown in Fig. 3, percent cell viability in the case of MNP-NIR treatment was 96.00 ± 6.25, 104.58 ± 8.40, and 96.35 ± 3.64 for CHME-5 human microglia cells, Human primary astrocytes and SKNMC human neuroepithelioma cells, respectively. The unaffected cell viability either during NIR-alone or MNPs-NIR exposure (Fig. 3) coincides with the insignificant temperature rise of the culture system (Fig. 2b). In fact, studies pertaining to the oncolytic effect of MNP-NIR exposure report that the cytotoxic effect is induced when the local temperature rise is greater than 42 °C [4,28]. The "no-effect" exhibition of cell viability in this study may also be because of the low light absorption by the natural endogenous cytochromes of cells upon short-term exposure, which may cause minimal temperature elevation accounting to the high cell survivability[19]. The unaffected percent cell viability potentiates the safe use of short-term MNP-NIR exposure for biomedical applications in the brain region.

While the MTT cell proliferation assay provides a general sense of cytotoxicity at a final time point after specific duration of treatment, analyzing the continuous growth behavior of cells over time may reveal kinetic effects of toxicity[23,24]. As such, astrocytes growth resistance/impedance was measured for 10 hours post-treatment using the electric cell-substrate impedance sensing method. This method measures the resistance (Ω) produced by growing cell monolayers over the electrodes and can detect changes in resistance to AC current flow that may occur with changes in the cell layer[47,48]. Primarily, growth resistance of each treatment group is compared with resistance of blank well containing culture media to obtain blank-normalized resistance value. As shown in Fig. 4, growth-resistance kinetics of astrocytes with all kinds of treatments showed a steady upward slope. The end point blank-normalized resistance (ΔΩ) value after 10 hrs of culture was 1.63, 1.49, 1.26, and 1.50 for NIR-alone treatment, MNPs-alone treatment,MNPs-NIR and untreated control exposure, respectively. Thus, the cell growth-resistances values were in the same range, which suggests a similar growth behavior of attached cells across all groups. The presence of MNPs on the electrode surface of the culture chip may result in a slower cellular attachment process; however, in a real in vivo scenario the effect on the cellular attachment process due to MNPs may be inconsequential. Moreover, the intrinsic AC properties of MNPs may also interfere with the AC current flow of the culture chip electrodes and subsequently may obstruct the cell growth-resistance behavior via the resistance measurement. These factors may have resulted in a relatively slower kinetic slope in MNPs-only and MNPs-NIR treated cells than the NIR-only and untreated control groups. A little up or down slope of the growth kinetics among groups is expected during the early hour due to variations in the cell attachment rate which can be influenced by many other factors such as initial cell density, cellular transitory metabolic slowdown due to treatment effects, etc.[47,48]. Thus, a longer monitoring of cell growth-behavent, impedance sensing method may reflect a healthy cellular status and factors such as cell attachment and metabolic pause can be nullified. Nonetheless, the astrocytes growth-resistance pattern obtained for the initial 10 hrs is in accordance with the MTT assay (Fig. 3) suggesting no harmful effect of short-term MNPs-NIR exposure on cell health.

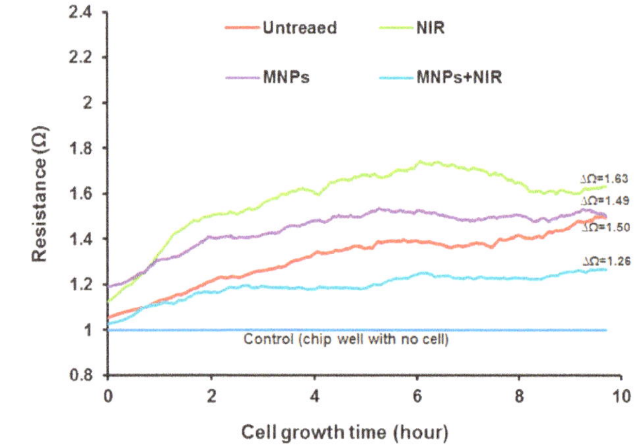

The end point blank-normalized resistance (ΔΩ) values were 1.63, 1.49, 1.26, and 1.50 for NIR-alone treatment, MNPs-alone treatment, MNPs-NIR, and untreated control exposure, respectively. Thus, the cell growth-resistances patterns were in the same range which suggests a similar growth behavior of attached cells across all groups.

Figure 4: Cell growth behavior measurements for human primary astrocytes over 10 hours of culture: Growth-resistance (Ω) kinetics of astrocytes with all kinds of treatments showed a steady upward slope.

Long term effect of MNPs-NIR treatment on dendrite and spine morphology (synaptic plasticity) of SKNMC cells was monitored. Spine morphology plays an important role in maximizing the effectiveness of the synaptic transmission in brain and to the periphery[17,49]. Treated or untreated cells were allowed to grow on cover slip for more than 72 hours such that homogenous elongation of dendritic and spinal projections can take place. Subsequently, cells were PBS-fixed and stained using the green-fluorescent membrane tracer 1, 1′-Dioctadecyl-3, 3,3′,3′-tetramethylindocarbocyanine perchlorate. This lipophilic dye uniformly labels lipid contents of plasma membrane via lateral diffusion. The stained cells on slides were now microscoped for confocal imaging. The 60X immersion objectives lens at 488 nm illusion and 2.5x electronic zoom could give us the required magnification to visualize dendritic spine of individual cells. Spine density were quantified using obtained confocal images using a well-established protocol[17,46,49] where ImageJ software is used to measure defined length of single cells and no. of spines present within that length is counted (spine density = number of spines/dendritic or cell length). As shown from confocal microscopy in Figs 5 and 6, healthy dendritic and spine morphology of SKNMC cells is evident for all treatments. NIR treated cells with or without MNPs showed a spinal density of 0.76 ± 0.13 and 0.77 ± 0.17 per μm², respectively, whereas the same in untreated cells were approximately 0.77 ± 0.20 per μm² (Fig. 6). Cells treated with MNPs only showed a spinal density of 0.76 ± 0.14 per μm²(Fig. 6). We have shown earlier, similar to this case (Fig. 5B), that MNPs alone did not affect the dendritic and spinal elongation17·46. Similarly, NIR-alone (Fig. 5C) and MNPs-NIR (Fig. 5D) treated cells showed dendritic and spinal morphology comparable to that of untreated controls (Fig. 5A). Thus, healthy synaptic plasticity was observed in all experimental groups. This result further substantiates the above observations that short-term MNPs-NIR exposures do not affect brain cell growth and behaviors and, as such, this novel, combined approach may be used for biomedical applications in brain regions.

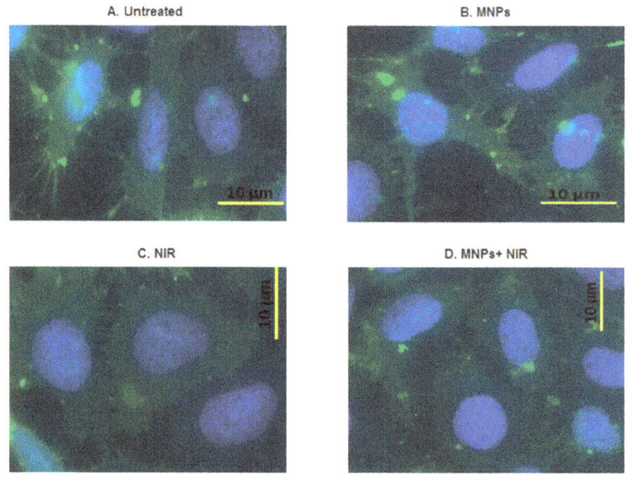

Confocal microscopy to evaluate the dendrite and spine morphology (synaptic plasticity) of SKNMC cells after 72 hours of NIR treatment in the absence (**C**) or presence (**D**) of MNPs in compare to control (**A**,**B**): Healthy dendritic and spine morphology of SKNMC cells is evident for all treatments. This suggests NIR phototargeting did not alter the neuronal synaptic plasticity and thus, similar to short-term effect (Figs 3 and 4), long term effect of MNPs-NIR phototargeting potentiate towards the safe use of this novel approach.

Figure 5

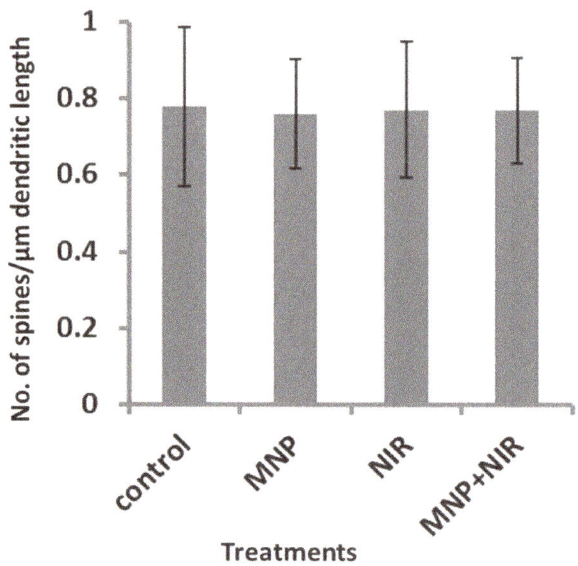

Figure 6: Spinal density (No. of spines/**μm** dendritic length) of SK-N-MC: spine density remains unchanged in all treatments (Untreated Control: 0.77 ± 0.20 per μm², MNPs: of 0.76 ± 0.14 per μm², NIR: 0.76 ± 0.13 per μm², and MNP + NIR: 0.77 ± 0.17 per μm²) which show coupling of transient near infrared photonic with magnetic nanoparticle have no adverse effect on brain cell growth and behaviors.

Prospective

We demonstrated that transient NIR irradiation in presence of MNPs is dissipation free and safe for brain cells. Our results suggest that MNPs-NIR phototargeting does not have adverse effect on the viability, growth behavior and plasticity of brain cells. Thus, selected power density of ~1.5 W/cm^2 and 2 minute time window for 808 nm NIR exposure in this report will open a referral point to explore higher power density laser for brain cells. This opens up a regime that can provide an exceptional opportunity for the safe use of this novel combinatorial approach for various biomedical innovations. Such as, transitory optoelectronic excitation-mediated molecular vibrations on MNPs surfaces, even for short NIR laser exposure, may change the charge distribution of MNPs surfaces and subsequently the ionic interaction of drugs can be broken. Thus controlled, on-demand drug release may be achieved. Discovery of NIR-sensitive materials such as hydrogels and liposomes may further advance the applicability of this approach where drugs bound MNPs can be encapsulated in liposomes or hydrogel layers. Brain targeting requires a faster reach of drug carriers with no or little exposure to RES for maximum bioavailability. External control over MNPs movement can be highly applicable for this purpose. Encapsulation can prevent drugs from digestion by direct reticuloendothelial system (RES) exposure and enzymztic activity of the peripheral circulation (blood), can remarkably improve the total drug loading capacity of the carrier by loading drugs on MNPs and encapsulating materials as well, and can enhance the drugs stability and prevent leaching in the blood circulation. Also, drug dissemination in healthy tissues can be prevented with maximized target bioavailability. A recent study by Tedford *et al.*[60] showed that transcranial and intraparenchymal exposure of 808 nm NIR laser can successfully penetrate 40 mm deep into the scalp, human skull and meninges. As such, short-term MNPs-NIR (with or without coating of NIR sensitive materials) phototargeting may be a new innovation for controlled, on-demand drug delivery in brain (Fig. 7) and for photobiomodulation to treat different psychological disorders. Moreover, while MNPs alone are used for MRI imaging; a "MNP-NIR sensitive fluorescent materials: core-shell" template can be an excellent use for multi-color imaging modalities. The NIR sensitive fluorescent materials may be liposomes, hydrogels, fluorescent carbon dots, or matrix of upconversion materials such as Erbium, Thulium, etc., which can emit different color fluorescent lights depending on NIR wavelength[64]. Nonetheless, sufficient research is required to establish permissible limits of different MNPs-NIR combination that have non-significant safety concerns before practical application on a real scenario.

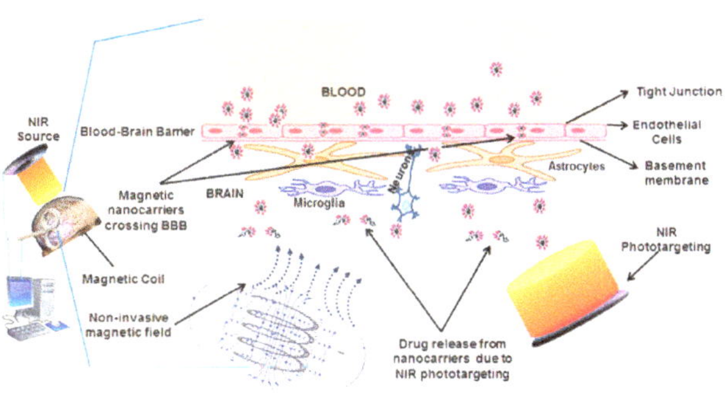

Also, while MNPs alone are used for MRI imaging; a "MNP-NIR sensitive fluorescent materials::core-shell" template can be an excellent use for multi-color imaging modalities. Combining fluorescent liposomes, hydrogels, carbon dots, or matrix of upconversion materials such as Erbium, Thulium, etc., can be used for this purpose because these materials can emit different color fluorescent lights depending on NIR wavelength.

Figure 7: Proposed mechanism for efficacious targeting, delivery, release, and monitoring of therapeutics to brain using transient NIR photonics with MNPs for noninvasive brain targeting: An *in silico*-controlled, non-invasive magnetic force will drive MNPs across BBB and simultaneously NIR phototargeting can cause release of associated drugs (transitory optoelectronic excitation-mediated molecular vibrations on MNPs surfaces, even for short NIR laser exposure, change the charge distribution of MNPs surfaces and subsequently the ionic interaction of drugs can be broken with subsequent drug release).

REFERENCES

1. Hofmann-Amtenbrink, M., Von Rechenberg, B. & Hofmann, H. In*Nanostructured Materials for Biomedical Application* (ed M. C. Tan) 119–149 (Transworld Research Network, 2009).
2. Sagar, V., Pilakka-Kanthikeel, S., Pottathil, R., Saxena, S. K. & Nair, M. Towards nanomedicines for neuroAIDS. *Rev Med Virol* 24, 103–124, doi: 10.1002/rmv.1778 (2014).
3. Santoso, M. R. & Yang, P. C. Magnetic Nanoparticles for Targeting and Imaging of Stem Cells in Myocardial Infarction. *Stem Cells Int 2016*, 4198790, doi: 10.1155/2016/4198790 (2016).
4. Parton, E., Palma, R. D. & Borghs, G. Biomedical applications using magnetic nanoparticles, 2008).
5. Unsoy, G. & Gunduz, U. Targeted silencing of Survivin in cancer cells by siRNA loaded chitosan magnetic nanoparticles. *Expert Rev Anticancer Ther*, 1–9, doi: 10.1080/14737140.2016.1184981 (2016).
6. Laurent, S., Saei, A. A., Behzadi, S., Panahifar, A. & Mahmoudi, M.Superparamagnetic iron oxide nanoparticles for delivery of therapeutic agents: opportunities and challenges. *Expert Opin Drug Deliv* 11, 1449–1470, doi: 10.1517/17425247.2014.924501 (2016).
7. Valdiglesias, V. *et al.* Are iron oxide nanoparticles safe? Current knowledge and future perspectives. *J Trace Elem Med Biol*, doi: 10.1016/j.jtemb.2016.03.017 (2016).
8. Long, N. V. *et al.* Biomedical Applications of Advanced Multifunctional Magnetic Nanoparticles. *J Nanosci Nanotechnol*15, 10091–10107 (2015).
9. Sagar, V., Atluri, V.S.R., Pilakka-Kanthikeel, S. & Nair, M. Magnetic nanotherapeutics for synaptic plasticity dysregulations during neuroAIDS and drug abuse. *Molecular Brain* 9(1), 57, doi: 10.1186/s13041-016-0236-0 (2015).
10. Nair, M., Kaushik, A., Jayant, R.D. & Sagar, V. Getting into the brain: potential of nanotechnology in the management of neuroAIDS. *Advanced Drug Delivery Reviews* 103, 202–217, doi: 10.1016/j.addr.2016.02.008 (2016).
11. Jayant, R.D. *et al.* Sustained release nanoART formulation

for the treatment of neuroAIDS. *International Journal of Nanomedicine*10, 1077–1093, doi: 10.2147/IJN.S76517 (2015).

12. Sagar, V. *et al.* Novel magneto electric nanodelivery of "microRNA mimic" across blood-brain barrier: Implications to cocaine modulation on HIV-associated neurocognitive disorders. *Journal of NeuroImmune Pharmacology* 9(1), 49 (2014).
13. Sagar, V. *et al.* Effect of Magneto-electric nanoparticle on deep brain motor coordination activity. *Journal of NeuroImmune Pharmacology* 10(S2), S99–S100 (2015).
14. Wiogo, H. T. R. *et al.* Insight into Serum Protein Interactions with Functionalized Magnetic Nanoparticles in Biological Media.*Langmuir* 28, 4346–4356, doi: 10.1021/la204740t (2012).
15. Benede, J. L., Chisvert, A., Giokas, D. L. & Salvador, A.Development of stir bar sorptive-dispersive microextraction mediated by magnetic nanoparticles and its analytical application to the determination of hydrophobic organic compounds in aqueous media. *J Chromatogr A* 1362, 25–33, doi: 10.1016/j.chroma.2014.08.024 (2006).
16. Ghose, S. K. *et al.* In *Developments in Earth and Environmental Sciences*. Vol. 7, 1–29 (Elsevier, 2007).
17. Sagar, V. *et al.* Therapeutical Neurotargeting via Magnetic Nanocarrier: Implications to Opiate-Induced Neuropathogenesis and NeuroAIDS. *J Biomed Nanotechnol* 11, 1722–1733 (2015).
18. Ding, H. *et al.* Enhanced blood-brain barrier transmigration using a novel Transferrin-embedded fluorescent magnetoliposome nanoformulation. *Nanotechnology*25, 055101–055101, doi: 10.1088/0957-4484/25/5/055101 (2014).
19. Shen, S. *et al.* Magnetic nanoparticle clusters for photothermal therapy with near-infrared irradiation. *Biomaterials* 39, 67–74, doi: 10.1016/j.biomaterials.2014.10.064 (2015).
20. Jain, T. K., Reddy, M. K., Morales, M. A., Leslie-Pelecky, D. L. &Labhasetwar, V. Biodistribution, clearance, and biocompatibility of iron oxide magnetic nanoparticles in rats. *Mol Pharm* 5, 316–327, doi: 10.1021/mp7001285 (2008).
21. Mura, S., Nicolas, J. & Couvreur, P. Stimuli-responsive nanocarriers for drug delivery. *Nat Mater* 12, 991–1003, doi: 10.1038/nmat3776 (2013).
22. Reineke, T. M. S.-R. Polymers for Biological Detection and Delivery. *ACS Macro Letters* 5, 14–18, doi: 10.1021/acsmacrolett.5b00862 (2016).
23. Medeiros, S. F., Santos, A. M., Fessi, H. & Elaissari, A. Stimuli-responsive magnetic particles for biomedical applications. *Int J Pharm* 403, 139–161, doi: 10.1016/j.ijpharm.2010.10.011 (2011).
24. Bansal, A. & Zhang, Y. Photocontrolled Nanoparticle Delivery Systems for Biomedical Applications. *Accounts of Chemical Research* 47, 3052–3060, doi: 10.1021/ar500217w (2014).
25. Kim, H., Chung, K., Lee, S., Kim, D. H. & Lee, H. Near-infrared light-responsive nanomaterials for cancer theranostics. *Wiley Interdiscip Rev Nanomed Nanobiotechnol* 8, 23–45, doi: 10.1002/wnan.1347 (2016).
26. Gu, L., Koymen, A. R. & Mohanty, S. K. Crystalline magnetic carbon nanoparticle assisted photothermal delivery into cells using CW near-infrared laser beam. *Scientific Reports* 4, 5106 (2014).
27. Wang, H. *et al.* Magnetic/NIR-responsive drug carrier, multicolor cell imaging, and enhanced photothermal therapy of gold capped magnetite-fluorescent carbon hybrid nanoparticles. *Nanoscale* 7, 7885–7895, doi: 10.1039/c4nr07335e (2015).
28. Zhu, C.-H., Lu, Y., Peng, J., Chen, J.-F. & Yu, S.-H. Photothermally Sensitive Poly(N-isopropylacrylamide)/Graphene Oxide Nanocomposite Hydrogels as Remote Light-Controlled Liquid Microvalves. *Advanced Functional Materials* 22, 4017–4022, doi: 10.1002/adfm.201201020 (2012).
29. Yan, B., Boyer, J. C., Habault, D., Branda, N. R. & Zhao, Y. Near infrared light triggered release of biomacromolecules from hydrogels loaded with upconversion nanoparticles. *J Am Chem Soc* 134, 16558–16561, doi: 10.1021/ja308876j (2012).
30. Fusco, S. *et al.* An integrated microrobotic platform for on-demand, targeted therapeutic interventions. *Adv Mater* 26, 952–957, doi: 10.1002/adma.201304098 (2014).
31. You, J. *et al.* Near-infrared light-sensitive liposomes for the enhanced photothermal tumor treatment by the combination with chemotherapy. *Pharm Res* 31, 554–565, doi: 10.1007/s11095-013-1180-7 (2014).
32. Tao, W., Wilkinson, J., Stanbridge, E. J. & Berns, M. W. Direct gene transfer into human cultured cells facilitated by laser micropuncture of the cell membrane. *Proceedings of the National Academy of Sciences of the United States of America* 84, 4180–4184 (1987).
33. Schneckenburger, H., Hendinger, A., Sailer, R., Strauss, W. S. &Schmitt, M. Laser-assisted optoporation of single cells. *J Biomed Opt* 7, 410–416, doi: 10.1117/1.1485758 (2002).
34. Palumbo, G. *et al.* Targeted gene transfer in eucaryotic cells by dye-assisted laser optoporation. *J Photochem Photobiol B* 36, 41–46, doi: 10.1016/S1011-1344(96)07335-6 (1996).
35. Liu, B. *et al.* Functional nanomaterials for near-infrared-triggered cancer therapy. *Biomater Sci* 4, 890–909, doi: 10.1039/c6bm00076b (2016).
36. Mohanty, S., Sharma, M. & Gupta, P. Laser-assisted microinjection into targeted animal cells. *Biotechnology Letters* 25, 895–899 (2003).
37. Hosokawa, Y. *et al.* Gene delivery process in a single animal cell after femtosecond laser microinjection. *Applied Surface Science*255, 9880–9884, doi: http://dx.doi.org/10.1016/j.apsusc.2009.04.111 (2009).
38. Stracke, F., Rieman, I. & Koenig, K. Optical nanoinjection of macromolecules into vital cells. *Journal of photochemistry and photobiology. B, Biology* 81,

136–142, doi: 10.1016/j.jphotobiol.2005.07.006 (2005).

39. Schinkel, H., Jacobs, P., Schillberg, S. & Wehner, M. Infrared picosecond laser for perforation of single plant cells. *Biotechnol Bioeng* 99, 244–248, doi: 10.1002/bit.21549 (2008).

40. Sordillo, L. A., Pu, Y., Pratavieira, S. o., Budansky, Y. & Alfano, R. R.Deep optical imaging of tissue using the second and third near-infrared spectral windows. *Journal of Biomedical Optics* 19, 056004–056004 (2014).

41. Tirlapur, U. K. & Konig, K. Targeted transfection by femtosecond lasec. *Nature* 418, 290–291, doi: 10.1038/418290a (2002).

42. Chu, M. *et al.* Near-infrared laser light mediated cancer therapy by photothermal effect of Fe_3O_4 magnetic nanoparticles.*Biomaterials* 34, 4078–4088, doi: 10.1016/j.biomaterials.2013.01.086 (2013).

43. Rudolf, H., Silvio, D., Robert, M. l. & Matthias, Z. Magnetic particle hyperthermia: nanoparticle magnetism and materials development for cancer therapy. *Journal of Physics: Condensed Matter* 18, S2919 (2006).

44. Maier-Hauff, K. *et al.* Efficacy and safety of intratumoral thermotherapy using magnetic iron-oxide nanoparticles combined with external beam radiotherapy on patients with recurrent glioblastoma multiforme. *Journal of Neuro-Oncology*103, 317–324, doi: 10.1007/s11060-010-0389-0 (2011).

45. Tang, Y. & McGoron, A. J. Combined effects of laser-ICG photothermotherapy and doxorubicin chemotherapy on ovarian cancer cells. *J Photochem Photobiol B* 97, 138–144, doi: 10.1016/j.jphotobiol.2009.09.001 (2009).

46. Pilakka-Kanthikeel, S., Atluri, V. S. R., Sagar, V., Saxena, S. K. &Nair, M. Targeted Brain Derived Neurotropic Factors (BDNF) Delivery across the Blood-Brain Barrier for Neuro-Protection Using Magnetic Nano Carriers: An *In-Vitro* Study. *PLoS One* 8, e62241 (2013).

47. Hondroulis, E., Zhang, Z., Chen, C. & Li, C.-Z. Impedance Based Nanotoxicity Assessment of Graphene Nanomaterials at the Cellular and Tissue Level. *Analytical Letters* 45, 272–282, doi: 10.1080/00032719.2011.633184 (2011).

48. Hondroulis, E., Liu, C. & Li, C. Z. Whole cell based electrical impedance sensing approach for a rapid nanotoxicity assay.*Nanotechnology* 21, 315103, doi: 10.1088/0957-4484/21/31/315103 (2011).

49. Atluri, V. S. *et al.* Vorinostat positively regulates synaptic plasticity genes expression and spine density in HIV infected neurons: role of nicotine in progression of HIV-associated neurocognitive disorder. *Mol Brain* 7, 37, doi: 10.1186/1756-6606-7-37 (2014).

50. Sun, Y.-k., Ma, M., Zhang, Y. & Gu, N. Synthesis of nanometer-size maghemite particles from magnetite. *Colloids and Surfaces A: Physicochemical and Engineering Aspects* 245, 15–19, doi: 10.1016/j.colsurfa.2004.05.009 (2004).

51. Liu, J., Wen, J., Zhang, Z., Liu, H. & Sun, Y. Voyage inside the cell: Microsystems and nanoengineering for intracellular measurement and manipulation. *Microsystems & Nanoengineering*1, 15020 (2015).

52. Saiyed, Z. M., Gandhi, N. H. & Nair, M. P. Magnetic nanoformulation of azidothymidine 5′-triphosphate for targeted delivery across the blood-brain barrier. *Int J Nanomedicine* 5, 157–166 (2010).

53. Shi, D., Sadat, M. E., Dunn, A. W. & Mast, D. B. Photo-fluorescent and magnetic properties of iron oxide nanoparticles for biomedical applications. *Nanoscale* 7, 8209–8232, doi: 10.1039/c5nr01538c (2015).

54. Carrey, J., Mehdaoui, B. & Respaud, M. Simple models for dynamic hysteresis loop calculations of magnetic single-domain nanoparticles: Application to magnetic hyperthermia optimization. *Journal of Applied Physics* 109, 083921, doi: 10.1063/1.3551582 (2011).

55. Vallejo-Fernandez, G. *et al.* Mechanisms of hyperthermia in magnetic nanoparticles. *Journal of Physics D: Applied Physics* 46, 312001 (2013).

56. Nair, M. *et al.* Externally controlled on-demand release of anti-HIV drug using magneto-electric nanoparticles as carriers. *Nat Commun* 4, 1707 (2013).

57. Kaushik, A., Jayant, R. D., Sagar, V. & Nair, M. The potential of magneto-electric nanocarriers for drug delivery. *Expert Opin Drug Deliv* 11, 1635–1646, doi: 10.1517/17425247.2014.933803 (2014).

58. Xie, B., Singh, R., Torti, F. M., Keblinski, P. & Torti, S. Heat localization for targeted tumor treatment with nanoscale near-infrared radiation absorbers. *Phys Med Biol* 57, 5765–5775, doi: 10.1088/0031-9155/57/18/5765 (2012).

59. Fomina, N., Sankaranarayanan, J. & Almutairi, A. Photochemical mechanisms of light-triggered release from nanocarriers. *Adv Drug Deliv Rev* 64, 1005–1020, doi: 10.1016/j.addr.2012.02.006 (2012).

60. Tedford, C. E., DeLapp, S., Jacques, S. & Anders, J. Quantitative analysis of transcranial and intraparenchymal light penetration in human cadaver brain tissue. *Lasers Surg Med* 47, 312–322, doi: 10.1002/lsm.22343 (2015).

61. Cassano, P. *et al.* Near-Infrared Transcranial Radiation for Major Depressive Disorder: Proof of Concept Study. *Psychiatry J* 2015, 352979, doi: 10.1155/2015/352979 (2015).

62. Barrett, D. W. & Gonzalez-Lima, F. Transcranial infrared laser stimulation produces beneficial cognitive and emotional effects in humans. *Neuroscience* 230, 13–23, doi: 10.1016/j.neuroscience.2012.11.016 (2013).

63. Jaque, D. *et al.* Nanoparticles for photothermal therapies. *Nanoscale* 6, 9494–9530, doi: 10.1039/c4nr00708e (2015).

64. Haro-Gonzalez, P. *et al.* Optical trapping of NaYF4: Er3+, Yb3+ upconverting fluorescent nanoparticles. *Nanoscale* 5, 12192–12199, doi: 10.1039/c3nr03644h (2013).

CHAPTER 14

MAGNETIC RESONANCE-GUIDED LASER INDUCED THERMAL THERAPY FOR GLIOBLASTOMA MULTIFORME

Sarah E. Norred and Jacqueline Anne Johnson

INTRODUCTION

Glioblastoma multiforme (GBM) is the most common type of malignant primary brain tumor in adults, with over 22,000 estimated diagnoses in 2012 [1]. GBM, often referred to as simply "glioblastoma," presents a unique clinical challenge in that these tumors generally have strong resistance to traditional therapies, can spread aggressively to other areas of the brain, and may be localized in susceptible areas such that the tumor may not be treated without causing damage to adjacent healthy tissue. In addition, the tumors themselves are characterized by heterogeneous areas of necrotizing tissue and peritumoral edema and often are asymptomatic until reaching a large size. As such, the prognosis is very poor, with a five-year relative survival rate no higher than 17% for patients aged between 20 and 44, and progressively lower for older patients [2].

Surgery is usually indicated as the first stage of treatment, along with chemotherapy and radiation therapies. Though near-complete resection of the tumor usually results in longer survival times, recurrence is very common. Many patients, such as those with deep-seated or recurrent tumors, may not be candidates for surgical debulking. In the last decade, laser induced thermal therapy (LITT) has been used as an alternative treatment for several types of brain tumors, including patients with recurrent glioblastoma who were not candidates for a second resection procedure [3]. LITT as a cancer treatment is a thermocoagulative therapy whereby laser irradiation is introduced into a tumor via percutaneous insertion of an optical fiber. The ablative heat generated causes necrosis in the tumor. This procedure has been used clinically in many tissues, including the brain [3–11], liver [12], lung [13], bone [14], and prostate [15].

In addition to having many flexible applications in other therapies, LITT is useful as a cancer treatment due to its minimally invasive debulking approach, repeatability, and compatibility with magnetic resonance imaging (MRI) and that it can be done concurrently with other treatments [4, 8]. It is furthermore useful specifically as a neurosurgical treatment as it produces lesions of a predictable size, does not use ionizing radiation, and can be performed with the patient awake [4, 8, 9]. This paper intends to review current clinical uses of LITT in the treatment of glioblastoma, as well as to evaluate LITT-based therapies, which may have application in the treatment of glioblastoma in the future.

CLINICAL USE OF LITT IN GLIOMAS

LITT has been used to treat many types of brain tumors, including metastatic brain tumors [7], astrocytoma, meningioma, ependymoma, chordoma, hemangioblastoma, and glioblastoma multiforme. It is as yet an uncommon therapy, though it is becoming increasingly popular as an alternative treatment to traditional resection and radiation therapies. Research is limited to areas that have available infrastructure for LITT application, though appropriate laser sources, image analysis, and thermal analysis software are each becoming more commercially available. Some of the research on the use of LITT in glioblastoma has been done concurrently with brain tumors of other types; a few studies have focused specifically on recurrent glioblastoma.

Citation: Sarah E. Norred and Jacqueline Anne Johnson, "Magnetic Resonance-Guided Laser Induced Thermal Therapy for Glioblastoma Multiformes," BioMed Research International, vol. 2014, Article ID 761312, 9 pages, 2014. doi:10.1155/2014/761312.

Laser Interaction with Brain Tissue

Prediction of thermal influence and lesion size caused by laser-tissue interaction is highly relevant to the treatment of glioblastoma as damage to healthy brain tissue can cause significant neurological deficits in patients. The laser interactions in LITT are predominantly thermal and dependent on the effects of applied laser power. Photons emitted from the optical fiber are absorbed by chromophores in the tumor, causing excitation and the subsequent release of thermal energy [16]. If the heat produced is maintained above a certain threshold temperature at which proteins begin to denature, the tissue irreversibly coagulates and the optical properties (most significantly, the absorbance properties) of the tissue change. It is important to carefully monitor the temperature induced by the applicator device, as too quick or large of an increase in temperature may cause carbonization—this increases the absorption of the tissue substantially, thus limiting the potential light penetration. Overheating may also induce vaporization in the target tissue, resulting in an explosive rupture one article has labeled the "popcorn effect" [17].

The types of lasers commonly used for LITT are either continuous-wave neodymium-doped yttrium aluminum garnet (Nd: YAG) lasers ($\lambda = 1064$ nm) or diode lasers ($\lambda = 800$–980 nm) which operate at a wide range of powers [4]. Nd: YAG lasers in particular are indicated for use in soft tissue with high blood perfusion, which demonstrate the highest penetration depths at wavelengths between 1000 and 1100 nm [6]. As brain tissue, especially white matter, has a high rate of perfusion [5], the common and relatively cost-effective Nd: YAG laser is an excellent choice for LITT in brain tumors. These wavelengths also lie within the near infrared (NIR) "therapeutic window," where scattering is significantly higher than absorption, allowing for optimal penetration into the tissue [16]. Diode lasers at a wavelength of 980 nm have been used in a few studies involving LITT in the brain, specifically chosen due to the local maximum in water absorption at this wavelength [8, 9]. Diode lasers also have relevance to certain nanoparticle therapies excited at these wavelengths.

In order to effectively estimate light penetration and thus lesion size, it is highly important to estimate the optical properties of the brain tissue, most importantly the tissue absorption coefficient. Though values vary for different brain tissues, in all cases, the absorption of coagulated tissue is generally greater than that of noncoagulated tissue [18].

Optical Fiber Applicators

Laser irradiation is introduced into the tumor via an optical fiber, usually of a diameter of about 600 μm, of which ~1 cm of the tip is exposed. This setup generates a uniform ellipsoid lesion along the axis of the fiber [8]. Previously, only this bare tip was used for LITT, resulting in significant overheating effects (carbonization and vaporization) at the tissue surface [11, 16]. In addition, the fiber itself can potentially be damaged in cases of extreme overheating. When this occurs, similar to the effects of carbonization and vaporization, the increase in local absorbance does not allow for good penetration, and coagulation is limited to the effects of thermal conduction.

Enclosing the fiber in a diffusion sheath allows for the power density to be reduced across the breadth of the tip sheath, allowing for a higher laser power to be used [8]. The sheath surface furthermore provides a nonstick surface such that the fiber may be moved within the coagulation tract easily. More important to recent research and to the use of LITT in glioblastoma is the inclusion of a "cooling catheter," a sheath-like device which cools the fiber with a constant stream of fluid (trials in the brain have used room-temperature water or saline) [8, 9] which circulates through the lumen of the sheath. The heat transfer induced by the cooling catheter lowers the temperature of both the fiber and the surrounding tissue, allowing the use of much higher power laser systems without inducing carbonization. See Figure 1.

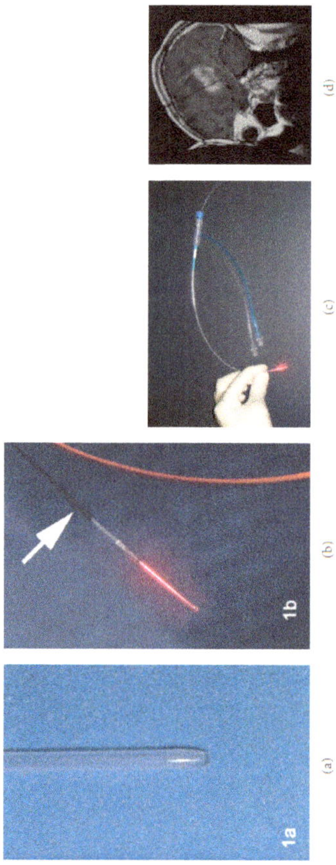

Figure 1: ((a), (b)) Visualase applicator tip used in initial clinical investigation of LITT in GBM patients [4]. (a) Fiber protection sheath. (b) Illuminated fiber tip, arrow notes magnetite particle coating used for visual reference during MR imaging [4]. (c) Most recent iteration of Visualase laser applicator, shown with cooling catheter attachment. (d) MR reference image showing the insertion of the entire length of the applicator [9].

Though only a small portion is used for LITT, the fiber itself is often very long as the laser source must be located outside of the MRI room when MR-guided imaging is used for the fiber placement. Specific paramagnetic markings added on the applicator allow for easy viewing of the location of the fiber tip when imaging the brain [9].

Perhaps most significant to the treatment of GBM is the relative ease at which multiple fibers may be used to attack large and irregularly shaped tumors, which may be used simultaneously or in tandem on a specific area. In principle, the power settings and thermal impact of each fiber can be controlled individually, allowing for highly tuned ablative treatment of tumors with irregular shape [8, 9] and potentially for those with heterogeneous optical properties. A practical concern when using this method in the brain, however, is appropriate spacing of the fiber entry sites to avoid collisions between bones anchors used to hold the fibers in place [9]. In this same case, the fiber tips were kept at least 1 cm apart while in the brain.

However, a complication was noted in one case of multiple fiber usage in a GBM tumor with a very large volume (~70 cm^3); LITT was attempted with three simultaneous lasers, causing postablation refractory edema that could not be controlled medically and thus required a hemicraniectomy. The investigators assert that control of edema volume should be a priority when large ablation volumes are planned. Though this case had the largest amount of fibers used in a single treatment, the same study had multiple cases where two fibers were used simultaneously without complications [9]. It should be noted, however, that in a similar study using similar thermal therapy and the same MR-guidance system, a patient with recurrent GBM received treatment with 3 fibers on a significantly smaller tumor but did not experience complications [8]. Furthermore, no significant postoperative complications were reported in any patients in a previous study of LITT in recurrent GBM using a Nd: YAG laser along with MR-guidance; this study included a number of patients with large tumor volumes (33.2 cm^3, 45.5 cm^3, and 79.8 cm^3) [4]. Though it may be difficult to draw salient conclusions from individual results across different tests, it may be inferred from the latter examples that patients need not necessarily be excluded from multifiber therapy on the basis of lesion size alone.

Fiber Placement and Thermal Imaging

As correct fiber placement is highly significant to the development of a treatment plan for GBM and the preservation of healthy neural tissue, it is important that all aspects of the therapy are compatible with the imaging technique used in the procedure. Imaging methods for LITT include ultrasound, computed tomography, fluoroscopy, and MRI [16]. In MRgLITT (magnetic resonance-guided laser induced thermal therapy), MRI information is used not only to develop a therapy plan and to accurately place the laser fiber (introduced into the brain using a placement catheter clamped by a plastic skull bolt), but also for the analysis of heating inside the brain tissue using MRTI (magnetic resonance temperature imaging).

An MR-guided technique is well suited for use in brain based tumors as MRI provides excellent definition and contrast in soft tissue, quick, real-time results, and the ability to monitor many different factors, including the local temperature of the coagulated tissues. Due to the often heterogeneous nature of glioblastoma tissue, which may present with peritumoral edema and necrotizing areas, accurate soft tissue modeling is highly necessary for treatment planning and qualitative clinical analysis of coagulation progress.

Systems for integrating laser equipment and MRI visualization for MRgLITT are commercially available. Visualase Thermal Therapy has been used in several Phase I clinical trials treating tumors in the brain and in at least two studies where glioblastoma has been included. Working at wavelengths from 800 to 1064 nm, the system is capable of controlling laser ablation and simultaneously performing analysis of real-time MRTI information in order to output a false-color temperature map and a "damage map" estimated by the Arrhenius damage equation. The system allows for tunable control points to be set such that the laser deactivates if a certain temperature threshold is reached. See Figure 2.

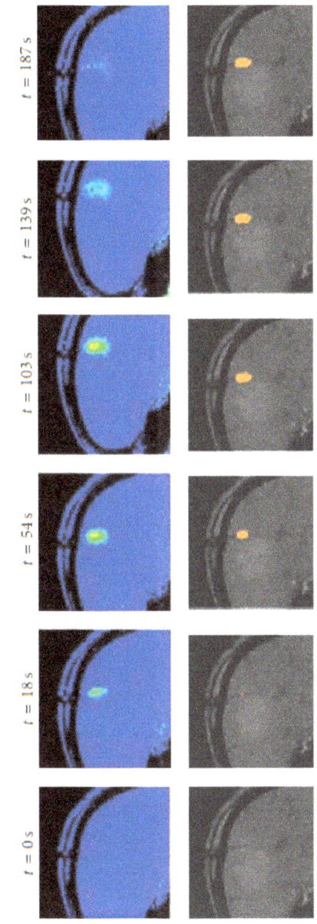

Figure 2: The commercial Visualase system allows the display of thermal distribution and an approximation of damage during the treatment [6]. Treatments including nanoparticle therapy would require much more complex modeling, as well as accurate estimations of nanoparticle concentration in the tissue.

Clinical Results

In the majority of available clinical studies where LITT has been used to treat glioblastoma, most patients who were candidates for standard surgical, chemotherapeutic, and radiotherapeutic treatments had already undergone these procedures. Though initially labeled as alternative or salvage therapy to be attempted after all other therapies have been maximized, there have been some cases where LITT has been used as a primary debulking procedure on the basis of surgical inaccessibility, including at least one patient with glioblastoma [9].

The earliest clinical trial specifically addressing LITT treatment of glioblastoma was in 2006, where 16 patients, all presenting recurrent GBM and all nonsurgical candidates, were administered LITT using a 1064 nm Nd: YAG laser source coupled into a quartz fiber (length 12 m, ID = 400 μm) [4]. The optical fiber used the diffusing tip to direct light emission into an ellipsoid profile. The fiber was positioned in the tumor using MR-guidance and a laser output of 6 W was selected. Thermal necrosis was induced in a ~2-3 cm diameter, with the axial length along the fiber being ~1 cm longer than the fiber itself.

The thermally affected area extended ~3-4 cm in diameter around the necrotized tissue. Only one fiber was used in each treatment, though some patients received repeated irradiations. No patient required postoperative intensive care.

The mean tumor volume treated was 21.6 ± 18.6 cm^3, with an average of 9.4 ± 3.3 kJ of energy delivered per tumor. Generally, tumors after irradiation exhibited a defined area of coagulation, which decreased in volume over time (e.g., see Figure 3). This is consistent with the results from a previous investigation by the same group [3]. The median overall survival time after diagnosis of recurrence was 9.4 ± 1.3 months, a substantial increase over the natural history survival prediction of< 5 months. The median survival time after the first portion of the study (10 cases) was 6.9 ± 1.7 months; the second portion of the study demonstrated an increase in survival time to 11.2 ± 2.0 months. Reasons for this increase include self-reported selection bias in choice of patients with higher Karnofsky Performance Status (KPS) and smaller tumor size [4], though future studies would later qualitatively confirm the presence of a steep learning curve in the surgical procedure [8, 9].

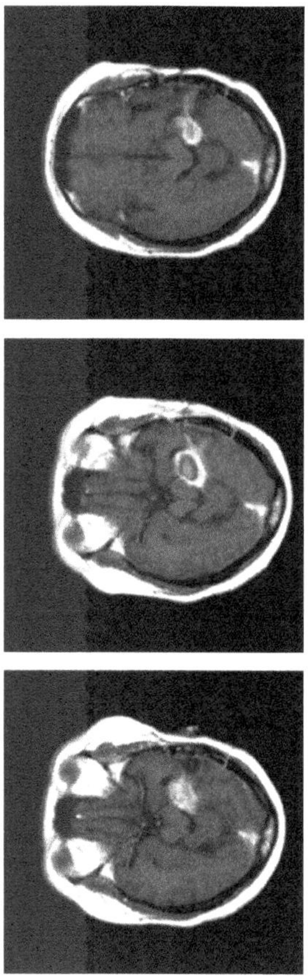

Figure 3: MR imaging of treatment progression of a recurrent GBM patient [4]: (a) Preoperative state and (b) 24 hours after LITT. A characteristic "ring" can be seen, indicating that necrosis has been achieved in the center of the tumor mass. (c) 17 months after LITT, tumor volume is reduced.

Two recently published studies have used the Visualase system and very similar surgical methods in patients with glioblastoma. Both studies used a 980 nm diode laser with a cooling catheter (Visualase). They used real-time MRTI feedback in order to monitor the thermal ablation zone and used software to set specific boundary conditions for temperature on the ablation area.

The first study examined glioblastoma in four patients that had recurrent tumors after surgical resection. The fiber was positioned with MR-guidance and LITT was administered at 10–15 W for 30–180 seconds. The average tumor volume was much smaller than in the previous GBM recurrence study, with the largest tumor treated at ~5.5 cm^3 [8]. The mean time of survival after LITT was 37 days. See Figure 4. A notable inference from this particular article notes the lack of correlation between the amount of tissue destroyed at the initial tumor site and the later recurrence of a tumor at that same site [8].

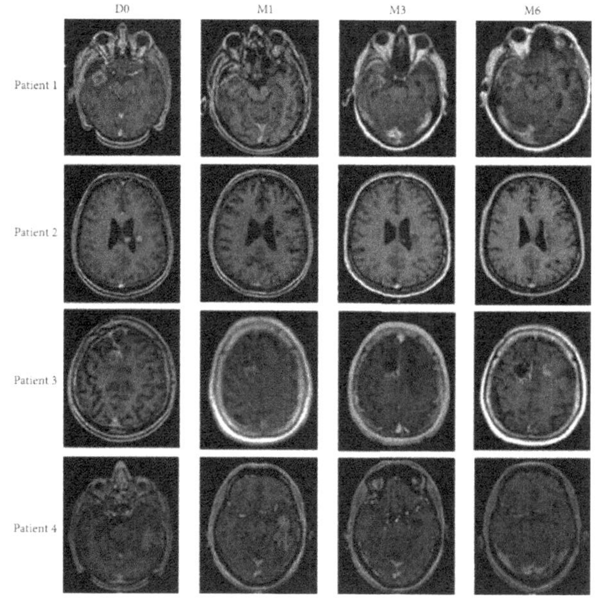

Figure 4: MR imaging of four patients with recurrent glioblastoma, following LITT on the day of treatment (D0), one month following (M1), three months following (3 M), and six months following (M6). Recurrence was observed in all patients, with a mean progression-free survival time of 37 days. Though LITT is suited for addressing many tissue types and tumor shapes/sizes, GBM masses are often diffuse in brain tissue. In order to discriminate between healthy and diseased tissue, highly specific therapies are desirable. Though more developed surgical methods may help in an immediate sense, reliance on simple hyperthermia outside the zones of ablative heat may not be sufficient for addressing microtumors seen in GBM.

The second study examined 20 patients with highly variable tumors in various stages of treatment and progression [9]. LITT was administered at an average of 11.0±1.4 W for an average of 13.9±10.7 minutes. If the catheter could be inserted in the long axis direction of a tumor, multiple ablations along the same entrance channel could be performed. The mean tumor size treated was 7.0 ± 9.0 cm^3. Mean survival time was not listed as the study was focused on the procedural aspects of the treatment. Investigators reflected on the presence of a steep learning curve in applying this treatment and recommended that this treatment be reserved for lesions less than 3 cm in diameter that are well circumscribed and noninfiltrating. However, the article does not provide a detailed list of tumor volumes and the investigators note that the size recommendation is based on experience and that most tumors treated were of a diameter less than 2 cm. Recently, [19] a first-in-humans Phase I clinical trial by Sloan et al. used the NeuroBlate system for recurrent glioblastoma multiforme. The NeuroBlate system, previously known as the AutoLITT system, in conjunction with MRI thermometry, reports the temperature of tissue and thermal dose, providing feedback to the surgeon to tailor tissue coagulation. The NeuroBlate software determines the likelihood of cell death in the monitored region. Ten patients who had an average age of 55 and average KPS score of 80 were treated. Three levels of dose were administered. An external review board judged the need for dose escalation as warranted by toxicity level. A drop in KPS score measures toxicity level. The mean tumor volume was s 6.8±5 cm^3. The median survival was 316 days. In conclusion, the NeuroBlate system represents new technology for controlled LITT in recurrent GBM.

Positron emission tomography (PET) has also been used recently to monitor the metabolic activity of the tumor before and after LITT treatments [10]. In a single case study, a 1064 nm laser was used to irradiate several points in a large glioblastoma in a 58-year-old patient. PET analysis determined that there was significantly less metabolic activity in the tumor area, in contrast to findings described by the contrasting area shown in the MRI. Future research may include PET scanning in order to better characterize the metabolic activity of the tumor before and after LITT treatment, as it may confirm assumptions derived from visual MR feedback and elucidate questions concerning the viability of local tissue after surgery [10].

FUTURE TECHNIQUES RELEVANT TO GLIO-BLASTOMA

The investigation of LITT as a treatment for glioblastoma is inherently limited by the application of local methods to a diffuse disease. In many respects, LITT by itself cannot be considered much more than palliative treatment when regarded through the lens of invasive and widely distributed cancers. In contrast, the development of nanoscale technologies as a method of specific, cell-by-cell attack for cancerous tissue is highly significant to the future of cancer therapies of all types. Though targeting methods are still under investigation, glioblastoma is especially suited to nanoscale treatments as several reliable targeting methods have been developed for this disease in vitro. Of particular interest is the large subtype of GBM tumors which overexpress epidermal growth factor receptor (EGFR), providing a feasible target for many kinds of nanoparticle therapies [20]. Other targeting methods include monoclonal antibody tagging, which may be specific to a wide range of cell types, including GBM-specific $CD133^+$ cancer stem-like cells [21].

Many preliminary studies have demonstrated that nanoparticles sensitive to irradiation in the NIR therapeutic range may be used for inducing hyperthermia in specific cells. LITT, then, may act as a method of delivering laser irradiation to nanoparticle-loaded tissues. In the context of use for cancer therapy, nanoparticles artificially increase the absorption in tissues, thus assisting the resultant heat and formation of necrosis when irradiated. In the case of targeted therapies, absorption is increased specifically in the targeted cell, thus preferentially heating that cell. This may be particularly useful for glioblastoma, which can develop tumor micromasses in a diffuse manner, including around arteries.

Preclinical Methods

Though not specifically focused on treatment of GBM, there have been several attempts in nanoparticle-assisted laser ablation in animal models, including an orthotropic canine model examining brain tumors, using gold shells surrounding silica cores or "gold nanoshells." Gold nanoshells are highly relevant to laser therapeutics as their plasmon resonance may be tuned to a value in the NIR spectrum (in this case, 780–800 nm). Gold is also highly inert in the body, with no risk of oxidation or toxicity. In this study, 144–150 nm gold nanoshells coated with polyethylene glycol (PEG) when injected intravenously were shown to selectively accumulate in tumors due to the enhanced permeability and retention effect seen in tumor vasculature [22]; even without antibody conjugation or other targeting methods, this method has traditionally shown good selectivity and accumulation in tumors over healthy tissue. A 15 W gallium arsenide diode laser (λ = 810 ± 20 nm) paired with a cooling catheter (Visualase Cooled Laser Application System) was used due to the proximity to the absorbance peak of the gold nanoshells and the minimization of absorbance in oxyhemoglobin and deoxyhemoglobin [23]. Canine transmissible venereal tumor (cTVT) cells were injected into the brain to form the tumor model. The Visualase system was used to interface with the MRI and monitor thermal changes in the brain. The laser was used at 3.5 W for 3 minutes on average. Irradiation was repeated in healthy tissue on the opposite hemisphere of the brain. The nanoparticles assisted the thermal ablation to achieve ablative temperatures of 65.8°C, where the normal tissue only experienced heating up to the ineffective hyperthermic temperature of 48.6°C. After a healing period, the tumors were harvested and analyzed for nanoparticle content, demonstrating good distribution in the ablated tissue.

A further study by the same group was conducted in the prostate in a canine model, where a similar technique was used to demonstrate higher volume of necrosis at the same exposure and laser power in a nanoparticle-laden tissue compared to native tissue. The same laser applicator and cooling catheter (Visualase) were used for MRgLITT. Gold nanoshells were injected directly into the parenchyma of a single hemisphere of the prostate, using the bilateral tissue portion as a control. The applicator was passed through both hemispheres, with laser settings of 3–3.5 W for an irradiation time of 2-3 minutes. Cross-sections of the dissected prostrate (see Figure5) demonstrated a much larger treatment area in the nanoparticle-laden tissue, with much more distinct and self-limited lesion boundaries [24].

This could indicate that some lasers, specifically those below 980 nm, may be better suited for nanomaterial excitation than for direct ablation itself. They may also be more suited for use in tissues with high perfusion due to their smaller absorptions of oxy- and deoxyhemoglobin. The exclusion of this particular fact in a choice of laser might be particularly inefficient for the brain, regardless of the speed of lesion development, considering that perfusion provides the largest challenge to heating efficiency in the brain, and thus high absorbance in blood tissue should be avoided.

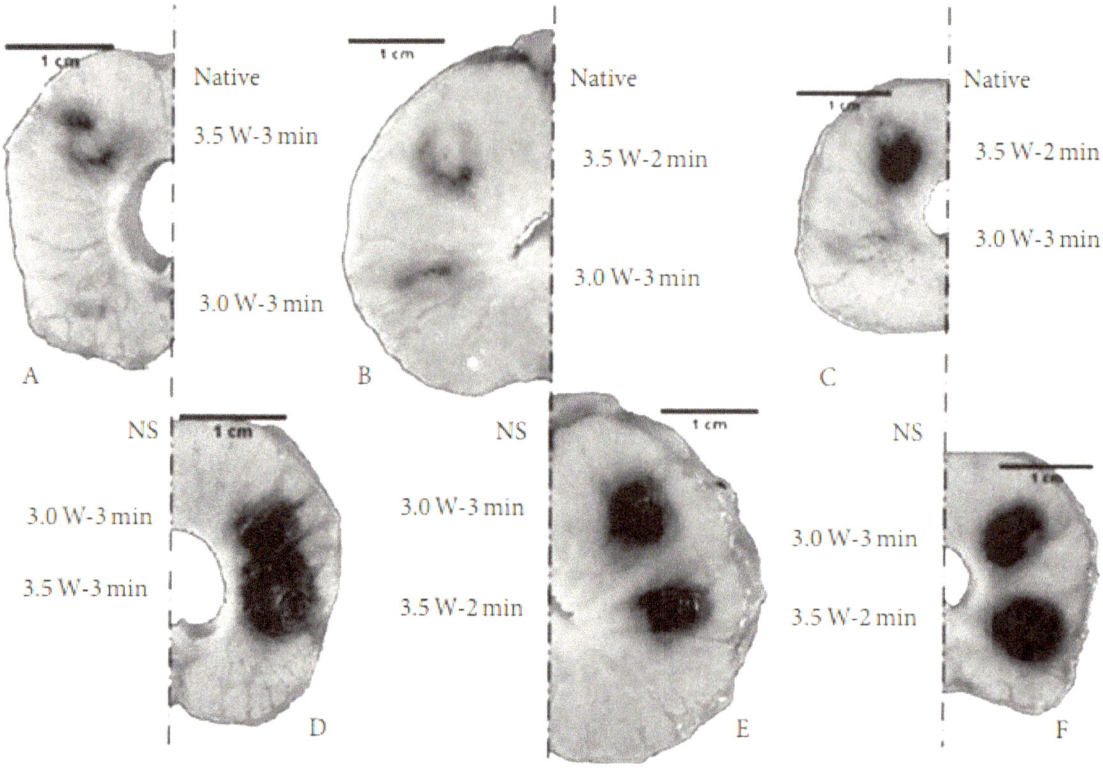

Figure 5: Dissection of formalin-fixed canine prostate tissue. Darkened areas indicate areas of thermal lesions induced at 3.0 or 3.5 W for 2 or 3 minutes. Sections A, B, and C indicate native tissue and sections D, E, and F indicate nanoparticle-laden tissue. 150 nm gold nanoshells were injected directly into the parenchyma of the prostate. It was found that lesion size grew with increasing power when applied to native tissue and that nanoshell application significantly increased the lesion size, but not with respect to power [23].

Particles accumulate in the tumor by virtue of the enhanced permeability and retention effect. When tumors accumulate nanoparticles this way, they can be efficiently ablated with high selectivity over surrounding healthy tissues [23, 24]. In the case of the 150 nm gold nanoshells used in the canine brain tumor model, TEM imaging demonstrated that tumors accumulated along capillary walls. Though fenestrated vasculature is characteristic of many GBM tumors, these therapies alone would likely not provide the cell-by-cell specificity necessary for avoiding healthy cells in a diffuse mass. Furthermore, accumulation via the enhanced permeability and retention effect may provide less distributed nanoparticle mass across the tissue, giving uneven resultant heating effects. However, this type of nanoparticle-assisted therapy allows a less ablative dose to translate into an artificially higher-heat ablative therapy in a self-limited lesion size [24].

In Vitro Methods

Gold nanorods and other gold nanostructures continue to be investigated in many in vitro applications, including in GBM-specific cell lines [25]. Gold nanorods are often regarded as superior to nanoshells (which may exceed 150 nm in diameter [23]) with regard to size and cross-sectional profile in tissue. According to Mie theory, the ratio of scattering to absorption increases with size, so smaller nanoparticles are preferred in processes such as LITT where absorption is greatly preferred [25].

Gold nanorods were used to induce death in a glioblastoma cell line (1321N1 human brain astrocytoma). The cells were incubated with the gold nanorods at varying concentrations in order to achieve this. It was determined that gold nanorods were internalized by tumor cells by confocal microscopy (see Figure 6) and transmission electron microscopy (TEM) and that overall viability of the cells decreased with exposure to large concentrations of nanorods for periods longer than 20 minutes. After irradiation with an 808 nm source at 1.2 W for 20 minutes, significant cell death was seen in the nanoparticle-laden cells, but not in either of the irradiation-free or nanoparticle-free controls [25].

Though gold nanostructures are by far the most investigated particle type with regard to NIR irradiation in biological application, new NIR-sensitive substrates continue to be produced that may provide a more cost-effective alternative to gold nanoparticles. Potential nanoparticle alternatives include specialized carbon nanotubes [21], which have been used to specifically target CD133+ GBM cells. Other NIR-sensitive particles include reduced graphene oxide particles [26] and low band gap polymers [27].

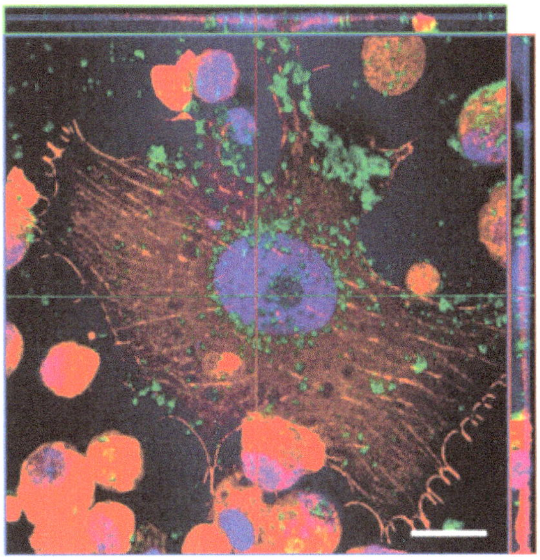

Figure 6: Confocal imaging of gold nanorod uptake in 1321N1 human brain astrocytoma cells after 12-hour exposure to 36 µg/mL. Green coloring corresponds to the gold nanorods. The image effectively demonstrates that nanorods can be internalized into glioblastoma cells, including in the nucleus. Future investigation may elucidate the impact of internalized particles compared to nonspecific tissue-accumulated particles with regard to specificity in LITT treatments [24].

CONCLUSIONS

Surgical debulking, by both operational resection and by LITT, does not yet provide a significant solution to mortality in GBM, see Table 1. However, LITT still proves to be a promising therapy for palliative care as it can significantly reduce the risk and discomfort caused by current resection surgeries—potentially providing a comparable extension of survival, but with a minimally invasive therapy that may be completed in a single day. LITT is also amenable to multiple treatments and to treatment of deep-seated and otherwise inaccessible tumors, which have limited alternate options for cytoreductive therapy.

Table 1: Gives a summary of average values in the treatment of GBM.

Ref.	Laser wavelength (nm)	Number of patients	Tumor volume treated (cm^3)	Energy delivered (kJ)	Survival time
[4]	1064	16	21.6 ± 18.6	9.4 ± 3.3	9.4 ± 1.3 months
[8]	980	4	5.5	0.3–2.7	37 days
[9]	980	20	7.0 ± 9.0	9.2	Not given
[19]	1064	10	6.8 ± 5.0	12.24 ± 9.7	316 days

In comparison to Gamma knife (GK) it is a more flexible technique, lending itself to treatment of the liver, lung, bone, and prostate as referenced earlier. The basic design of GK lends itself to treatment of the brain only and also requires the placement of a stereotactic head frame, which is painful. Deep-seated tumors can be treated by GK but not tumors on the periphery. Abnormal cells are not removed by the GK technique but are only damaged; therefore the physician must monitor the disease from weeks to months whereas LITT can be monitored in real time.

It is evident that nanotechnology will shape much of future cancer therapies. LITT may be of great use to these therapies if it can introduce laser irradiation into areas not otherwise accessible, such as those seen in deep-seated and diffuse GBM. It is already evident that the type of laser used for this therapy may not necessarily be dependent on the amount of conductive heat that it may generate alone—immediate conclusions may be that longer ablation times with nanoparticle additions give superior results than shorter times which may produce less predictable thermal damage.

In these nanotherapeutic models, the value of the cooling catheter used in the laser applicator is particularly evident—the cooling of tissue medial to the laser applicator avoids the immediate coagulation usually seen at this point, resulting in a significant limitation in penetration depth. It can be postulated that further advances in cooling the laser applicator could be of great value to nanoparticle-enhanced ablation treatments, allowing much deeper optical penetration into the tumor. This could be uniquely useful to GBM in that deeper optical penetration could provide better access to diffuse cell structures.

CONFLICT OF INTERESTS

The authors declare that there is no conflict of interests regarding the publication of this paper.

REFERENCES

1. Glioblastoma Multiforme, 2013, http://www.aans.org/Patient%20Information/Conditions%20and%20Treatments/Glioblastoma%20Multiforme.aspx.
2. Brain and Spinal Cord Tumors in Adults, 2012, http://www.cancer.org/cancer/braincnstumorsinadults/detailedguide/index.
3. H.-J. Schwarzmaier, F. Eickmeyer, W. von Tempelhoff et al., "MR-guided laser irradiation of recurrent glioblastomas," Journal of Magnetic Resonance Imaging, vol. 22, no. 6, pp. 799–803, 2005.
4. H. J. Schwarzmaier, F. Eickmeyer, W. von Tempelhoff et al., "MR-guided laser-induced interstitial thermotherapy of recurrent glioblastoma multiforme: preliminary results in 16 patients," European Journal of Radiology, vol. 59, no. 2, pp. 208–215, 2006.
5. H. J. Schwarzmaier, I. V. Yaroslavsky, A. N. Yaroslavsky, V. Fiedler, F. Ulrich, and T. Kahn, "Treatment planning for MRI-guided laser-induced interstitial thermotherapy of brain tumors—the role of blood perfusion," Journal of Magnetic Resonance Imaging, vol. 8, no. 1, pp. 121–127, 1998.
6. Carpentier, R. J. McNichols, R. J. Stafford et al., "Real-time magnetic resonance-guided laser thermal therapy for focal metastatic brain tumors," Neurosurgery, vol. 63, supplement 1, no. 1, pp. ONS21–ONS29, 2008.
7. Carpentier, R. J. McNichols, R. J. Stafford et al., "Laser thermal therapy: real-time MRI-guided and computer-controlled procedures for metastatic brain tumors," Lasers in Surgery and Medicine, vol. 43, no. 10, pp. 943–950, 2011.
8. Carpentier, D. Chauvet, V. Reina et al., "MR-guided laser-induced thermal therapy (LITT) for recurrent glioblastomas," Lasers in Surgery and Medicine, vol. 44, no. 5, pp. 361–368, 2012.
9. P. R. Jethwa, J. C. Barrese, A. Gowda, A. Shetty, and S. F. Danish, "Magnetic resonance thermometry-guided

laser-induced thermal therapy for intracranial neoplasms: initial experience," Neurosurgery, vol. 71, supplement 1, pp. 133–145, 2012.

10. N. Galldiks, W. von Tempelhoff, D. Kahraman et al., "11C-methionine positron emission tomographic imaging of biologic activity of a recurrent glioblastoma treated with stereotaxy-guided laser-induced interstitial thermotherapy," Molecular Imaging, vol. 11, no. 4, pp. 1–7, 2012.

11. T. Kahn, M. Bettag, F. Ulrich et al., "MRI-guided laser-induced interstitial thermotherapy of cerebral neoplasms," Journal of Computer Assisted Tomography, vol. 18, no. 4, pp. 519–532, 1994.

12. S. Clasen and P. L. Pereira, "Magnetic resonance guidance for radiofrequency ablation of liver tumors,"Journal of Magnetic Resonance Imaging, vol. 27, no. 2, pp. 421–433, 2008.

13. T. J. Vogl, N. N. N. Naguib, T. Lehnert, and N.-E. A. Nour-Eldin, "Radiofrequency, microwave and laser ablation of pulmonary neoplasms: clinical studies and technical considerations—review article,"European Journal of Radiology, vol. 77, no. 2, pp. 346–357, 2011.

14. F. Streitparth, U. Teichgräber, T. Walter, K. D. Schaser, and B. Gebauer, "Recurrent osteoid osteoma: interstitial laser ablation under magnetic resonance imaging guidance," Skeletal Radiology, vol. 39, no. 11, pp. 1131–1137, 2010.

15. D. A. Woodrum, L. A. Mynderse, K. R. Gorny, K. K. Amrami, R. J. McNichols, and M. R. Callstrom, "3.0T MR-guided laser ablation of a prostate cancer recurrence in the postsurgical prostate bed," Journal of Vascular and Interventional Radiology, vol. 22, no. 7, pp. 929–934, 2011.

16. R. J. Stafford, D. Fuentes, A. A. Elliott, J. S. Weinberg, and K. Ahrar, "Laser-induced thermal therapy for tumor ablation," Critical Reviews in Biomedical Engineering, vol. 38, no. 1, pp. 79–100, 2010.

17. H. J. Schwarzmaier, F. Eickmeyer, V. U. Fiedler, and F. Ulrich, "Basic principles of laser induced interstitial thermotherapy in brain tumors," Medical Laser Application, vol. 17, no. 2, pp. 147–158, 2002.

18. N. Yaroslavsky, P. C. Schulze, I. V. Yaroslavsky, R. Schober, F. Ulrich, and H.-J. Schwarzmaier, "Optical properties of selected native and coagulated human brain tissues in vitro in the visible and near infrared spectral range," Physics in Medicine and Biology, vol. 47, no. 12, pp. 2059–2073, 2002.

19. Sloan, M. Ahluwalia, J. Valerio-Pascua et al., "Results of the neuroblate system first-in-humans phase I clinical trial for recurrent glioblastoma clinical article," Journal of Neurosurgery, vol. 118, no. 6, pp. 1202–1219, 2013.

20. Westermark, "Glioblastoma—a moving target," Upsala Journal of Medical Sciences, vol. 117, no. 2, pp. 251–256, 2012.

21. H. Wang, S. H. Chiou, C. P. Chou, Y. C. Chen, Y. J. Huang, and C. A. Peng, "Photothermolysis of glioblastoma stem-like cells targeted by carbon nanotubes conjugated with CD133 monoclonal antibody," Nanomedicine, vol. 7, no. 1, pp. 69–79, 2011.

22. Y. Matsumura and H. Maeda, "A new concept for macromolecular therapeutics in cancer chemotherapy: mechanism of tumoritropic accumulation of proteins and the antitumor agent smancs," Cancer Research, vol. 46, no. 12, part 1, pp. 6387–6392, 1986.

23. J. A. Schwartz, A. M. Shetty, R. E. Price et al., "Feasibility study of particle-assisted laser ablation of brain tumors in orthotopic canine model," Cancer Research, vol. 69, no. 4, pp. 1659–1667, 2009.

24. J. A. Schwartz, R. E. Price, K. L. Gill-Sharp et al., "Selective nanoparticle-directed ablation of the canine prostate," Lasers in Surgery and Medicine, vol. 43, no. 3, pp. 213–220, 2011.

25. T. Fernandez Cabada, C. S. de Pablo, A. M. Serrano, P. Guerrero Fdel, J. J. Olmedo, and M. R. Gomez, "Induction of cell death in a glioblastoma line by hyperthermic therapy based on gold nanorods,"International Journal of Nanomedicine, vol. 7, pp. 1511–1523, 2012.

26. J. T. Robinson, S. M. Tabakman, Y. Liang et al., "Ultrasmall reduced graphene oxide with high near-infrared absorbance for photothermal therapy," Journal of the American Chemical Society, vol. 133, no. 17, pp. 6825–6831, 2011.

27. M. Macneill, R. C. Coffin, D. L. Carroll, and N. H. Levi-Polyachenko, "Low band gap donor-acceptor conjugated polymer nanoparticles and their NIR-mediated thermal ablation of cancer cells,"Macromolecular Bioscience, vol. 13, no. 1, pp. 28–34, 2012.

CHAPTER 15

LASER-DRIVEN ION BEAMS

G. A. Pablo Cirrone, Giacomo Cuttone, Francesco Romano, Francesco Schillaci, Valentina Scuderi, Antonino Amato, Giacomo Candiano, Michele Costa, Giuseppe Gallo, Giuseppina Larosa, Georg Korn, Renata Leanza, Rosanna Manna, Mario Maggiore, Valentina Marchese, Daniele Margarone, Giuliana Milluzzo, Giada Petringa and Antonella Tramontana

Keywords: Laser-driven ion, ELIMED beamline, medical applications, magnetic transport elements, dosimetry, electromagnetic pulse

INTRODUCTION

The concept of coherent acceleration, that can be advised as the starting idea of the new particle acceleration approach, based on the high-intensity laser matter interaction, was introduced for the first time in 1957 by Veksler [1]. Ever since, enormous theoretical, as well as experimental progress, has been achieved in this field until the year 2000, which can be considered the *annus mirabilis* of laser-driven ion acceleration. In 2000 three experiments [2,3,4] reported on the generation of intense and energetic (up to 58 MeV) proton beams from solid targets irradiated with high-intensity (from 3×10^{18} W/cm^2 to 3×10^{20} W/cm^2) laser pulses.

The typical experimental set-up used in these experiments consists of a high power laser beam impinging on a thin solid target; thanks to the laser-matter interaction charged particles are accelerated in both directions with respect to the laser-irradiated target surface. For the first time, protons were emitted as a rather collimated beam, along the target normal direction rather than in the isotropic and low-brilliance emission typical of previous experiments [5,6].

So far different acceleration regimes, for instance target normal sheath acceleration (TNSA) [7], the radiation pressure acceleration [8], "Light sail" acceleration [9], "Laser piston" acceleration [10], "Break-out afterburner" acceleration [11], have been studied and several experimental results on laser driven ion energies [12,13,14], obtained mainly within the TNSA regime, have been reported in literature. The peculiarities of the proton emitted in the forward direction (*i.e.* in the opposite direction with respect to the laser-irradiated target surface) in these new experiments were extremely promising suggesting that laser-driven proton/ion acceleration might represent a new opportunity in the particle acceleration field and generating also a huge interest in both fundamental research as well as in the possible multidisciplinary applications.

Actually, laser accelerated beams show different and, in some cases, extreme characteristics with respect to the conventional ones. A very high peak current, a broad energy spectrum, a wide angular distribution and a rather small transverse and longitudinal emittance are the main features of a typical laser-driven ion beam generated in the so-called TNSA (Target Normal Sheath Acceleration) acceleration regime, which represents the most known and experimentally investigated laser-induced acceleration process.

In this framework, a collaboration between the INFN-LNS (Nuclear Physics Laboratory, Catania, Italy) and the ASCR-FZU (Institute of Physics of the Czech Academy of Science), in charge for the ELI-Beamlines facility implementation, has been established in 2012. The main aim of the collaboration, named ELIMED (ELI-Beamlines MEDical applications), is to investigate the feasibility to design and realize a beam transport line for optically accelerated beams to be used for multidisciplinary and medical applications. In 2013 the ELI-Beamlines Institute started the realization of a facility in Prague, where one of the experimental halls, called ELIMAIA (ELI Multidisciplinary Applications of laser-Ion Acceleration), will host the beam line dedicated to the ion acceleration and multidisciplinary applications. The ELIMED beam line, which will be developed by LNS-INFN, will represent the section of the ELIMAIA activity addressed to the transport, handling and dosimetry of the laser-driven ion beams allowing the achievement of stable, controlled and reproducible beams that,

Citation: G. A. Pablo Cirrone , Giacomo Cuttone, Francesco Romano, Francesco Schillaci, Valentina Scuderi, Antonino Amato , Giacomo Candiano, Michele Costa, Giuseppe Gallo, Giuseppina Larosa , Georg Korn, Renata Leanza, Rosanna Manna , Mario Maggiore , Valentina Marchese , Daniele Margarone, Giuliana Milluzzo, Giada Petringa and Antonella Tramontana, Laser-Driven Ion Beams, doi:10.3390/app5030427.

in the future, will be available for all the users interested in multidisciplinary and medical applications of such innovative technology.

OVERVIEW OF THE ELIMAIA BEAM LINE

As already pointed out, the ELIMED main goal is the realization of a beam transport line (from the laser-target interaction chamber to the final sample irradiation point) and a number of dosimetric endpoints that will enable the users to apply laser-driven ion beams (protons, alpha-particles and carbon/lithium ions) in multidisciplinary fields. The technical solution proposed for the realization of the ELIMED beam line is based on a modular system, which allows adapting and combining the devices depending on the different phases of the project and the different user's requirements. Three different sections of the ion beam line have been identified for its implementation: the Ion Beam Collection and Diagnostics; the In-Vacuum Ion Beam Selection, Transport and Diagnostics and the Dosimetry and sample Irradiation systems.

Laser-driven ion beams are characterized by high intensities, several species with multiple charge states, broad energy spectra and energy-dependent angular distributions. Therefore, in order to make these beams suitable for multidisciplinary purposes, the main task of the beam transport line will be to control the particle energy and angular distributions as well as the flexibility and reproducibility of the delivered beam spot size and dose distribution. The requirements needed to perform accurate dosimetric measurements and radiobiological irradiations for pre-clinical studies imply that the transported beam must fulfill specific characteristics. Indeed, these preparatory pre-clinical studies, which are based on *in vivo* irradiations on mice, have to be done with a level of accuracy on delivered dose close to the one characterizing human tumor treatments. The specific characteristics of the transported beams are mainly related to the beam flatness and symmetry at the irradiation point, which have to be within 3%.

In particular, the purpose of the collecting and focusing section is to reduce the initial angular divergence of the particle beam accelerated from the target improving the transmission and the selection efficiency of the entire transport system. The focusing element will consist of a set of permanent magnet quadrupoles (PMQs) [15,16,17] based on a hybrid Halbach array in order to maximize the magnetic field strength, quality and uniformity by keeping the system as compact as possible. A quadrupole set is necessary to obtain a focusing effect on both transversal planes, as well as for a suitable matching with selection system acceptance. The PMQs system will be placed few cm downstream the target. Since the field strength is fixed for each magnet, the PMQs system will be provided with mechanical stages allowing modifying the relative distance of each quadrupole along the beam direction (*i.e.* the longitudinal axis) and, thus, the focal point position as a function of the beam energy. The moving system will be provided with a remote control.

The energy selection system is the critical element of the whole beam line. The beam accelerated by the laser-target interaction, after the collecting and focusing section, might still show a broad energy spectrum. Therefore, it is important to realize a system able to select ion beams with controlled energy spread and kinetic energy. The energy selector system (ESS) proposed will be placed along the beam transport line, downstream from the PMQs system. The ESS will consist of tunable resistive conductors allowing varying the performances depending on the ion species to be selected. It will be based on a resistive coil system with alternating field gradient. A set of resistive quadrupoles placed downstream from the ESS will allow recovering both vertical and radial focusing of the beam after the selection. Two steering magnets (in both vertical and horizontal directions) will allow for control of the beam direction after the selection and correcting of any possible beam misalignments.

The beam transport elements, as it has been established in the signed contract, will be characterized and tested using conventional accelerated and laser-driven ion beams. An experimental hall at the INFN-LNS has been entirely dedicated to the ELIMED project in order to perform all the tests required for the ELIMED beam line using proton beams delivered by the Tandem and by the Superconducting Cyclotron (CS) (LNS, Catania, Italy).

After the main beam transport elements, diagnostics elements specifically optimized to measure the beam emittance for laser-driven ions will be used for the on-line characterization of the collected beam. Complementary diagnostics systems such as Silicon Carbide (SiC), Diamond or Secondary Emission Monitor (SEM) detectors will be provided. This detector will be used in Time of Flight (ToF) configuration, providing a spectral analysis of the accelerated and transported ions.

A precise knowledge of the absolute dose, delivered by the incoming radiation, is crucial in several applications, as for instance the hadrontherapy ones, where the dosimetric accuracy is a key prerequisite. Indeed, for clinical accelerator-driven proton beams, the dose uncertainties at a reference depth in water with calibrated plane-parallel ionization chambers are 2.3%, according to the Code of Practice for Dosimetry IAEA TRS-398 of 2006. Nevertheless, the pulse properties of laser-driven ion beams differ significantly from those commonly provided by medical accelerators in pulse duration, peak current and correspondingly pulse dose rate and energy spectrum. Thus, detectors must be dose-rate independent allowing measuring very intense and short pulses without suffering from saturation effects and guarantying the required accuracy also for high intense beam pulse. Furthermore, it is mandatory to develop on-line beam monitoring and relative dosimetry systems, suitable to operate with high intense beam pulse and strong electromagnetic noise, ensuring a negligible beam perturbation. In order to fulfill all the requirements new detection devices and techniques must be designed and developed. The solution proposed to perform relative and absolute dose measurements implies different alternative approaches, as for instance SEM detectors [18], CR39 solid-state track detectors [19], Gafchromic films [20], transmission ionization chamber for high intense beams [21] and Faraday Cups (FC) [22].

Figure 1 shows a schematic layout of the ELIMED beam line where the elements composing the three different sections are clearly visible in their preliminary design.

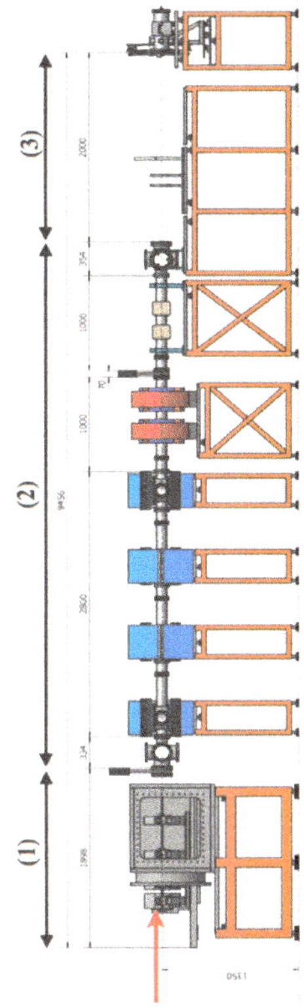

Figure 1: Scheme of the ELIMED beam line showing the three different sections: (**1**) ion beam collection and diagnostics; (**2**) in-vacuum ion beam selection, transport and diagnostics; (**3**) dosimetry and sample irradiation. Distances are in mm.

Monte Carlo (MC) simulations will play a key role for the realization of the ELIMED beam line transport elements and for the development of the detectors. Indeed, MC results allow accurate designing of some key elements, particularly for the in-air section of the beam line, and to predict the particle transport, as well as the fluence and dose at given points of interest. Moreover, radioprotection studies can be carried out in order to design appropriate shielding solutions. For the simulation of ELIMED beam line the MC toolkit GEANT4 (GEometry ANd Tracking) [23] will be used. GEANT4 is one of the most versatile and widespread codes used today for particle tracking, as well as being widely used for different physical applications. It is a C++ object oriented toolkit allowing the simulation of particle interactions with matter. It provides advanced functionalities for all the typical domains of detector simulation: geometry and material modelling, description of particle properties, physical processes, tracking events and run management, user interface and visualization. Initially developed for High Energy Physics experiments (HEP) simulation, GEANT4 is now widely used also for multidisciplinary applications and, in particular, medical physics. It allows the tracking of any charged and uncharged particle relevant for radio diagnostics and radiation therapy.

The ELIMED collaboration already designed and realized a beam transport line prototype composed by a quadrupole-based focused-system and by an energy selector allowing to transport and select up to 30 MeV protons.

The design and realization of the three different sections of the ELIMED beam line, as well as the development of the diagnostics/dosimetry detectors and the MC simulations, will be carried out in parallel. The beam line, after dedicated tests with conventional accelerator beams and laser-driven beams, will be finally assembled at ELI-Beamlines at the end of 2017.

ELEMENTS OF THE BEAM LINE PROTOTYPE

The Focusing System Prototype

The permanent magnet quadrupoles (PMQs) prototype, designed and realized at the INFN-LNSS, consists of two PMQs of 80mm length and two PMQs of 40 mm length with an active bore of 20 mm and 1 mm thick shielding pipe, placed inside the bore for the protection of the magnets. The magnetic field gradient is about 100 T/m [17]. The system has been designed to improve the transmission and the selection efficiencies of a magnetic energy selector prototype based on four permanent magnet dipoles with alternating field gradient (see section 3.2 for a detailed description).

The PMQs are based on a hybrid Halbach cylindrical array with eight sectors, as shown in Figure 1. The poles are set at 45° with respect to the horizontal axis and are attached to four iron sections, almost saturated, used as supporting structure as well as magnetic flux guides. The poles have a rectangular main body (13 × 14 mm^2) with two smaller pieces close to the bore, which allow increases of the field and, hence, the gradient inside the bore itself. The T-like magnets between two poles are modeled as three independent squared pieces (10 × 10 mm^2). The magnetic features of each section of the quadrupole have been evaluated from the BH curve of the materials, NdFeB N50 for the magnets and iron XC10 for the other four parts. Each magnet has its own direction for the permanent magnetization (see red arrows in Figure 2).

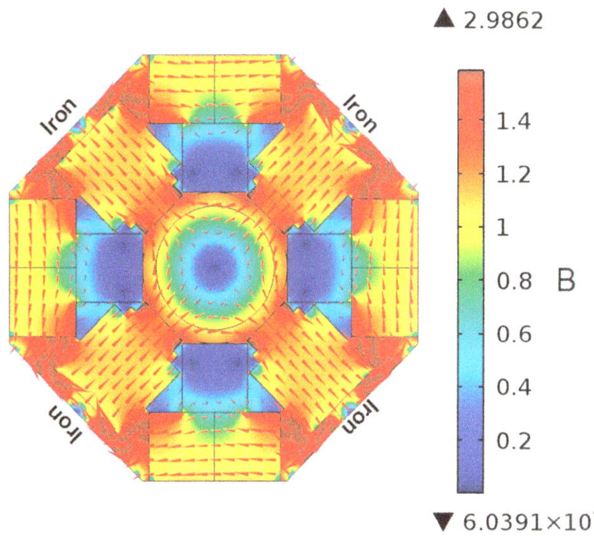

Figure 2: The layout of the PMQs system prototype based on hybrid Halbach design.

Design based on hybrid Halbach design with standard rectangular magnetic blocks. Field maps are simulated with 0.5 mm step and scaled according to the measurements provided by the Sigma Phi Company (Vannes, France) during the manufacturing process [17].

The system can be used for focusing and collecting protons up to 30 MeV, although, according to the simulation results, the best performances can be achieved for energies lower than 20 MeV. An example of particle tracing performed using the code for optics calculation Trace Win [24], in the PMQs system optimized for the 5 MeV proton injections in the energy selector prototype is shown in Figure 3. The ESS is simulated as a three collimator system in line: collimator 1 and 2, with a 10-mm diameter aperture, are used to remove from the beam particles with large divergence, the slit is a rectangular aperture of 1 × 10 mm and is responsible for the energy selection and the selected energy resolution. The focusing system ensures a transmission efficiency of about 1%–2% for the whole beam line.

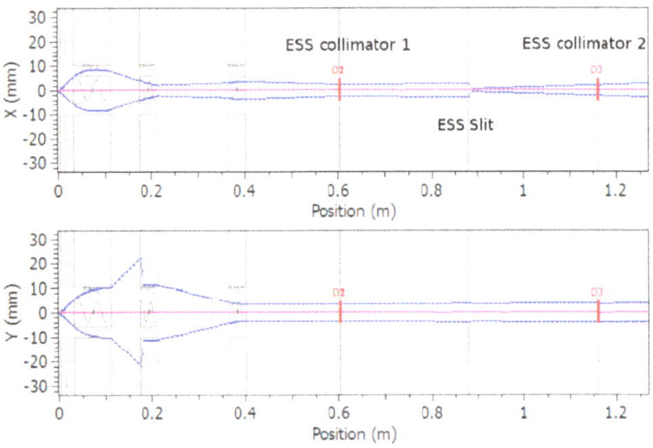

Figure 3: Horizontal envelop (**upper panel**) and vertical envelop (**lower panel**) of the beam line for a monochromatic beam (5 MeV) with a spot size of 100 µm and 170 mrad half angle divergence.

The PMQs prototype has been tested with 10 MeV proton beam delivered by the Tandem accelerator at the LNS. Figure 4shows the comparison between the 10 MeV proton beam spot distribution predicted by the beam optics simulations and the experimental data taken using Gafchromic film, EBT3 type. As one can see from the comparison, a good agreement between the simulation and the measured spot size has been obtained.

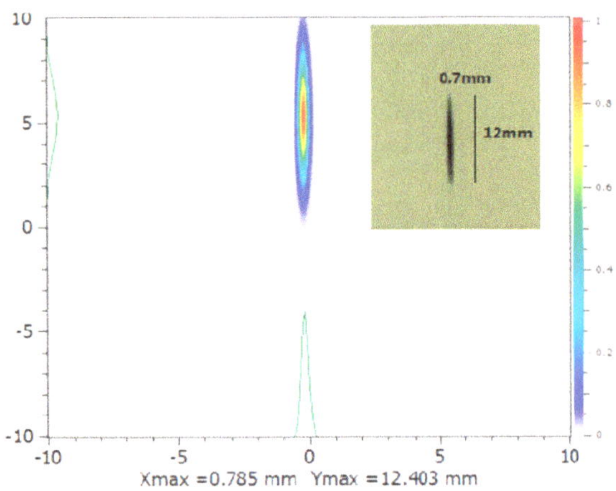

Figure 4: Comparison between the simulation results and the experimental measurements (inset) for the 10 MeV proton beam spot along the PMQs system.

A new system, characterized by higher field uniformity and capable of focusing ions with energies up to 60 MeV/n, has been recently designed for the ELIMED section of the ELIMAIA beam line. It will consist of 5 PMQs with the characteristics described in the Table 1. This is the lowest proton energy of interest for medical applications, for instance, eye melanoma are treated worldwide with 60 MeV proton beams. This is the reason why transport and selection of 60 MeV/n ions represents our first goal in the feasibility studies related to the applications of laser driven ion beams for therapeutic purposes. After few years of operation, different acceleration regimes will be investigated at the ELIMAIA facility reaching higher energies and, hence, extending the studies at least up to 250 MeV, which is the maximum energy used in proton therapy.

Table 1: The ELIMED PMQs system features.

n° of PMQ	Geometric length	Field gradient	Bore diameter
1	160 mm	101 T/m	30 mm
2	120 mm	99 T/m	30 mm
2	80 mm	94 T/m	30 mm

The layout of the PMQs is shown in Figure 5. It is based on a standard trapezoidal Halbach array surrounded by two external hybrid arrays made of rectangular block and iron. The inner array is mainly responsible for the field quality of the quadrupole, the external arrays are necessary to increase the volume of the permanent magnet material to reach the required field gradient.

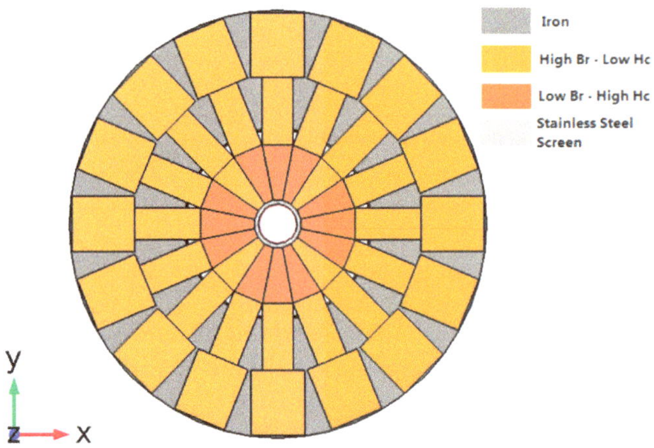

Figure 5: ELIMED PMQs layout.

The Energy Selector System Prototype

A prototype of an energy selector system (ESS), designed and developed at the INFN-LNS a few years go [25], is based on four permanent magnet dipoles with alternating fields, similar to a bunch compressor chicane [26]. Each dipole is 88 mm long; the drift space between the central and peripheral dipoles is 85 mm whereas a 10 mm drift distance separates the central dipoles. In such drift space, the particles with different energies are dispersed along the radial plane and a moving slit is used to select ions within a certain energy range. Figure 6 shows a scheme of the ESS prototype together with the parameters for the calculation of the path length of a proton inside the magnetic chicane, for the calibration equation and for the acceptance of the system.

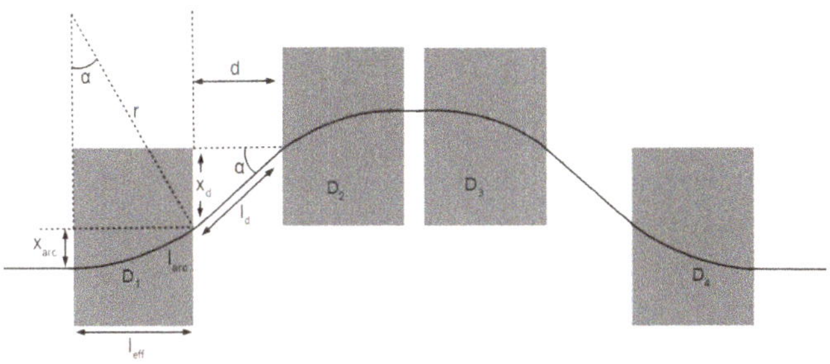

Figure 6: Scheme of the ELIMED ESS prototype.

The ESS prototype has been successfully calibrated using accelerator-driven proton beams with FWHM energy spread of 0.5% in the energy range between 2.5 and 5 MeV at the Laboratori Nazionali di Legnaro (LNL) in Padova and between 4.5 and 12 MeV at the LNS in Catania. Figure 7 shows an example of the experimental results of the slit positions as a function of the energy measured in the energy range available at the LNL, along with a comparison with theoretical predictions. Based on the parameters shown in Figure 6, the equation describing the slit position as a function of the energy can be derived:

$$T = \sqrt{\left(\frac{qx(\Delta x - \tan\alpha_2)}{\frac{(1-\cos\alpha_1)}{B_1} + \frac{(1-\cos\alpha_2)}{B_2}}\right)^2 + M_0^2 c^4} - E_0 \quad (1)$$

where q, M_0 and E_0 are charge, mass and rest energy of the ion beam, respectively, Δx is the slit position, α_i and B_i are the bending radius and the magnetic field of the first and the second dipoles, respectively. As shown in Figure 7 (lower panel) the Equation (1) can evaluate the ion energy in a very good agreement with the experimental measurements (upper panel).

The energy resolution of the system can be estimated from the derivative of Equation (1) as shown in Equation (2).

$$E_{res} = \left(\frac{dT}{dx}\right) \times \Delta x = \frac{\left(\frac{qc}{\frac{1-\cos\alpha_1}{B_1} + \frac{1-\cos\alpha_2}{B_2}}\right) \times \sqrt{(E_0+T)^2 - M_0^2 c^4}}{E_0 + T} \times \Delta x \quad (2)$$

According to the scaling factor, the experimental results indicate that the system can select protons with an energy resolution of about 30% at 30 MeV ensuring a better resolution for lower energies.

The experimental results have also been compared with the GEANT4 predictions showing a rather good agreement and allowing a fine-tuning of the simulation parameters.

The ESS prototype has been also characterized with laser-driven proton beams at the TARANIS laser facility (Queen's University, Belfast, UK). During the test experiment two energies, 4.5 and 7 MeV, have been selected using the selection system. Energy spreads, fluence, angular divergence, spot sizes of the proton beam coming out from the energy selection system have been measured for both energies selected. Preliminary results show energy spreads of about ±7% and ±8% for 4.5 and 7 MeV, respectively. The measured average fluence is 5×10^6 p/cm^2 when a 4.5 MeV proton beam is selected and about 10^6 p/cm^2 in the case of 7 MeV protons. The results of this experiment will be published elsewhere.

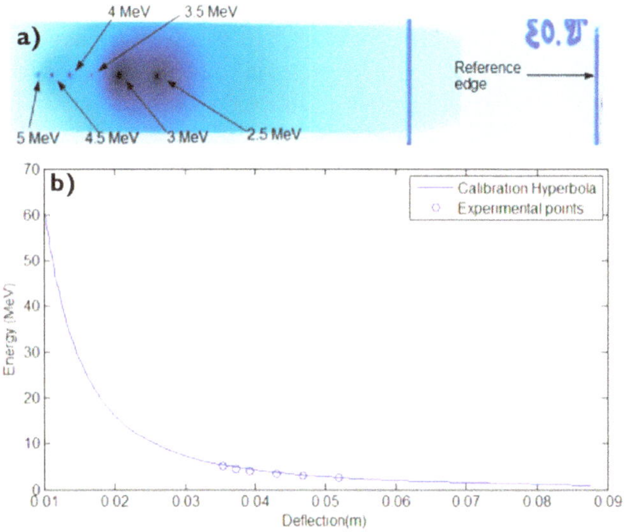

Figure 7: (**a**) Experimental measurements of the slit position as a function of the energy, using accelerator-driven proton beam in the energy range between 2.5 and 5 MeV and (**b**) comparison with the calibration Equation (1) of the energy selection system.

The predicted transmission efficiency of the ESS prototype is quite low (few particles out of ten thousands). Nevertheless, the PMQs system coupled to the ESS will allow increased efficiency up to 1%–2% and also selecting beam more precisely since the energy selection is particularly sensitive to the particle divergence. The beam optics simulation of the whole system, composed by the PMQs and ESS prototypes, is shown in Figure 8.

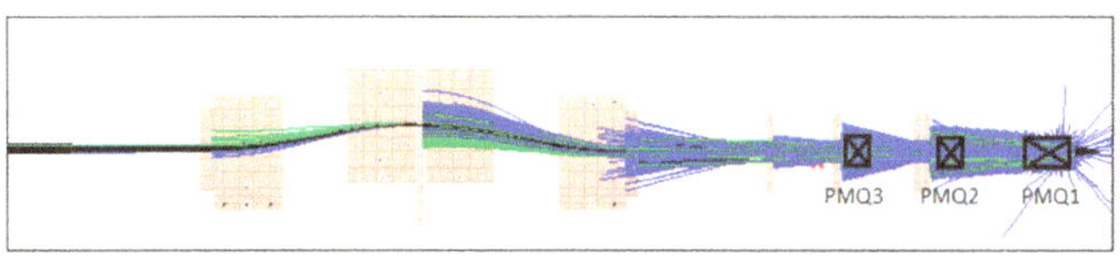

Figure 8: The results of the full beam optics simulation for 5 MeV proton selection (black trajectories). Blue traces represent lower energies and green trajectories represent higher energies. A proton beam is transported from the right to left direction.

A new energy selection system dedicated to the ELIMED section of the ELIMAIA beam line will be designed by the end of the year. It will be based on four electromagnets and will allow the selection of proton and carbon beams up to a maximum energy of 60 MeV/n with a maximum energy spread of 30%.

On-line Beam Diagnostics and Dosimetry

The pulse properties of optically accelerated ion beams differ significantly from those commonly provided by conventional accelerators in pulse duration, peak current and correspondingly pulse dose rate and energy spectrum. Thus, among obvious properties, such as operational stability, the development of innovative techniques for diagnostics and dosimetry represents a crucial step towards multidisciplinary applications of laser-driven beams with the required uncertainty.

Typical laser-driven beam specifications report proton burst duration of the order of 0.1–1 ns, with intensity ranging from 10^{10} to 10^{12} p/burst. Therefore, the detectors developed for the on-line beam monitoring systems and for the relative dosimetry have to be dose-rate independent, in order to be able to measure very intense and short pulses without saturation effects and suitable to operate in presence of a strong electromagnetic pulse (EMP). Different alternative approaches will be followed in the development of detectors for on-line non-invasive beam current monitoring.

Different diagnostics detectors will be placed, in particular, after the focusing and the selection systems. A detector, based on the well-known pepper pot method, will be specifically designed and optimized for single shot emittance measurement of the laser-driven proton beam after the PMQs system. On the other hand, particle identification and fluence/current measurements will be performed using Silicon Carbide (SiC), CVD diamond detectors and secondary particle emission detectors (SEM), based on the secondary electron emission from a thin metallic foil hit by ion beams. These radiation-hard detectors, suitable to work in harsh radiation environments, will be used in ToF mode for plasma ion intensity and energy distribution measurements in the energy range between few MeV/n up to 60 MeV/n. They will be placed at different distances from the target along the beam-line, in particular, after the PMQs system and after the energy selection device at the end of the in-vacuum section of the beam-line. The high temporal resolution and the fast response of SiC and CVD detectors, together with the high signal-to-noise ratio characteristics, already tested in the previous experimental campaign at PALS and at TARANIS laser facilities, will allow determining the energy of the identified selected ions with resolution of about 20% for the maximum energy selected (*i.e.* 60 MeV/n).

So far, no protocol for absolute dosimetry for optically accelerated ion beams has been established. In order to fulfill this task, a reliable and accurate dosimetric characterization of laser-driven charged particle beams has to be performed. Therefore, devices and procedures to develop a calibration method for absolute dose evaluation have to be implemented.

The device proposed to perform absolute dose measurements will consist of a Faraday Cup (FC) specifically designed to measure the beam charge with high accuracy and extract the absolute dose for pulsed laser-driven beams. Indeed, thanks to the linear response with respect to the beam intensity and the dose rate independent signal, a FC is expected to allow charge collection with high level of accuracy, also in case of high pulsed and extremely intense beams.

A FC prototype [27] for absolute dosimetry with high-pulsed ion beams has been recently designed and realized at the INFN-LNS within the ELIMED project. The FC technical design has been inspired to similar detectors already developed for ion-beam dosimetry [22, 28], adopting innovative geometrical solutions in order to further improve the overall charge collection efficiency [29].

In the absolute dose measurement using a FC, the main uncertainty sources are related to the precise measurements of the total charge carried by the beam, of the proton beam energy spectrum and of the effective beam area, needed to extract the fluence distribution. Each of these quantities has to be separately and accurately evaluated. An issue related to the total charge measurements is represented by the emission of secondary particles, usually electrons, produced by the interaction of the protons with the entrance foil and the cup material itself. The emitted secondary electrons may be either collected or leave the cup, affecting the collected charge measurements. The secondary electrons produced in the vacuum window former may lead to a total charge underestimation, while those produced within the cup material, if produced with sufficient energy, may leave the collecting cup and cause a charge overestimation. In order to maximize the charge collection measurement accuracy by increasing the probability of capturing the secondary electrons emitted, shape, sizes, materials and electric field characteristics have been modeled and determined, with the help of the GEANT4 MC simulations and electrostatic ion trajectory software, such as the COMSOL FEM (Finite Element Method) [30] and the Simion FEA (Finite Element Analyse). These studies aimed to increase the efficiency and reliability of the FC design as a charged particle current detector. Figure 9 shows the final FC prototype configuration with all the geometrical details. The FC prototype technical details and the simulation results on the electric field generated can be found in ref. [27].

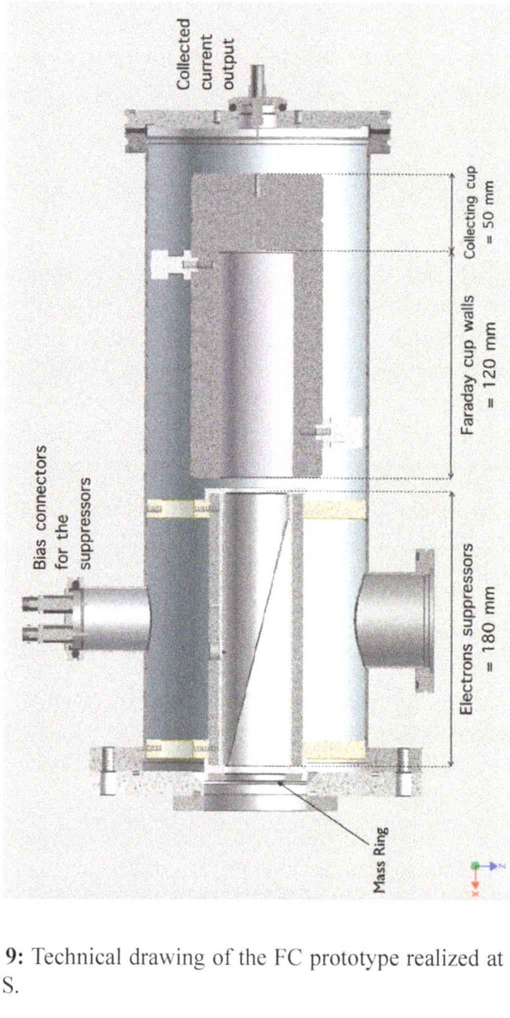

Figure 9: Technical drawing of the FC prototype realized at the INFN-LNS.

Typical FC detectors have cylindrically symmetric components and employ coaxial electrostatic fields to recapture the ejected electrons. In the FC prototype, the electric field results by the combined effect of two coaxial electrodes. The external electrode is a metallic hollow cylinder and the internal one, follows a similar design presented by Thomas *et al.* [29], and is a peculiar beveled cylinder. The addition of this element breaks the electrostatic coaxial symmetry of the field, provided solely by the external electrode, by introducing a relevant transverse electric field component, as it can be seen in Figure

10, which allows recapturing secondary electrons improving the deflection of the secondary electrons emitted by both the entrance window and the cup.

In order to investigate the FC prototype response, an experimental campaign has been performed using the 62 MeV proton beams delivered by the superconducting cyclotron (CS) at the CATANA protontherapy facility [31] at the INFN-LNS (Catania, Italy). A complete characterization of the FC prototype charge measurement varying the bias applied to the suppression electrodes, the proton beam incident current and the dose delivered has been also performed. The FC prototype has been placed about 9 cm downstream from the final collimator along the beam-line, a 20 mm diameter collimator has been used for the measurements. The FC operating vacuum was 10^{-5} mbar. The FC current signal as a function of the collecting time was measured using of the electrometer Keithley 6517A.

In order to compare the secondary electron suppression efficiency using the external electrode alone and the two coaxial electrodes together, the characterization of the FC prototype charge collection as a function of the applied voltage has been performed using both configurations. The voltage has been varied from −4000 Volt to 4000 Volt. Each charge measurement has been performed delivering a given dose of 15 Gy. The dose released has been monitored using the online transmission ionizations chambers. Figure 11 shows some preliminary results of the collected charge as a function of the applied voltage. Each charge measurement has been obtained integrating the current signal for the time interval required to deliver 15 Gy.

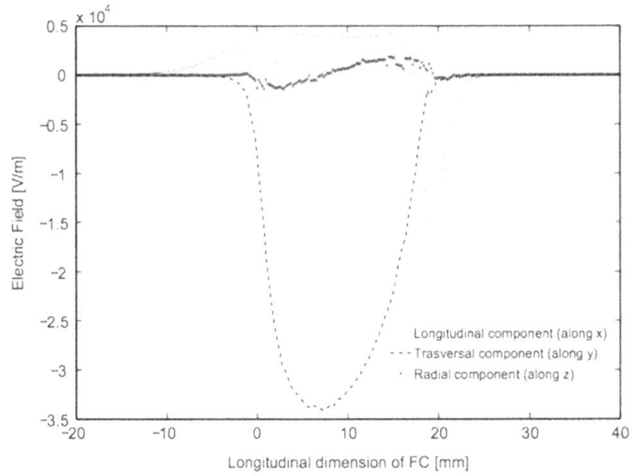

Figure 10: Transversal, longitudinal and radial components of the FC prototype electric field as a function of the FC longitudinal size, obtained using the COMSOL software (COMSOL, Brescia, Italy).

As one can see, a typical flat voltage response plateau in the negative voltage range between −2400 V and −400 V can be observed; on the other hand, for positive voltages a collected charge dependence increasing with the voltage applied is evident. The charge collection response observed in the FC prototype is in agreement with the typical behavior of standard FCs used for absolute dosimetry [22,28]. The analysis of the charge measurement data using the internal beveled electrode is still on going. Nevertheless, based on a preliminary comparison the electric field resulting from the two coaxial electrodes improves the electron suppression efficiency.

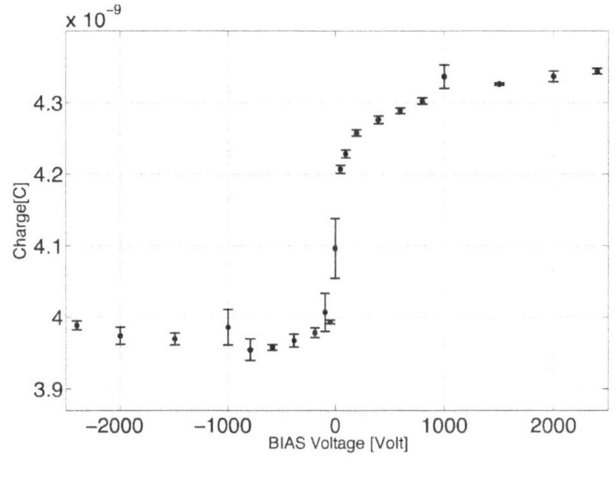

Figure 11: Charge collected by the FC prototype as a function of the bias applied to the external electrode.

Recently, characterizations of the electromagnetic pulse (EMP) effect on the FC prototype signal has been carried out at the PALS laser facility, in Prague (Czech Republic) and at the TARANIS laser facility at the Queen's University in Belfast (UK). Typically, a high power laser-matter interaction and the subsequent plasma production are accompanied by the emission of an intense EMP, which propagates inside and outside the interaction chamber [32] interfering with the diagnostic devices and affecting their response. Therefore, particularly for a dosimeter, it is crucial to perform a systematic study, in terms of frequencies and amplitude of the EMP generated by the laser pulse, to perform accurate dose evaluations. Indeed, the EMP produces a not-negligible background signal that, if not carefully characterized, does not allow a precise charge measurement [32].

The FC prototype has been tested by using a laser-driven proton beam during an experiment [27] performed at the PALS laboratory (Prague, Czech Republic), where a 2 TW laser system (about 1 KJ of energy delivered on target and a time duration of 300 ps) is available. The FC was mounted inside the experimental hall at approximately 2 and 5 m from the interaction chamber for a characterization of the EMP propagating outside of the chamber and directly connected to the interaction chamber at 30° in the backward direction with respect to the target normal axis. The FC signal was registered with a 2 GHz Le Croy Digital Oscilloscope. Figure 12 shows a typical FC signal registered during the experiment.

As one can see from the inset in Figure 12, the EMP is characterized by frequency components from few hundreds of MHz up to 2 GHz. Based on the frequency analysis, the significant noise signal results from the EMP propagation inside the interaction chamber [33, 34] and seems to be predominant with respect to the charge signal, affecting the measurement of the total collected charge. Nevertheless, the cumulative integration of the FC signal, as shown in Figure 13, reveals the presence of a charge contribution that, although very small, can be separated by the huge EMP noise contribution [27]. Indeed, even though the EMP signal is present, a null final cumulative value is expected due to their identical positive and negative contributions to the integral. On the other hand, for those signals registered when the FC is mounted on the interaction chamber (*i.e.* along the plasma emission direction) a signal generated by the incoming charged particles accelerated from the target is expected to be present together with the typical EMP contribution. Therefore, as it can be observed in Figure 13, the cumulative of those signals deviates from a null final integral showing a decreasing-negative trend, which indicates a continuous negative contribution to the integrated signal.

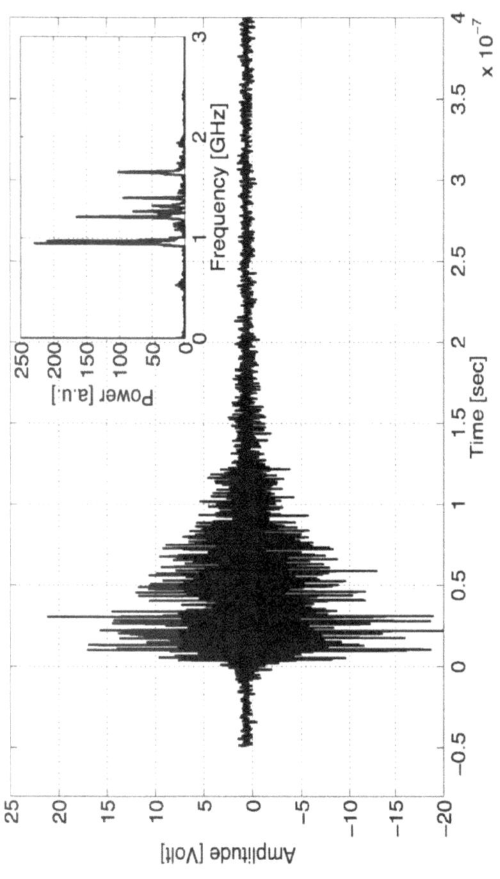

Figure 12: A FC signal registered during the PALS experiment. The inset shows the Fast Fourier transformation of the signal.

This result indicates the necessity to perform a thorough study to allow the separation of the noise contribution from the physical signal. The use of appropriate electronic circuits, as filter amplifiers, and also proper EMP shielding will allow removing or reducing the higher frequency contributions related to the EMP.

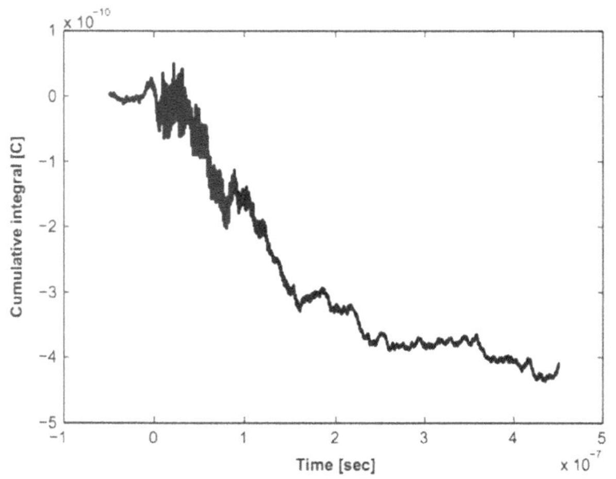

Figure 13: The cumulative integration of the FC signal.

The analysis of the data taken during the experimental campaign performed at the TARANIS laser facility at the Queen's University in Belfast (UK) is currently on going and the results will be published elsewhere.

CONCLUSIONS

In this paper an overview of the ELIMED beam line section, that will be installed at the ELI-Beamlines facility in Prague, has been presented. In particular, the prototypes of the quadrupoles and of the energy selector (ESS) have been shown and some results of the tests performed with both accelerator-driven and laser-driven beams have been discussed. The focusing properties of the quadrupole prototype have been tested at the INFN-LNS in Catania with 10 MeV proton beams, confirming the performances predicted by the beam optics simulations. The energy selector prototype has been calibrated at INFN-LNS and INFN-LNL with proton beams in the energy range between 2.5 and 12 MeV, demonstrating that this system can select protons with an energy resolution of 30% at 30 MeV, ensuring a better resolution at lower energy. Moreover, the ESS has been also characterized with laser-driven proton beams at the TARANIS facility in Belfast (UK), obtaining energy spreads of 7% and 8% selecting, respectively, 4.5 and 7 MeV proton energies. For both devices, respectively involved in the focusing and selection of the laser-driven ion beams, a feasibility study has been carried out in order to realize the final versions to be assembled in 2017 along the ELIMAIA beam line in Prague. The feasibility study of the new quadrupoles has been already completed and the characteristics have been described in detail in this paper. The design of the final version of the ESS is still in progress.

Concerning the diagnostics, different solutions will be adopted and implemented for the ELIMAIA beam line in Prague, involving the use of SiC, CVD diamonds and SEM detectors, used in ToF mode for proton energy distribution measurements in the energy range between few MeV/n up to 60 MeV/n. Some of them have been tested during previous experimental campaigns at PALS and TARANIS to characterize them in the laser environment. In the same environment, a prototype of a FC, built with innovative solutions for the internal electrode, has been also tested in order to characterize the EMP and analyze its frequency components. These studies are of crucial importance since the EMP, typically present in the laser-driven ion facilities, can affect the measurements as it is predominant over the detector signals. In particular for the FC, the obtained results would support adopting specific solutions to disentangle such EMP contributions from the physical signal, for instance filter amplifiers and/or dedicated EMP shielding. A dosimetric characterization of the FC has been also carried out with accelerator-driven beams at INFN-LNS. In particular, the FC response, as a function of the applied voltage to the electrode for the electron suppression, has been measured for voltages ranging between −4000 V and 4000 V. A typical flat response has been observed between −2400 V and −400 V, as confirmed by other authors.

Most of the activities related to the prototypes have been carried out aiming to preliminarily study and characterize the physical properties of laser-driven ion beams and optimize the final versions of the main transport elements, as well as the detectors. The ELIMED beam line section will be delivered and assembled in Prague, where a first user's facility dedicated to multidisciplinary studies will offer the possibility to study the biological properties and the potentialities of laser-driven ion beams with well controlled systems. By means of the transport devices and the diagnostics/dosimetric systems, precise dosimetric measurements and accurate cell sample irradiations will be possible at the ELIMAIA beam line, giving the possibility to study the characteristics of these beams from the clinical point of view and, therefore, provide a deeper understanding of their peculiarities for a possible future use in the medical field.

ACKNOWLEDGMENTS

This work has been performed within the ELIMED activities supported by the INFN (Italian Institute for Nuclear Physics), the MIUR (Italian Ministry of Education, Research and University) and by the ELI-Beamlines Contract no S 14-187, under Laser Gen (CZ.1.07/2.3.00/30.0057), under the Ministry of Education, Youth and Sports of the Czech Republic (ELI-Beamlines reg. No. CZ.1.05/ 1.1.00/02.0061), the institution Fyzikalni ustav, AV CR, v.v.i and under the project, co-financed by the European Social Fund and the state budget of the Czech Republic.

CONFLICTS OF INTEREST

The authors declare no conflict of interest.

REFERENCES

1. Veksler, V. The principle of coherent acceleration of charged particles. *Sov. J. Atom. Energy* 1957, *2*, 525–528.
2. Clark, E.L.; Krushelnick, K.; Davies, J.R.; Zepf, M.; Tatarakis, M.; Beg, F.N.; Machacek, A.; Norreys, P.A.; Santala, M.I.K.; Watts, I.; *et al.* Measurements of Energetic proton transport through Magnetized Plasma from Intense Laser Interactions with Solids. *Phys. Rev. Lett.* 2000, *84*, 670–673.
3. Maksimchuk, A.; Gu, S.; Flippo, K.; Umstadter, D.; Bychenkov, Y.V. Forward ion acceleration in thin films driven by high-intensity laser. *Phys. Rev. Lett.* 2000, *8*.
4. Snavely, R.A.; Key, M.H.; Hatchett, S.P.; Cowan, T.E.; Roth, M.; Phillips, T.W.; Stoyer, M.A.; Henry, E.A.; Sangster, T.C.; Singh, M.S.; *et al.* Intense high-energy proton beams from petawatt-laser irradiation of solids. *Phys. Rev. Lett.* 2000, *85*.
5. Gitomer, S.J.; Jones, R.D.; Begay, F.; Ehler, A.W.; Kephart, J.F.; Kristal1, R. Fast ions and hot electrons in the laser-plasma interaction. *Phys. Fluids* 1986, *29*, 2679–2688.
6. Fews, A.P.; Norreys, P.A.; Beg, F.N.; Bell, A.R.; Dangor, A.E.; Danson, C.N.; Lee, P.; Rose, S.J. Plasma ion emission from highintensity picosecond laser pulse interactions with solid targets. *Phys. Rev. Lett.* 1994, *73*.
7. Macchi, A.; Borghesi, M.; Passoni, M. Ion acceleration by superintense laserplasma interaction. *Rev. Mod. Phys.* 2013, *85*.
8. Robinson, A.P.L.; Zepf, M.; Kar, S.R.; Evans, G.; Bellei, C. Radiation pressure acceleration of thin foils with circularly polarized laser pulses. *New J. Phys.* 2008, *10*.
9. Macchi, A.; Veghini, S.; Pegoraro, F. Light sail acceleration reexamined. *Phys. Rev. Lett.* 2009, *103*.
10. Esirkepov, T.; Borghesi, M.; Bulanov, S.V.; Mourou, G.; Tajima, T. Highly efficent relativistic-ion generation in the laser-piston regime. *Phys. Rev. Lett.* 2004, *92*.
11. Yin, L.; Albright, B.J.; Bowers, K.J.; Jung, D.; Fernández, J.C.; Hegelich, B.M. Three-dimensional dynamics of breakout afterburner ion acceleration using high-contrast short-pulse laser and nanoscale targets. *Phys. Rev. Lett.* 2011, *107*.
12. Schwoerer, H.; Pfotenhauer, S.; Jäckel, O.; Amthor, K.-U.; Liesfeld, B.; Ziegler, W.; Sauerbrey, R.; Ledingham, K.W.D.; Esirkepov, T. Laser-plasma acceleration of quasi-monoenergetic protons form microstructured targets. *Nature* 2006,*439*, 445–448.
13. Henig, A.; Steinke, S.; Schnürer, M.; Sokollik, T.; Hörlein, R.; Kiefer, D.; Jung, D.; Schreiber, J.; Hegelich, B.M.; Yan, X.Q.; *et al.* Radiation-pressure acceleration of ion beams driven by circularly polarized laser pulses. *Phys. Rev. Lett.*2009, *103*.
14. Zeil, K.; Kraft, S.D.; Bock, S.; Bussmann, M.; Cowan, T.E.; Kluge, T.; Metzkes, J.; Richter, T.; Sauerbrey, R.; Schramm, U. The scaling of proton energies in ultrashort pulse laser plasma acceleration. *New J. Phys.* 2010, *12*.
15. Halbach, K. Physical and optical properties of rare earth cobalt magnets. *Nucl. Instrum. Methods* 1981, *187*, 109–117.
16. Mihara, T.; Iwashita, Y.; Kumada, M.; Spencer, C.M. Superstrong Adjustable Permanent Magnet for a Linear Collider Final Focus. In Proceedings of the 12th Linear Accelerator Conference, Lubeck, Germany, 16–20 August 2004.
17. Schillaci, F.; Maggiore, M.; Cirrone, G.A.P.; Cuttone, G.; Giove, D. Errors and optics study of a permanent magnet quadrupole system. *JINST* 2015, *10*.
18. Kramer, D.; Dehning, B.; Holzer, E.B.; Feriolia, G.; Stockner, M. Simulation and measurements of secondary electron emission beam loss monitors for LHC. *Nucl. Phys. B* 2007, *172*, 246–249.
19. Szydlowskia, A.; Badziakb, J.; Fuchsc, J.; Kubkowskab, M.; Parysb, P.; Rosinskib, M.; Suchanskab, R.; Wolowskib, J.; Anticic, P.; Mancicc, A. Application of solid-state nuclear track detectors of the CR-39/PM-355 type for measurements of energetic protons emitted from plasma produced by an ultra-intense laser. *Radiat. Meas.* 2009, *44*, 881–884.
20. Devic, S.; Seuntjens, J.; Sham, E.; Podgorsak, E.B.; Schmidtlein, C.R.; Kirov, A.S.; Soares, C.G. Precise radiochromic film dosimetry using a flat-bed document scanner. *Med. Phys.* 2005, *32*.
21. Boag, J.W.; Wilson, T. The saturation curve at high ionization intensity. *Br. J. Appl. Phys.* 1952, *3*.
22. Cambria, R.; Hérault, J.; Brassart, N.; Silari, M.; Chauvel, P. Proton beam dosimetry: A Comparison between the Faraday Cup and an ionization chamber. *Phys. Med. Biol.* 1997, *42*, 1185–1196.
23. Geant4. Available online: http://geant4.cern.ch (accessed on 17 August 2015).

24. TraceWin 2.8.2.4. Available online: http://irfu.cea.fr/Sacm/logiciels/index3.php (accessed on 17 August 2015).
25. Scuderi, V.; Jia, S.B.; Carpinelli, M.; Cirrone, G.A.P.; Cuttone, G.; Korn, G.; Licciardello, T.; Maggiore, M.; Margarone, D.; Pisciotta, P.; *et al.* Development of an energy selection system for laser-driven proton beam applications. *Nucl. Instrum. Methods .Phys. Res. Sect. A* 2014, *740*, 87–93.
26. Castro, P. Beam trajectory calculations in bunch compressors of TTF2. DESY Technical Note 01–03; April; 2003.
27. Cirrone, G.A.P.; Romano, F.; Scuderi, V.; Amato, A.; Candiano, G.; Cuttone, G.; Giove, D.; Korn, G.; Krasa, J.; Leanza, R.; *et al.* Transport and dosimetric solutions for the ELIMED laser-driven beam line. *Nucl. Instrum. Methods Phys. Res. Press* 2015.
28. Cuttone, G.; Raffaele, L.; Tonghi, B.L.; Rovelli, A.; Sabini, M.G.; Egger, E.; Kacperek, A.; Brai, M.; Bartolotta, A.; Teri, G.; *et al.* First Dosimetry Intercomparison Results for the CATANA project. *Phys. Med.* 1999, *15*, 121–130.
29. Thomasa, J.D.; Hodgesa, G.S.; Seelyb, D.G.; Morozb, N.A.; Kvale, T.J. Performance enhancement study of an electrostatic Faraday Cup detector. *Nucl. Instrum. Methods Phys. Res. A* 2005, *536*, 11–21.
30. COMSOL. Available online: http://www.comsol.it (accessed on 17 August 2015).
31. Cuttone, G.; Cirrone, G.A.P.; di Franco, G.; la Monaca, V.; Nigro, S.L.; Ott, J.; Pittera, S.; Privitera, G.; Raffaele, L.; Reibaldi, A.; *et al.* CATANA protontherapy facility: The state of art of clinical and dosimetric experience. *Eur. Phys. J. Plus* 2011, *126*, 1–7.
32. Mead, M.J.; Neely, D.; Gauoin, J.; Heathcote, R.; Patel, P. Electromagnetic pulse generation within a petawatt laser target chamber. *Rev. Sci. Instrum.* 2004, *75*, 4225–4227.
33. Park, S.H.; Jeong, Y.U.; Lee, K.T. Faraday cup measurements of a laser-induced plasma for a laser-proton acceleration.*J. Korean Phys. Soc.* 2006, *49*, 342–346.
34. Marco, M.D.; Pfeifer, M.; Krousky, E.; Krasa, J.; Cikhardt, J.; Klir, D.; Nassisi, V. Basic features of electromagnetic pulse generated in a laser-target chamber at 3-TW laser facility PALS. *J. Phys. Conf. Ser.* 2014, *508*.

CHAPTER 16

RECENT DEVELOPMENTS IN LASER DRIVEN INERTIAL FUSION

M. Ghoranneviss and A. Salar Elahi

INTRODUCTION

Controlled generation of fusion energy for power stations received an essential turning point by interaction of picosecond laser pulses of powers above terawatts with plasmas resulted in ultrahigh acceleration of plasma layers with a thickness of dielectric increased skin depths [1]. These plasma blocks contained directed energetic ions with extremely high ion current densities opening a new way of ignition of fusion by direct laser generation of a fusion flame in uncompressed solid fuel [2]. This reaction as a kind of side-on ignition was considered before as the following described hydrodynamic Chu-model but only the new experimental results of the plasma blocks provided the necessary conditions for a plane geometry interaction for fusion of deuterium-tritium. Fast ignition was the alternative first formulated by Tabak et al. [3] and elaborated [4] following the discussions about the highest deuterium-tritium (DT) fusion gains based on shock-free spherical ideal compression at thermal ignition [5–7]. The gains of more than 10^{12} fusion neutrons were significantly higher than all earlier attempts with exploding pushers [8], where shocks led to strong degrading deviations from the ideal adiabatic implosion. The ideal adiabatic compression led to the highest gains at volume burn and the discovery of volume ignition [9], confirmed as "Wheeler modes" [10] and then summarized [8]. The discovery of this shock-free compression was reached systematically by varying the experimental parameters in parallel to computations, which, in contrast to those of exploding pushers, arrived at the highest gains as predicted [9], and the result [5] of which was called "Yamanka compression." Gains above 10^{13} neutrons were achieved [11] and as well as compression to 2000 times that of the solid state, where it was crucial that laser beam smoothing [12] be used [13]. Further improvement to gains led to gains of 2×10^{14} neutrons [14] all with direct drive and volume ignition [8] in contrast to about 1000 times lower gains with indirect drive spark ignition [15]. What was most disappointing was the rather low maximum temperature T at the highest compression [11] of only about = 3 × 106 K, while the theory of spherical compression and thermal ignition said it should have reached higher temperatures. Nevertheless, the spherical compression and thermal ignition scheme was followed up by building the $4 Billion laser facility NIF (National Ignition facility) under the initial leadership of Campbell (2006) followed by Moses et al. (2006) leading close to the historic first controlled exothermic nuclear fusion reactions [16], while simultaneously it was considered by Tabak et al. [3] to overcome the problem of the unexpected low temperature at compressions [11] by "fast ignition." The scheme at the first discussions (2006) was to use the spherical plasma compression to more than 1000 times the solid state density and to produce an additional heating of the center area by an additional laser pulse of picoseconds duration with an energy in the range above 10^5 J. In order to produce laser pulses of this short duration and power above terawatt (TW) and up to more than petawatt (PW), the chirped pulse amplification CPA [17, 18] was invented for solid state lasers. Another method is the dye laser pulse amplification by KrF lasers known as the Szatmari-Schäfer method [19, 20]. Since the controlled generation of clean, safe, virtually inexhaustive, and everywhere accessible fusion energy at sufficiently low cost in power stations using nuclear fusion reactions is an extremely important goal to overcome the problems inherent in the present energy sources, an enormous budget has been

Citation: M. Ghoranneviss and A. Salar Elahi, "Recent Developments in Laser Driven Inertial Fusion,", vol. 2014, Article ID 802054, 14 pages, 2014. doi:10.1155/2014/802054.

and will further be spent into research to solve this problem. A summary of this work can be found in a special issue of the IAEA based journal "Nuclear Fusion" which shows how many of the new physics problems have been solved and break-even has almost been reached. The main effort has been devoted to the reaction of DT at temperatures between 10 and 100 Million degrees centigrade where the reacting fuel, a high temperature plasma, has to be confined by magnetic fields (MCF). A review of the forthcoming ITER project and the complementary stellarator program is given in this volume by G. Janeschitz and R. Wolf. So far, the highest fusion gain has been achieved in JET (Joint European Torus) where a magnetically confined plasma torus (D:T ratio of 40:60), heated by 21 MW of neutral beam and 3 MW of electromagnetic wave power, has, for 2 seconds, produced 16 MW of fusion power representing a fusion gain of 66% [21, 22]. The extensive work on IFE and its success, with fusion gains not far below break-even, is traditionally based on lasers with pulses in the ns range and, for the last few years, on the new direction of fast igniters with PS laser pulses as reviewed in the following Section 2. These broad stream IFE developments are modified by a few new experiments which arrived at results significantly different form the broad stream observations. These anomalous results were reported by Sauerbrey, by Jie Zhang et al, and by Badziak et al. and could be understood only by realizing that theses authors used laser pulses of very high quality with suppression of prepulses. The main motivation for this explanation was guided by the knowledge of earlier computations shown in the following Figure 1. A consequence is that a modification of the fast ignition is possible, which may lead to a very simplified and therefore lower cost fusion reactor using plasma blocks with extremely high ion beam current densities. More details of these new aspects are described in Section 3. Sauerbrey's experimental discovery of ultrahigh laser acceleration of plasma by nonlinear forces will be presented in section 4. Analysis of acceleration measurements with doppler effect will be presented in section 5. Also, Application to a radical new laser driven fusion of solid density fuel without nuclear radiation will be presented in section 6. Conclusions will be presented in section 7.

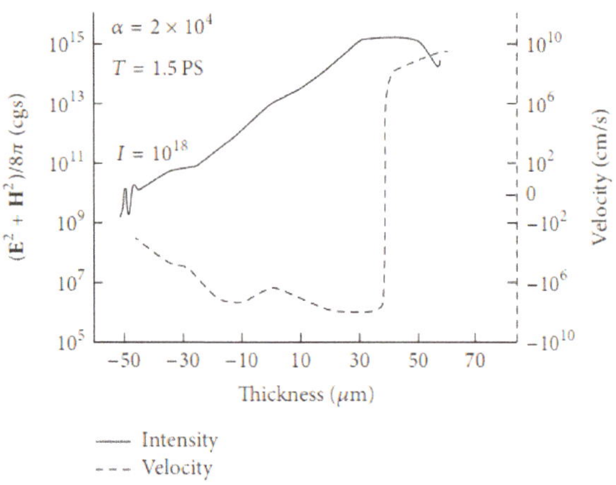

Figure 1: Generation of blocks of deuterium plasma moving against the neodymium glass laser light (positive velocities v to the right) and moving into the plasma interior (negative velocities) at irradiation by a neodymium glass laser of 1018 W/cm^2 intensity onto an initially 100 eV hot and 100 µm thick bi-Rayleigh profile with minimum internal reflection. The electromagnetic energy density $(E^2 + H^2)/(8\pi)$ is shown at the same time of 1.5 PS after start of the constant irradiation.

FUSION ENERGY BY LASERS

It was underlined at the time of the discovery of the laser that irradiation ignition for fusion reactions may be considered using laser radiation. This is based on the fact that Planck radiation of 1 keV temperature has an emission of 1017 W/cm2, an intensity which is well available from lasers. In the following, we describe how some developments to these aims went along three different lines which finally merged in schemes with mutual support.

Volume Ignition

These studies started with the laser fusion computations using the adiabatic self-similarity model as it was used initially in the computations by Basov and Krokhin, Dawson, Engelhard, and Hora. If an energy E_o is uniformly deposited into a volume V_o (with radius R_o) of DT of maximum density *no* related to the solid state density *ns*, with a break-even energy E_{BE}, at an initial zero expansion velocity and an optimum temperature of 17 keV, the subsequent adiabatic expansion results in a fusion gain *G* (DT fusion energy per E_o, with a break-even energy E_{BE} of 6 MJ):

$$G = \left(\frac{E_o}{E_{BE}}\right)^{1/3} \left(\frac{n_o}{n_s}\right)^{2/3} \quad (1)$$

$$= \text{const } n_o R_o. \quad (2)$$

Formula (1) is based on computations of 1964, published by Hora and Pfirsch in 1970, and the algebraically identical formula (2) by Kidder in 1974. When in these computations the alpha self-heat, the X-ray reabsorption, and the fuel depletion were included, the volume ignition was discovered resulting in very high gains up to 2000 and low ignition temperatures down to keV and less at compressions to higher than 1000 times the solid state density. This volume ignition was later confirmed by Kirkpatrick and John Wheeler, by Lackner, by Colgate et al., and by others. The most detailed computations by Xiantu He et al. and Martinez-Val et al. included reheat by neutrons and followed the thermal nonequilibrium between electrons, ions, and the Planck-radiation during the very short times of the ignition process such that the very high fusion gains were confirmed. Initial experiments clearly showed that at laser interaction with irradiated solid spheres most of the laser energy uniformly went into a spherical core fully following the self-similarity expansion (a part of the laser energy went into a nonlinear (ponderomotive) force driven fast plasma ablation). It is remarkable that the later measured highest DT neutron generation at direct laser drive exactly followed the self-similarity model. Any shock generation had to be avoided (stagnation free Yamanaka compression) in contrast to the shock waves at the fast pusher with much lower gains. The highest gain fusion reactions followed volume burn on the way to volume ignition where the physics is "robust" based on the ideal adiabatic dynamics as it was underlined by Lackner and Colgate et al.

Radiation Ignition and Spark Ignition

The motivation for this line came from two arguments. First, it was the enthusiasm that the laser irradiation may arrive at similar conditions of radiation ignition in solid or moderately compressed DT fuel as envisaged from the beginning by John Nuckolls based on the radiation ignition or propagating thermonuclear burn in uncontrolled reactions. The second argument was that the gains of (1) were terribly low before the very high gains for volume ignition were discovered. The ingenious approach to solve both questions was the introduction of the spark ignition scheme not long after 1960. The question of directly igniting solid DT by a laser pulse was answered negatively when computations indicated that the fusion flame needs to have an energy flux density of

$$F > F^* = 10^8 \text{ J/cm}^2, \tag{3}$$

or an irradiation corresponding to an ion current density

$$j > j^* = 10^{10} \text{ Amp/cm}^2. \tag{4}$$

The condition for beam fusion (4) was by more than five orders of magnitude out of possibilities for ion beams. For achieving condition (3), Nuckolls proposed the spark or central core ignition where the laser radiation (or after conversion into X-rays as "indirect drive") compressed the DT in a very sophisticated way. Isobaric compression should result in a very hot but low density central core whose fusion reaction was then to ignite a very high gain fusion detonation wave into the surrounding low temperature high density mantle. The reaction in the core was shown to be an ideal volume ignition and the energy flux density for driving the fusion detonation wave was in the range of $f = 7 \times 10^8$ J/cm^2, fulfilling condition (3). The detailed computation based on otherwise known realistic conditions arrived at generation of 1000 MJ fusion energy produced by a laser pulse of 10 MJ. The problem remains as how to achieve such exclusive temperature and density profiles and how to achieve the necessary orbital symmetry. The study of this spark ignition by indirect drive laser fusion is the aim of the big laser NIF at the Lawrence Livermore National Laboratory. It should be mentioned that parallel to this spark ignition scheme, a pure volume ignition scheme (Section 2.1) leads up to 200 times higher fusion energy per input laser energy by using few MJ ns laser pulses with volume compression to a few thousand times the solid state density and ignition temperatures of 1 keV or even less. This may be considered at least as a very conservative solution for laser driven IFE with the technical task to reduce the size and costs of lasers by many orders of magnitudes. This solution may be possible by physics similar to the reduction of the size of transistors.

Fast Ignition

When spherical laser irradiation did compress polyethylene-like polymers to 2000 times the solid state density, only a temperature of about 300 eV was measured. In order to reach fusion conditions it was considered to deposit an additional laser pulse of PW power and PS duration at the centre of the compressed plasma. These pulses were just becoming available by chirped pulse amplification (CPA) or the Schäfer technique. However, when performing the first experiments with these pulses, all kinds of relativistic effects were measured: generation of 100 MeV electrons, GeV ions, 20 MeV gammas, exotic nuclear reactions, pair production, and so forth. This was against the initial aim to deposit the pulse energy at the center of the 1000 times solid precompressed plasma for spark ignition. Modifications of the fast igniter line were developed of which we mention the laser generation of 5 MeV intense proton beams for depositing energy into the center of recompressed DT fuel for spark ignition. Another modification by Nuckolls and Wood opens the possibility of the initial aim to ignite nearly uncompressed solid DT fuel. Based on the before mentioned PW-PS laser pulses in 1000 times recompressed plasma, Nuckolls et al. expected that 5 MeV electron beams can be generated with such an intensity that the before mentioned 100 MJ fusion energy may be produced by a 10 kJ laser pulses. The requirement to produce these fusion gains above 10,000 in a fully controlled way is then fulfilled by using "a large mass of low density DT compressed fuel." The advantage to use the precompression to only 10 times the solid state DT fuel is explained and also how even lower precompression is of an advantage. The use of uncompressed solid state fuel was elaborated in the same sense by using ion beams instead of electrons based on a new scheme where a laser driven plasma block (or piston) ignites the fusion flame. This scheme is described in the next section.

LASER DRIVEN ION BEAMS FOR CONTROLLED IGNITION OF SOLID DT

With TW-PS laser pulses a very anomalous new phenomenon was measured, in contrast to the broad stream of observations. This new phenomenon is based on the few interaction measurements which avoid relativistic self-focusing in contrast to the usual experiments where self-focusing results in all kinds of relativistic effects; see above. The essential condition is to use laser pulses with extreme suppression of prepulses, needing a contrast ratio of 10^8 or better. The measured and theoretically understood and numerically reproduced nonlinear force driven plasma blocks with space charge neutralized ion current densities surprisingly reached 10^{11} Amps/cm^{2}fulfilling condition (4) for igniting nearly uncompressed solid DT similar to the Nuckolls-Wood scheme with electron beams.

Unique Interaction without Relativistic Self-Focusing

We summarize the essential aspects of these new phenomena. The key question is whether there are conditions under which the interaction of a focused laser beam at the surface of a solid target in vacuum follows the conditions of a plane wave interaction described in one dimension or whether the laser beam, as in most of the usual cases, produces a pregenerated plasma in front of the target, performing relativistic self-focusing and undergoing a shrinking of the laser beam to diameters of a wavelength with subsequent enormous intensities resulting in the relativistic effects. The plane wave interaction was studied by hydrodynamic computations including nearly all realistic and general plasma properties of which one of the numerous cases is shown in Figure 1. Within about a PS interaction of a 10^{18} W/cm^2 neodymium glass laser pulse on a deuterium target with an initially very low reflection (bi-Rayleigh) density profile, the nonlinear (ponderomotive) force is generating blocks of plasma of about 15 vacuum wave lengths thickness moving with velocities exceeding 10^9 cm/s against the laser light and a similar block moving into the target interior.

It was many years later only that these plane geometry conditions were available in experiments, showing such velocities gained by irradiation of similar excimer laser intensities at less than a PS duration, as measured by Sauerbrey. The resulting accelerations of 10^{20} cm/s^2 were in full agreement with the expectation from the nonlinear force interaction. Another key experiment was that by Zhang et al. where 100 fs TW laser pulses of about 30-wave lengths diameter hit a target and the X-ray emission was measured. The laser pulses produced very much lower X-rays than usually known from other experiments with the same intensities. The uniqueness of this experiment was recognized and checked by the procedure that additionally a lower intensity similar pulse was irradiated on the target at varying times from 10 to 100 PS before the main pulse. At short time preirradiation, no change of the low X-ray emission was seen, but as soon as the prepulse time reached 70 PS and more, suddenly the very high X-ray emission appeared as known from all the usual main stream experiments. The later given explanation was evident: it was thanks to the clean laser pulse technique where the contrast ratio (ratio of suppression of any prepulse) for the main pulse was 10^8. When the 70 PS prepulse was incident, a plasma plum was generated of a depth about two times the focus diameter. This was sufficient for the main pulse to shrink to about one wave length diameter by relativistic self-focusing so that the very high X-ray intensities resulted as in the usual cases, Figure 2. The skin layer acceleration (SLA) of the plasma block (Figure2(b)) was just what Sauerbrey observed earlier confirming in retrospect that his 350 fs TW laser pulses were sufficiently clean. The pulses were produced by the Schäfer method amplifying the 350 fs dye laser pulses through an activated KrF laser medium not needing gratings or pulse compression as with Mourou's chirped pulse amplification (CPA). Sauerbrey measured (by Doppler effect) an acceleration A in a carbon plasma front moving against the laser being produced by a 350 fs TW KrF laser pulse at 3.5×10^{17}W/cm^2 of

$$A_{\text{exp}} = 10^{20}\ \text{cm/s}^2. \tag{5a}$$

This corresponds to an electric field $E^2 = 2.9 \times 10^{15}$ erg/cm^3 and a densiy $n_i m_i$ of the accelerated plasma layer of 5.4×10^{-3} g/cm^3 at the critical density $n_i = 1.6 \times 10^{21}$ cm^{-3} for C^{+6} ions. The nonlinear force for the simplified plane geometry is

$$f_{\text{NL}} = -\frac{(\partial/\partial x)\left(\mathbf{E}^2 + \mathbf{H}^2\right)}{8\pi} = n_i m_i A$$
$$= -\left(\frac{1}{16\pi}\right)\left(\frac{\omega_p}{\omega}\right)^2\left(\frac{d}{dx}\right)\mathbf{E}^2. \tag{5b}$$

Assuming for simplification $dx = \Delta x = 10$ μm and a swelling S=2 (the experiments of Badziak et al. for PS pulses resulted in $S = 3.5$) we find the theoretical value:

$$A_{\text{NL}} = 1.06 \times 10^{20}\ \text{cm/s}^2. \tag{5c}$$

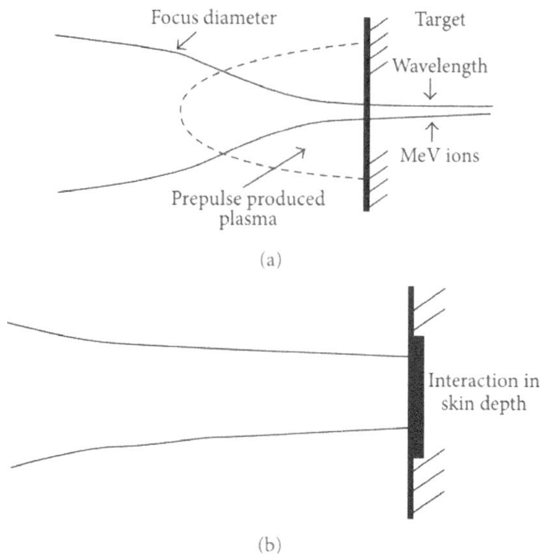

Figure 2: Laser interaction with a target, (a) if the laser pulse produced a plasma plume in front of the target by a prepulse causing shrinking of the beam to less than one wave length diameter due to relativistic self-focusing and (b) avoiding the prepulse for plane geometry interaction within the skin layer.

Both experiments were a clear confirmation of the plane wave plasma interaction, in agreement with the plane wave interaction theory and computation, Figure 1, and under exclusion of self-focusing. The anomaly was fully realized by the experiments by Badziak et al. from ion emission with high contrast ratio PS-TW laser pulses. This mutually confirmed the high contrast ratio of the excimer laser pulses as well as the very clean techniques of the other experiments. Irradiation of TW-PS neodymium glass laser pulses on copper targets should have led to Cu^{+13} ions of 22 MeV energy due to relativistic self-focusing, but only 0.5 MeV ions were detected. Furthermore, it was strange that the number of the fast ions did not change at all when the laser intensity was varied by a factor 30. The energy of the fast ions was fully linear with the laser intensity as expected from a nonlinear force acceleration. The measured X-ray emission leading to a quiver energy swelled dielectrically by a factor three fully satisfactorily agreed with the measured ion energies from a plane wave interaction which happened within the skin depth. The volume of the skin layer was not intensity dependent. This confirmed the observation of the intensity independence of the number of fast ions. These experiments were repeated with gold targets confirming all the details of the skin layer mechanism including the generation of the plasma block, Figure 3, moving into the target as experimentally confirmed from thin irradiated films and confirming an ion current density in the blocks above:

$$j > 10^{10}\ \text{A/cm}^2. \tag{6}$$

The measurements confirmed also that the PS-TW interaction resulted only in one narrow ion beam accelerated against the laser light as expected from the nonlinear force acceleration while the use of longer laser pulses as usually resulted in several groups of ions moving into a wide angle against the laser light. Now we report some details including recent numerical results for deuterium plasmas. The computations are based on the fact that before the PS main pulse, a less than 50 PS earlier acting laser prepulse produces a preplasma layer of thickness L_{pre} at least several times smaller than the laser focal spot diameter d_f. The main laser pulse interacts most intensively with the plasma in the skin layer near the surface of the critical electron density $n_{ec} = m_e\omega^2/4\pi e^2$ (ω is the laser frequency) and the geometry of the interaction is almost planar ($L_{pre} \ll d_f$). The high plasma density gradient in the interaction region produces nonlinear ponderomotive forces acting, at the laser beam incidence perpendicular to the target surface, nearly parallel to the target normal. The force density f_{NL} can be expressed as the one-dimensional negative gradient of the electromagnetic energy density of the laser field given by its (dielectric modified) electric and magnetic vectors E and H:

$$f_{NL} = -\frac{(\partial/\partial x)\left(\mathbf{E}^2 + \mathbf{H}^2\right)}{8\pi}. \qquad (7)$$

The gradients of the energy density near the critical surface result in two opposite nonlinear forces which break the plasma and drive two thin (~few λ) plasma blocks towards the vacuum and towards the plasma interior, Figure 3, respectively (λ is the laser wavelength). The density of the plasma blocks is high (the ion density $n_i \approx n_{ec}/z$, where z is the ion charge state) but the electron temperature is fairly moderate at sub relativistic laser intensities. Thus, the Debye length, λ_D, is small ($\lambda_D \ll \lambda$) inside the block and ions, being closely attached to electrons, move together with the electron cloud driven directly by the ponder motive force. Since $n_i \sim 10^{21}$–10^{22} cm^{-3} at $\lambda \sim 0.3$–1 μm, even at moderate ion velocities $V_i \sim 10^7$ –10^8 cm/s, the ion current densities can be very high (~ 10^9 –10^{10} A/cm2 or higher). The computations used the genuine two-fluid model and were performed for a 20-μm hydrogen, inhomogeneous plasma layer of initial density increasing in the direction of the laser beam propagation. Both the linear plasma density profile and the profile described by the function $n(x) = 2 \times n_{ec}$ $\{1 + \exp[(x - x_c)/05 \cdot L_n]\}^{-1}$ were considered as initial values. For both profiles the boundary plasma densities were chosen in such a way that the critical plasma density n_{ec} was placed near the middle of the layer. The initial inhomogeneity of the plasma was characterized by the (relative) plasma density gradient scale length, L_n/λ, where $L_n = n_e\,(\partial n_e/\partial x)^{-1}$. The initial temperatures of both electrons and ions were assumed to be of 30 eV. Most of the calculations were carried out for laser pulses of a Gaussian shape and for a laser wave length $\lambda = 1.05$ μm. Our numerical studies were focused on the influence of the initial plasma inhomogeneity and the laser pulse parameters (intensity, duration) on the plasma characteristics and, particularly, on the current densities and velocities of ion fluxes produced by the laser-plasma interaction.

The following results are for the nonlinear initial plasma density profile (the results for the linear profile were qualitatively similar). Figure 4 represents spatial distributions of the ion velocity, the ion current density, the ion density, and the electron temperature of plasma for the three-time intervals after the beginning of the laser-plasma interaction. We see the generation of the under dense plasma block moving against the laser (negative velocities and current densities) and the over dense plasma block behind the critical surface moving in forward laser direction (positive velocities and current densities) in accordance with the simple physical picture.

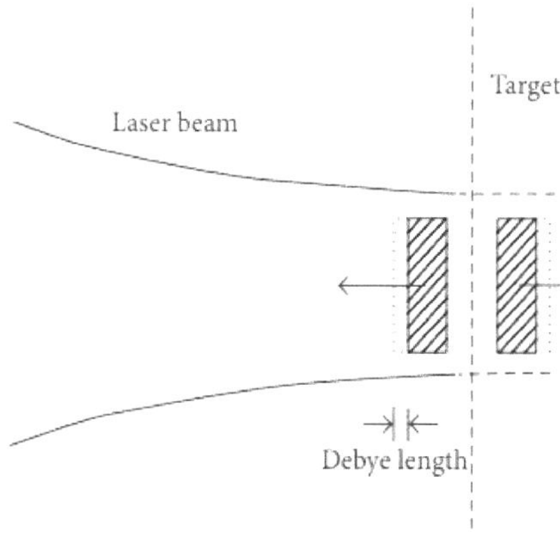

Figure 3: Scheme of skin depth laser interaction where the nonlinear force accelerates a plasma block against the laser light and another block towards the target interior. In front of the blocks are electron clouds of the thickness of an effective Debye length of less than 500 nm.

The plasma density profile is significantly disturbed near the critical surface and the electron temperature is the highest there. The effect of the plasma density gradient scale length on the maximum ion current densities and the maximum ion velocities of backward- and forward-emitted ion fluxes can be seen in Figure 5. For both fluxes there exist optimum values of $L_n/$ but they are located in essentially different regions of L_n/λ: the highest current densities and velocities for the backward flux are attained at the small density gradients and for the forward flux at the high density gradients.

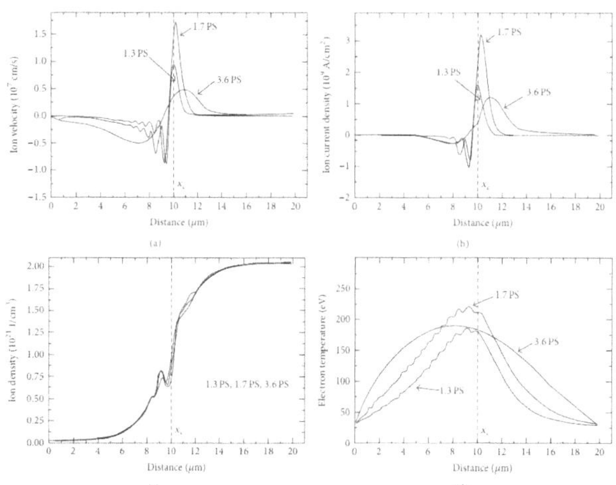

Figure 4: Spatial distributions of the ion velocity (a), the ion current density (b), the ion density (c), and the electron temperature (d) of plasma for various times measured from the beginning of the laser-plasma interaction. $\tau_L = 1$ PS, $I_L = 10^{16}$ W/cm^2, $L_n/\lambda = 2.5$.

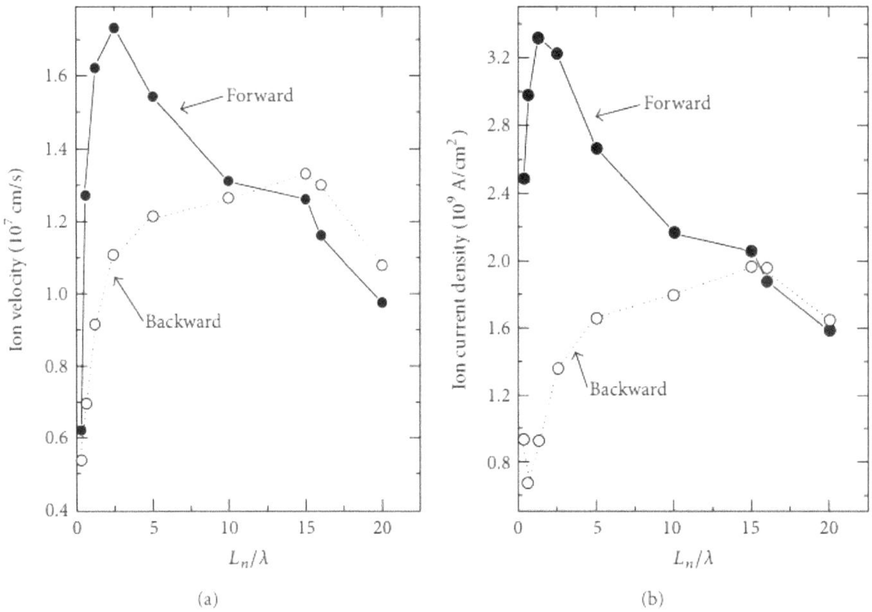

Figure 5: The dependencies of the maximum ion velocities (a) and the maximum ion current densities (b) on the plasma density gradient scale length. $\tau_L = 1$ PS, $I_L = 10^{16}$ W/cm^2.

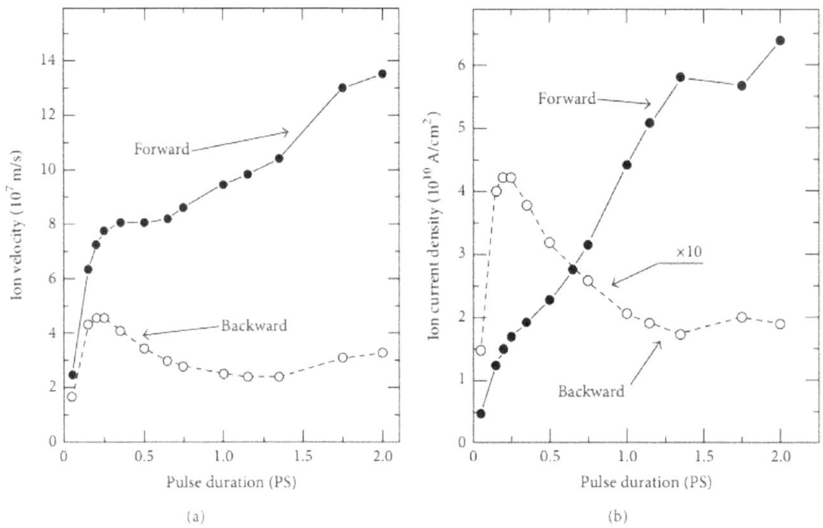

Figure 6: The maximum ion velocities (a) and the maximum ion current densities (b) as a function of the laser pulse duration. $I_L = 10^{17}$ W/cm^2, $L_n/\lambda = 1$.

The influence of laser pulse duration, τ_L, on characteristics of ion fluxes is illustrated in Figure 6. For the forward emitted ions, the maximum ion velocity and the maximum ion current density continuously increase with an elongation of the laser pulse. Contrary to that, in the case of backward emitted ions, distinct maxima of these quantities occurs at τ_L sub-PS ranges. One of the characteristic features of the dependencies presented in Figures 5 and 6 are the essential differences in the run of these dependencies for forward and backward-emitted ion fluxes. These differences can be understood better when we consider the dielectric swelling factor $S = 1/|n|$ which is a measure of increase in the electromagnetic energy density of the laser field in plasma in relation to its vacuum value (n is the plasma refractive index). The necessary condition for the acceleration of plasma (ions) by the ponder motive force in the backward direction is $S>1$ and, usually, the higher S, the higher the backward-directed ponderomotive force. The swelling factor depends not only on the plasma density gradient but also on parameters of a laser pulse and particularly on its intensity and the pulse duration. As it results from our numerical simulations, the swelling factor (its maximal value) decreases when the plasma density gradient λ/L_n, the intensity, or the pulse duration increase (within the ranges of these parameters shown in the figures). For instance, at $L_n/\lambda = 1$, $IL = 10^{17}$ W/cm^2, the swelling factor decreases from $S\approx 2$ for $\tau_L = 0.2$ PS down to $S \approx 1.3$ for $\tau_L = 2$. Thus, at high values of the above three parameters (λ/L_n, I_L, τ_L), the forward-directed force from the usual light pressure predominates the backward-directed one and, as a result, the velocities and current densities of forward-emitted ions are considerably higher than the ones for backward-emitted ions. The decreasing swelling factor is also the main reason for the faster increase in v_f and j_f (than v_b and j_b, resp.) when the laser intensity increases. The block generation was reproduced also by an essentially different genuine two-fluid computation, Figure 7. In conclusion, the numerical simulations based on the two-fluid hydrodynamic plasma model confirmed both qualitatively and quantitatively the idea of production of ultrahigh- currentdensity ion beams with the use of ponderomotive forces induced at the skin layer interaction of a short laser pulse of subrelativistic intensity with a thin inhomogeneous plasma layer. The results of the computations were found to be consistent with the simplified theory and measurements.

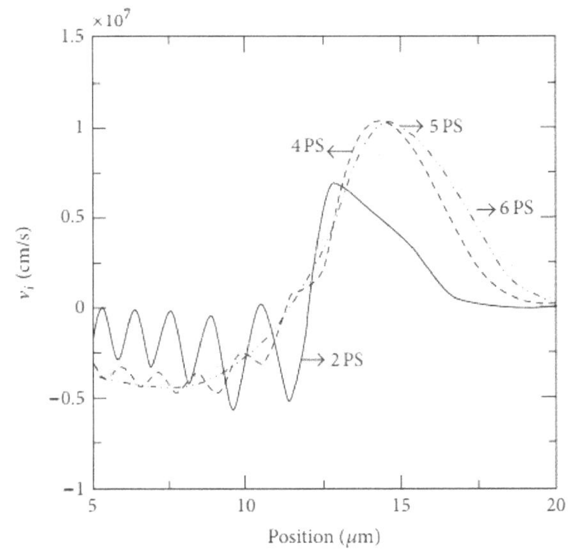

Figure 7: Ion velocity profiles at times 2, 4, 5, and 6 PS after irradiation of a 4 PS neodymium glass laser pulse on a deuterium plasma of initial density of a linearly increasing ramp (abscissa in micrometer of the depth, with critical density at 13) of 30 eV temperature, confirming the generation of an ablating plasma block (negative velocity) and a compressing plasma block (positive velocity).

Application for Controlled Ignition

Since the ion current densities, (6), in the laser driven plasma blocks from the skin layer or corona area of the irradiated plasma fulfilled the condition (4) for ignition of solid DT, we evaluated what has been achieved experimentally with respect to the energy flux density F, (3). The measurements included values of F up to 10^6 J/cm^2. If the ignition of DT has to use ion energies of 80 Kev for the optimum resonance conditions this gives an upper limit for the laser intensity while simultaneously a minimum of F is necessary. The values of $F^* = 10^8$ J/cm^2 may be not the final answer. It has been discussed before that the detailed evaluation of the interpenetration of the energetic blocks with the cold solid DT may result in 20 times lower values

(i) If the increased collision frequency based on the quantum correction is included,

(ii) If the modification of the stopping power by collective effects is included, and

(iii) If the double layer caused reduction of the thermal conduction is taken into account.

But even with the pessimistic aim to work with F^* values above 108 J/cm^2, this can be reached with short wave length lasers, for example, the double frequency of KrF

The remaining problem is only whether the interaction mechanisms of the plasma block of sufficient thickness is comparable to the interaction process as in the case of the electron beam in the Nuckolls-Wood scheme; otherwise, there is the similarity to the spark ignition at the interaction of the hot core with the surrounding plasma at spark ignition. For achieving the necessary thickness of the block, one has to produce it by irradiating a solid thin DT layer of spherical shape where the compressing part is then ballisticly moving into a focus for the interaction with the main solid DT fuel. During this motion, there is an increase of the density of the fast shell. Sine a minor heating is unavoidable apart from the nonlinear force driven directed motion, the shell is a little expanding thermally. But in view of the spherical shrinking of the shell, the final interaction occurs with the same high ion current density needed for conditions (4) however in a block of increased thickness needed for the ignition. This ballistic increase was mentioned before for the proton ignition generated by PW-PS laser pulses in precompressed DT fuel. It was shown that the ion current densities in plane geometry can be 1000 times higher than known before from measurements without the plasma block acceleration. With this result alone the scheme of proton fast ignition should reach the conditions of high gain fusion reactions. If then the ballistic layer acceleration is used to be focused into the main fuel, the conditions for ignition again will be improved. In this case indeed the ion energy for the ignition is not in the range of 80 keV as in the before mentioned ignition of solid DT fuel, but the ion blocks have to be relativistic with ion energies in the 5 MeV range. The details of this block acceleration have been evaluated numerically. In all our computations we used the genuine two fluid hydrodynamics including collisions. It is remarkable that the very different collision-less treatment with PIC methods resulted in very similar results at comparable conditions. The plasma block interaction for highly relativistic laser intensities was studied also in view of fusion ignition and further applications. Application of plasma block acceleration for improving the proton-fast-ignition fusion scheme with 1000 times higher ion current densities from the block acceleration was shown in the Figure 8.

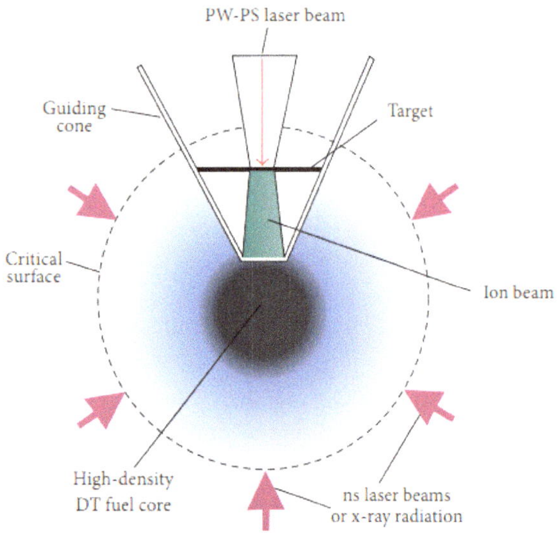

Figure 8: Application of plasma block acceleration for improving the proton-fast-ignition fusion scheme with 1000 times' higher ion current densities from the block acceleration.

SAUERBREY'S EXPERIMENTAL DISCOVERY OF ULTRAHIGH LASER ACCELERATION OF PLASMA BY NONLINEAR FORCES

Sauerbrey's [21] measurement of ultrahigh acceleration of plasma blocks by lasers turned out to be basically a nonthermal interaction process for which clarification was needed and developed in due course, as the problems were of a rather complicated nature. First we give the final short-cut result before going into details. The final result in hydrodynamic plasma theory is based on the force density f in a plasma to be determined by the gas-dynamical pressure $p = 3n_p kT/2$ where n_p is particle density, k is Boltzmann's constant, and T is the temperature and by the present electric and magnetic fields E and H:

$$\mathbf{f} = -\nabla p + \mathbf{f}_{NL}. \qquad (8)$$

For fields of a laser of frequency ω defining a complex optical constant n in the plasma, the nonlinear force with the unity tensor 1 is

$$\mathbf{f}_{NL} = -\left(\nabla \cdot \left[\mathbf{EE} + \mathbf{HH} - 0.5\left(\mathbf{E}^2 + \mathbf{H}^2\right)\mathbf{1} + \left(1 + \frac{(\partial/\partial t)}{\omega}\right)\left(\mathbf{n}^2 - 1\right)\mathbf{EE}\right]\right)(4\pi)^{-1} - \left(\frac{\partial}{\partial t}\right)\mathbf{E} \times \frac{\mathbf{H}}{4\pi c} \qquad (9)$$

which is dominant over the gas dynamic force (also called the thermokinetic force $f_{th} = -\nabla p$) if the quiver energy of the electrons in the laser field is larger than the energy of thermal motion (Hora 1969; 1985; see (8.87) and (8.88) of Hora [23]). For simplified one-dimensional geometry and perpendicular laser irradiation, the force (2) can be reduced to the time averaged value [24]:

$$\mathbf{f}_{NL} = -\frac{(\partial/\partial x)\left(\mathbf{E}^2 + \mathbf{H}^2\right)}{8\pi} = -\frac{\left(\omega_p/\omega\right)^2 (\partial/\partial x)\left(E_v^2/\mathbf{n}\right)}{16\pi}, \qquad (10)$$

where E_V is the amplitude of the electric field of the laser in vacuum. The second expression corresponds to the formulation of the ponderomotive force in electrostatics while the first expression includes identities from the later Maxwellian theory with the magnetic field as shown before [23]. Computations for plane geometry interaction with inclusion of the nonlinear force interaction, with thermal laser absorption by collisions and equipartition processes in the dynamically developing optical plasma properties for interaction of neodymium glass laser irradiation of 10^{18} W/cm^2 intensity on deuterium having an initial doubleRayleigh density profile (see Hora [25], Figures 9 and 10), arrived at a velocity distribution and an electromagnetic energy density as shown in Figure 1 after 1.5 PS interaction time. The laser was irradiating from the right hand side and a plasma block was moving against the laser light and another one into the deeper target. The velocity at this very short time at the closest part to the laser was more than 10^9 cm/s. This corresponds to an average acceleration of more than 5×10^{20} cm/s^2

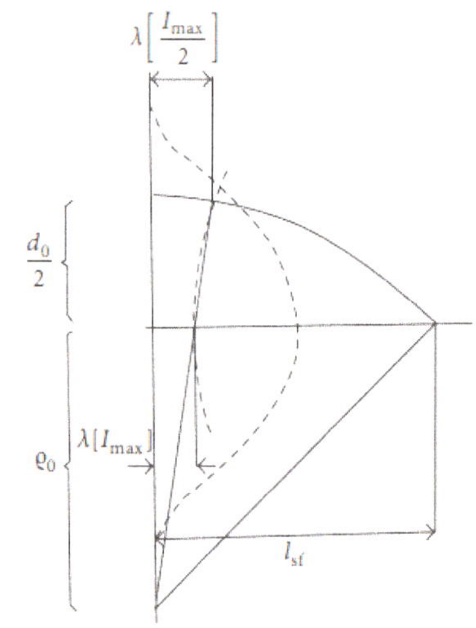

Figure 9: Relativistic self-focusing [26]: relativistic change of electron mass results in shorter optical propagation at higher intensity than at lower intensity. Relativistic self-focusing of a laser beam from the left with a dashed Gaussian intensity profile and radius $d_o/2$ moves into plasma right from the vertical line. Due to the relativistic change of the electron mass at the quiver motion, the effective wave length within the plasma is dielectrically shorter at higher intensity then at lower causing a concave bending of the initially plane wave front with shrinking of the beam to less than wave length diameter [26, 31, 32].

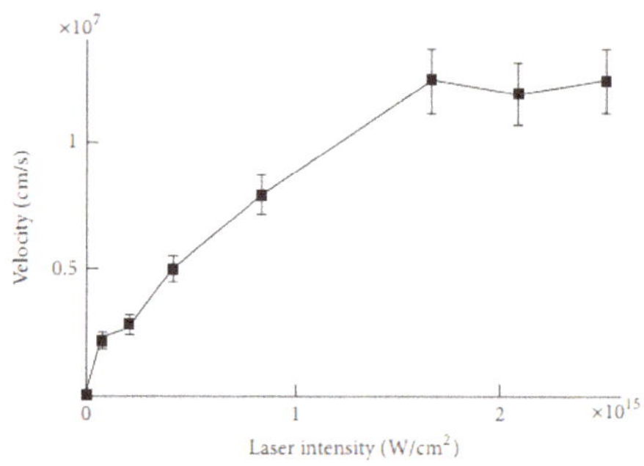

Figure 10: Intensity dependence of the velocity of the plasma front

from the Doppler shift of the reflected 700 fs KrF laser pulses from Al target [37].

These results of the computation were initially published in 1978 (see [25]) but it took a long time [21] before an experimental confirmation of these ultrahigh accelerations was measured. The reason was not only the question of how to produce the PS laser pulses up to the range of terawatt (TW) power, but there was the difficulty of relativistic self-focusing [26]. Each laser prepulse produced a plasma plume where any very intense laser beam which was relativistically squeezed to less than wave length diameter, Figure 9, producing very high intensities resulting in emission of highly charged ions to energies far above MeV and extremely intense very short wave length X-rays. These extreme mechanisms due to relativistic self-focusing had to be suppressed for verifying the plane geometry of the early computations and also to provide the conditions that the ultrahigh acceleration could be detected directly by Doppler effect. A confirmation of these conditions in the measurements of Sauerbrey [21] and the nonthermal processes was supported by following experiments [27,28], after similar measurements led to the conclusion that relativistic self-focusing was suppressed and shown from further related experiments [1, 2, 29]. This was in drastic contrast to the usual experiments with PS-TW and few petawatt (PW) laser pulses studied by the broad stream of experiments after Cowan et al. [30] observed high acceleration of ions and electrons, pair production, gamma rays with subsequent nuclear transmutations, and so forth. The characteristic for the very rare experiments of Sauerbrey [21], Zhang et al. [27], and Badziak et al. [28], see Hora et al. [2], is the use of the extremely high contrast ratio above 10^8 for the suppression of the relativistic self-focusing for achieving the enormous suppression of laser prepulses before the arrival of the main pulse on the targets. The reason for this was a specific need with the unique KrF laser pulses used by Sauerbrey [21] where subpicosecond laser pulses from dye lasers propagated in an inverted KrF laser gas for pulses reaching nearly TW power [19, 20]. The KrF excimer wave length of 248 nm, in contrast to the longer laser wave lengths of the usually applied lasers, was disturbing the experiments by the amplified spontaneous emission (ASE) radiation where the short wave length was preionizing the target too much. It was therefore highly necessary to reduce the ASE to a sufficient degree by suppression of any prepulse using plasma mirrors and other techniques for this type of lasers [20]. Sauerbrey therefore used lasers with a cutoff of the prepulses by a factor above 10^8 (contrast ratio) and had an interaction of the plane one-dimensional wave front of the interacting laser avoiding any self-focusing. The resulting plane wave front geometry provided then for the very first time the conditions for which the interaction was calculated before (Figure 1). Sauerbrey measured an acceleration of the plane plasma fronts perpendicular to the target of 10^{20} cm/s^2 against the laser as immediately given from the Doppler shift. This was in full agreement with the nonlinear force acceleration as computed before. Variations with respect to experimental accuracy were of minor nature for comparison in view of the significant fact that these ultrahigh accelerations were 100,000 times higher than the thermal pressure acceleration with the largest NIF (National Ignition facility) laser (Park et al. 2010) using nanosecond laser pulses.

ANALYSIS OF ACCELERATION MEASUREMENTS WITH DOPPLER EFFECT

Compared with the Doppler experiments with KrF lasers, the continuation with solid state lasers indicated a number of complexities which still have to be studied. It was possible [33] that a suppression of relativistic self-focusing for generation of plane plasma blocks as known from the numerous experiments with solid state lasers by Badziak et al. [34, 35] was possible by the proof of very low emission of X-rays [36] following the method of Zhang et al. [27]. Doppler experiments were performed with KrF PS laser pulses [37–39] which are now discussed in view of the later reached clarification between thermal and nonlinear force acceleration processes. Figure 10 shows the measured plasma block velocities from the Doppler effect at varying KrF laser intensities for 700 fs laser pulses with sufficiently high contrast ratio to suppress the ASE. The laser pulse of 2.6×10^{15} W/cm^2 intensity produced a velocity of $1,25 \times 10^7$ cm/s of the plasma at irradiation of aluminium [37]. This corresponds to an acceleration of 1.6×10^{19} cm/s^2 . The question is then how fast the laser energy can be converted into plasma motion. At this intensity, the quiver energy of electrons

$$\varepsilon_{osc} = \frac{e^2\mathbf{E}_v^2}{(2m_e\omega^2)} \tag{11}$$

using the electron mass me is 31.2 eV with the electric field amplitude E_V of the laser in vacuum for the KrF laser. This is sufficient to produce plasma electrons in the target within about a laser period by the electrons impacting in the target and with other aluminium ions to be ionized. The generation of an inhomogeneous plasma density in the interaction region can well be assumed to be within this very short time to produce a dielectric swelling S=3 which determines the electric field E by dielectric properties given by the temporary and spatially change during the: interaction dynamics by the absolute value of the refractive index n:

$$E^2 = \frac{E_v^2}{\mathbf{n}} \qquad S = \frac{1}{\mathbf{n}}. \tag{12}$$

This number S can be larger or smaller. The experience from the evaluation of the experiments of Badziak et al. [28] arrived at a swelling of a value of 3 [29]. The example of Figure 1 of the numerical velocity of deuterium with an assumed initial Rayleigh density profile had a swelling above S = 15. Swelling by S=3 produced a quiver energy of the electrons in the considered case for aluminium within fs duration had an energy of 93 eV and may then have produced high ionization Z of the aluminium. The dynamics of the plasma will then be determined by the nonlinear force (3) acting on the electron cloud again within the time of quiver motion to an acceleration whose dynamics will be determined by the inertia of the ion cloud due to electrostatic attraction in the same way as in the example of Figure 1 in agreement with the experiments of Sauerbrey [21]. Taking full ionization Z = 13, the front of the fastest ions arrives at a velocity of 1.2 × 107 cm/s. In view of a possible higher number S or lower number Z, it is sufficient at this stage only to see that the 700 fs laser interaction is sufficient to produce the measured acceleration. This is in contrast to any thermal mechanism, where first the quiver energy of the electrons has to be changed by collisions into random motion determining a temperature which then by a rather long equipartition time leads to heat the ions to move with the measured high plasma velocity. Similar acceleration velocities from Doppler experiments by Veres et al. [38] agree with the considered results from aluminium where the intensity dependence in comparison with the experiments of Badziak et al. [28] and the difference between irradiation of polystyrene and of aluminium may be interpreted along the discussed results of Foldes at al. [¨ 37]. At oblique incidence of lasers on the target, the nonlinear force is directed perpendicular to the plasma surface (see Section 11 of [23]) and the laser field energy density is not drastically modified. The acceleration of the ions is happening directly by the conversion of the laser field energy into the nonlinear force driven electron cloud driving the ion cloud by electrostatic attraction. This is in contrast to thermal absorption and heating which takes a much longer time than the interaction of the PS laser pulses. Computations at plasma surfaces with similar initial Rayleigh density profiles as in Figure 1 were performed, however, using a later developed genuine two-fluid hydrodynamic computation scheme. This code was developed by using the Euler equations of electrons and for ions separately combined by the electric fields given by the Poisson equation. These internal electric fields in the plasmas were eliminated in Schluter's [¨ 40] two-fluid hydrodynamics where the Euler equations were added to arrive at a one-fluid plasma equation of motion (1) used in the computations of Figure 1, and where the difference of the of the Euler equations led to a generalized Ohm's law for the plasmas.This elimination of the electric fields was correct for homogeneous, uniform plasmas similar to metals where the electric conductivity is similar to plasmas and where any electric field is decaying in much shorter times than femtoseconds (fs). For dimensions where the Debye length is much shorter than the considered length dimensions, the electric fields can well be ignored too. If these fields did appear in extraterrestrial plasmas as known by Alfven [41] and his associates, it was mentioned in a book review about this work by Kulsrud [42] that "these fields are intuitively not clear." In contrast to these general views in plasma theory, the genuine two-fluid hydrodynamics showed the strong appearance of these electric fields especially in the highly inhomogeneous plasmas at very intense laser interaction ([43, 44]; see Sections 8.8, 8.9, and 10.7 of [23]). It was then most interesting to see how these electric fields modify the results of Figure 1. After the genuine two-fluid code was applied to study the experiments of Badziak et al. [35] using genuine two-fluid hydrodynamics [45, 46] and other details of the nonlinear force dominating the laserplasma interaction (Cang et al. 2006), a numerical evaluation was achieved for the initially Rayleigh plasma density profiles similar to Figure 1 [47], where high swelling factors, S, were again the result. It turned out that the net ion motion was not much changed, Figure 11, compared with Figure 1 due to the fact that there was a dominant nonthermal conversion of laser energy into the macroscopic ion motion by the nonlinear force. The very general code with inclusion of collisions for thermal absorption and equipartition for ion heating were highly delayed and not strongly affecting the plasma dynamics determined by the ions for the very high laser intensity interactions during pulses in the PS range

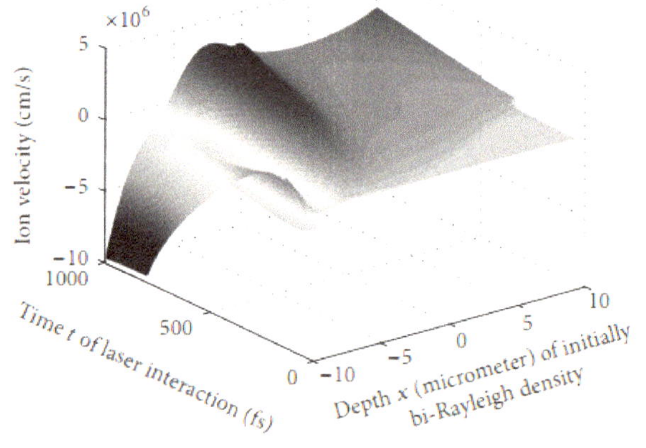

Figure 11: Ion velocity within an initially bi-Rayleigh $\alpha = 9.59 \times 10^3$ cm^{-1} deuterium plasma with initially zero velocity and with 100 eV initial temperature located between $x = 10$ and $+10$ μm at neodymium glass laser irradiation from the left hand side of 2×10^{16} W/cm^2 intensity during the time between 0 and 750 fs [47].

APPLICATION TO A RADICAL NEW LASER DRIVEN FUSION OF SOLID DENSITY FUEL WITHOUT NUCLEAR RADIATION

What was important with the ultrahigh acceleration was that extremely high current densities in the highly directed space charge neutral plasma blocks arrived at 10^{11} Amps/cm^2 or more. This is again more than a million times higher than accelerators could provide for ion beam fusion and permitted a comeback of the reaction of solid state, uncompressed or modestly compressed, fusion fuel by side-on ignition of a fusion flame. This was absolutely impossible with the first side-on ignition calculations for solid density fusion fuel (Chu 1971), [48] but this has changed now with the >PWPS laser pulses [18, 30]. It is potentially possible for energy production in power stations to achieve gains of 10,000 similar to the Nuckolls and Woods [49] scheme using PSlaser produced very high density 5 MeV relativistic electron beams in analogy to the here treated nonlinear force driven plasma blocks based on ions. The theory of Chu (1971) had to be updated [50] by the then not known effects of the thermal inhibition factor and by the collective effect of the stopping power (Gabor 1953). For laser fusion of deuterium tritium (DT), extremely clean PS laser pulses with a contrast ratio above 10^8 may drive the controlled reactions in power stations with pulses in the range of few dozens of PW power (Hora 2009). These are close to technical realization. What was very surprising is that the reaction of hydrogen and the boron isotope 11 (HB11) is less than ten times only more difficult than the DT fusion. This will generate less radioactivity in the entire reaction and in the waste than burning coal, per energy production [51, 52]. In contrast to the need of extremely high fuel compression in the usual thermally ignited laser-fusion schemes, the side-on ignition is simplifying the process and it can be expected that power production can be at considerably lower cost than present lowest cost sources, as today's cost in Eurocents/kWh: nuclear fission 2, coal 5, wind 8, and photovoltaics 38. For the next exploration of the side-on ignition of laser fusion with nonlinear force driven plasma blocks, the initial computations (Hora 2009; Hora et al. 2009) [51] are now generalized to use the genuine two-fluid model [43, 44] in order to study details of shock generation and very high electric field dynamics in the extremely inhomogeneous plasma in the fusion flame fronts. This is also for preparation of specific experiments with PWPS lasers of sufficient contrast to explore the revolutionary new scheme. Figure 12 shows results of the ion density of the fusion flame when developing into solid density DT fuel after a PS laser pulse initiated the fusion flame. It is very interesting to see that the local ion density in the thin flame front moves with a velocity of 1.55×10^8 cm/s and the density in the flame front is four times higher than the DT. This is an automatic result of the genuine two-fluid computation and agrees with the Rankine-Hugoniot theory of shock generation.

Conclusions

In conclusion, interaction of TW-PS laser pulses with plasma results in a skin layer mechanism for nonlinear (ponderomotive) force driven two-dimensional plasma blocks (pistons). This mechanism relies on a high contrast ratio for suppression of relativistic self-focusing. Space charge neutral plasma blocks are obtained with ion current densities larger than 10^{10} Amp/cm^2. Using ions in the MeV range results in 1000 times higher proton or DT current densities than the proposed proton fast igniter requires. This should result in better conditions of this fast ignitor scheme. The ballistic focusing of the generated plasma blocks and then short time thermal expansion increases their thickness but keeps the high ion current densities. As shown here, this approach then provides conditions that are very favorable for efficient fast ignition of a fusion target. If successful, this approach to fast ignition could significantly simplify operation of an IFE plant, allowing very attractive energy production costs. What is evident is the measurement of the ultrahigh acceleration of plasmas at interaction with subpicosecond high intensity laser pulses if relativistic self-focusing is avoided by using very high contrast ratios for suppression of prepulses. The high contrast was initially necessary with KrF laser pulses to avoid amplified spontaneous emission (ASE). The generation of the ultrafast plane highly directed plasma blocks and avoiding relativistic self-focusing was confirmed also by X-ray emission measurements [27] and the properties of the directed ion blocks based on the number of the fast ions limited to the dielectric strongly increased skin depth. [28]. The confirmation of the acceleration by the nonlinear (generalized ponderomotive) force in agreement with computed plane geometry conditions, Figure 1, [2, 25] demonstrated a nonthermal process where the laser energy was nearly completely converted into macroscopic acceleration of the space-charge neutral electron cloud in the plasma and the electrostatically attached ion cloud. These results are basically confirmed but may need more information about detailed properties. The use of the ultrafast plasma blocks for side-on ignition of uncompressed (or modestly compressed) solid fusion fuel following the early computations of Chu (1971) are based on hydrodynamic theory leading to ignition of high gain fusion with next available high contrast laser pulses of dozens of petawatt-picosecond laser pulses for DT fuel but even for HB11 fuel where less radioactivity is generated than from burning coal per gained energy (Hora et al. 2009) [51]. One problem is whether the hydrodynamic treatment for fusion is sufficient and whether not details of particle interpenetration [53] need to be included as known from Particle In Cell (PIC) computations as discovered by Wilks et al. (1991). A comparison between results from hydrodynamics and PIC evaluations for the range of interesting laser intensities (Limpouch et al. 2006) showed a rather good agreement. Laser generation at thin foils with high contrast ratio was studied in a number of experiments (see, e.g., [22]) and may need to be reconsidered as in similar cases [29] where the interaction by thermal processes need a comparison with nonlinear force effects.

CONFLICT OF INTERESTS

The authors declare that there is no conflict of interests regarding the publication of this paper.

REFERENCES

1. H. Hora, J. Badziak, F. P. Boody et al., "Effects of ps and ns laser pulses for giant ion source," Optics Communications, vol. 207, no. 1–6, pp. 333–338, 2002.
2. H. Hora, J. Badziak, M. N. Read et al., "Fast ignition by laser driven particle beams of very high intensity,"Physics of Plasmas, vol. 14, no. 7, Article ID 072701, 2007.
3. M. Tabak, J. Hammer, M. E. Glinsky et al., "Ignition and high gain with ultrapowerful lasers," Physics of Plasmas, vol. 1, no. 5, pp. 1626–1634, 1994.
4. E. M. Campbell, High Intensity Laser-Plasma Interaction and Applications to Inertial Fusion and High Energy Density Physics [Doctor of Science thesis], University of Western Sydney, New South Wales, Australia, 2005.
5. C. Yamanaka and S. Nakai, "Thermonuclear neutron yield of 10^{12} achieved with Gekko XII green laser,"Nature, vol. 319, no. 6056, pp. 757–759, 1986.
6. E. Storm, "Press Conference Lawrence Livermore National Laboraotory," 1986.
7. E. Storm, J. D. Lindl, E. M. Campbell, et al., "Progress in laboratory high-gain ICF: prgress for the future Livermore," LLNL Report 47312, 1988
8. H. Hora, H. Azechi, Y. Kitagawa et al., "Measured laser fusion gains reproduced by self-similar volume compression and volume ignition for NIF conditions," Journal of Plasma Physics, vol. 60, no. 4, pp. 743–760, 1998.
9. H. Hora and P. S. Ray, "Increased nuclear fusion yields of inertially confined DT plasma due to reheat,"Zeitschrift Für Naturforschung A, vol. 33, pp. 890–894, 1978.
10. R. C. Kirkpatrick and J. A. Wheeler, "Physics of DT ignition in small fusion targets," Nuclear Fusion, vol. 21, no. 3, pp. 389–401, 1981.
11. H. Azechi, T. Jitsuno, T. Kanabe et al., "High-density compression experiments at ILE, Osaka," Laser and Particle Beams, vol. 9, no. 2, pp. 193–207, 1991.
12. Y. Kato, K. Mima, N. Miyanaga et al., "Random phasing of high-power lasers for uniform target acceleration and plasma-instability suppression," Physical Review Letters, vol. 53, no. 11, pp. 1057–1060, 1984.
13. H. Hora, "Smoothing and stochastic pulsation at high power laser-plasma interaction," Laser and Particle Beams, vol. 24, no. 3, pp. 455–463, 2006.
14. J. M. Soures, R. L. Mccrory, C. P. Vernon, et al., "Direct-

drive laser-fusion experiments with the OMEGA, 60-beam, >40 kJ, ultraviolet laser system," Physics of Plasmas, vol. 3, pp. 2108–2112, 1996.

15. J. D. Lindl, "The Edward Teller medal lecture: the evolution toward indirect drive and two decades of progress toward ignition and burn," in Edward Teller Lectures: Laser and Inertial Fusion Energy, H. Hora and G. H. Miley, Eds., pp. 121–147, Imperial College Press, London, UK, 2005.
16. S. H. Glenzer, E. Moses, B. J. MacGowan, et al., "Demonstration of ignition radiation temperatures in indirect-drive inertial confinement fusion hohlraums," Physical Review Letters, vol. 106, Article ID 085004, 5 pages, 2011.
17. D. Strickland and G. Mourou, "Compression of amplified chirped optical pulses," Optics Communications, vol. 56, no. 3, pp. 219–221, 1985.
18. G. Mourou and T. Tajima, "Ultraintense lasers and their applications," in Inertial Fusion Science and Applications, V. R. Tanaka, D. D. Meyerhofer, and J. Meyer-ter-Vehn, Eds., pp. 831–839, Elsevier, Paris, France, 2002.
19. S. Szatmari and F. P. Schäfer, "Simplified laser system for the generation of 60 fs pulses at 248 nm," Optics Communications, vol. 68, pp. 196–201, 1988.
20. S. Szatmari, "High-brightness ultraviolet excimer lasers," Applied Physics B, vol. 58, pp. 211–223, 1994.
21. R. Sauerbrey, "Acceleration in femtosecond laser-produced plasmas," Physics of Plasmas, vol. 3, no. 12, pp. 4712–4716, 1996.
22. M. Kaluza, J. Schreiber, M. I. K. Santala et al., "Influence of the laser prepulse on proton acceleration in thin-foil experiments," Physical Review Letters, vol. 93, no. 4, Article ID 045003, 2004.
23. H. Hora, Plasemas at High Temperature and Density, Springer, Heidelberg, Germany, 1991.
24. F. F. Chen, "Physical mechanisms for laser-plasma parametric instabilities," in Laser Interaction and Related Plasma Phenomena, H. J. Schwarz and H. Hora, Eds., vol. 3, pp. 291–313, Plenum Press, New York, NY, USA, 1974.
25. H. Hora, Physics of Laser Driven Plasmas, John Wiley & Sons, New York, NY, USA, 1981.
26. H. Hora, "Theory of relativistic self-focuing of laser radiation in Plasmas," Journal of the Optical Society of America, vol. 65, no. 8, pp. 882–886, 1975.
27. P. Zhang, J. T. He, D. B. Chen et al., "X-ray emission from ultraintense-ultrashort laser irradiation,"Physical Review E, vol. 57, pp. 3746–3752, 1998.
28. J. Badziak, A. A. Kozlov, J. Makowski et al., "Investigations of ion streams emitted from plasma produced with a high-power picosecond laser," Laser and Particle Beams, vol. 17, no. 2, pp. 323–329, 1999.
29. H. Hora, "Skin-depth theory explaining anomalous picosecond-terawatt laser plasma interaction II,"Czechoslovak Journal of Physics, vol. 53, no. 3, pp. 199–217, 2003.
30. T. E. Cowan, M. D. Perry, M. H. Key et al., "High energy electrons, nuclear phenomena and heating in petawatt laser-solid experiments," Laser and Particle Beams, vol. 17, no. 4, pp. 773–783, 1999.
31. L. Cicchitelli, H. Hora, and R. Postle, "Longitudinal field components for laser beams in vacuum,"Physical Review A, vol. 41, no. 7, pp. 3727–3732, 1990.
32. T. Häuser, W. Scheid, and H. Hora, "Theory of ions emitted from a plasma by relativistic self-focusing of laser beams," Physical Review A, vol. 45, no. 2, pp. 1278–1281, 1992.
33. X. Yang, G. H. Miley, K. A. Flippo, and H. Hora, "Energy enhancement for deuteron beam fast ignition of a precompressed inertial confinement fusion target," Physics of Plasmas, vol. 18, no. 3, Article ID 032703, 2011.
34. J. Badziak, S. Głowacz, S. Jabłłoński et al., "Production of ultrahigh ion current densities at skin-layer subrelativistic laser-plasma interaction," Plasma Physics and Controlled Fusion, vol. 46, no. 12, pp. B541–B555, 2004.
35. J. Badziak, S. GŁowacz, S. JabŁoński, P. Parys, J. WoŁowski, and H. Hora, "Generation of picosecond high-density ion fluxes by skin-layer laser-plasma interaction," Laser and Particle Beams, vol. 23, no. 2, pp. 143–147, 2005.
36. K. Flippo, private communication, 2011.
37. B. Földes, J. S. Bakos, K. Gál et al., "Properties of high harmonics generated by ultrashort UV laser pulses on solid surfaces," Laser Physics, vol. 10, no. 1, pp. 264–269, 2000.
38. G. Veres, G. Kocsis, E. Rácz, and S. Szatmári, "Doppler shift of femtosecond laser pulses from solid density plasmas," Applied Physics B: Lasers and Optics, vol. 78, no. 5, pp. 635–638, 2004.
39. B. Földes and S. Szatmári, "On the use of KrF lasers for fast ignition," Laser and Particle Beams, vol. 26, no. 4, pp. 575–582, 2008.
40. Schlüter, "Dynamik des plasmas—I: grundgleichungen, plasma in gekreutzten feldern," Zeitschrift für Naturforschung A, vol. 5, pp. 72–78, 1950.
41. H. Alfven, Cosmic Plasma Dordrecht, Reidel, 1981.
42. R. Kulsrud, "Book review: Hannes Alfven," Physics Today, vol. 34, no. 4, p. 56, 1983.
43. P. Lalousis and H. Hora, "First direct electron and ion fluid computation of high electrostatic fields in dense inhomogeneous plasmas with subsequent nonlinear laser interaction," Laser and Particle Beams, vol. 1, no. 3, pp. 283–304, 1983.
44. H. Hora, P. Lalousis, and S. Eliezer, "Analysis of the inverted double layers produced by nonlinear forces in a laser-produced plasma," Physical Review Letters, vol. 53, no. 17, pp. 1650–1652, 1984.
45. S. Głowacz, J. Badziak, S. Jabłonski, and H. Hora, "Numerical modelling of production of ultrahigh-current-density ion beams by short-pulse laser-plasma interaction," Czechoslovak Journal of Physics, vol. 54, no. 3, pp. C460–C467, 2009.
46. S. Glowacz, H. Hora, J. Badziak, S. Jablonski, Y. Cang,

and F. Osman, "Analytical description of rippling effect and ion acceleration in plasma produced by a short laser pulse," Laser and Particle Beams, vol. 24, no. 1, pp. 15–25, 2006.

47. R. Sadighi-Bonabi, E. Yazdani, Y. Cang, and H. Hora, "Dielectric magnifying of plasma blocks by nonlinear force acceleration with delayed electron heating," Physics of Plasmas, vol. 17, no. 11, Article ID 113108, 5 pages, 2010.

48. J. L. Bobin, "Nuclear fusion reactions in fronts propagating in solid DT," in Laser Interaction and Related Plasma Phenomena, H. Schwarz and H. Hora, Eds., vol. 4, pp. 465–494, Plenum Press, New York, NY, USA, 1974.

49. J. L. Nuckolls and L. Woods, "Future of inertial fusion energy," in Proceedings of the International Conference on Nuclear Energy Systems ICNES Albuquerque, H. S. Park and B. Remington, Eds., 2002, Astrophyhsics and Space Science.

50. H. Hora, B. Malekynia, M. Ghoranneviss, G. H. Miley, and X. He, "Twenty times lower ignition threshold for laser driven fusion using collective effects and the inhibition factor," Applied Physics Letters, vol. 93, no. 1, Article ID 011101, 3 pages, 2008.

51. H. Hora, G. H. Miley, M. Ghoranneviss, B. Malekynia, N. Azizi, and X.-T. He, "Fusion energy without radioactivity: laser ignition of solid hydrogen-boron (11) fuel," Energy and Environmental Science, vol. 3, no. 4, pp. 479–486, 2010.

52. Y. Li, "Nuclear power without radioactivity," in Highlights in Chemical Technology, vol. 7, Royal Chemical Society, London, UK, 2010.

53. H. Hora, "Interpenetration burn for controlled inertial confinement fusion by nonlinear forces,"Atomkernenergie, Kerntechnik, vol. 42, no. 1, pp. 7–10, 1983

CHAPTER 17

PALEOMAGNETIC MEASUREMENTS

Hirokuni Oda, Chuang Xuan and Yuhji Yamamoto

Keywords: Superconducting rock magnetometer, Pass-through measurement, Sensor response Deconvolution, Laser interferometry geomagnetic excursion, U-channel Sediment long-core sample

INTRODUCTION

The development of pass-through superconducting rock magnetometers (SRM) has enabled continuous high sensitivity measurement of remanent magnetizations of sediments (e.g., Dodson et al. 1974; Goree and Fuller 1976; Weeks et al. 1993), leading to rapid accumulation of high-resolution paleomagnetic and environmental magnetic records. The continuous records acquired on pass-through SRMs, especially those with u-channels (Tauxe et al. 1983), have revolutionized paleomagnetism through reconstruction of paleomagnetic field at unprecedented resolution and scale (e.g., Valet and Meynadier 1993; Guyodo and Valet 1999; Valet et al. 2005; Roberts 2006; Ohno et al. 2008; Channell et al. 2009; Roberts et al. 2013) and greatly contributed to the paleomagnetic data archive based on sediments (Xuan and Channell2009; Brown et al. 2015).

Continuous measurements on pass-through SRMs, however, are subjected to smoothing and distortion due to the convolution of sensor responses, controlled by the geometry of pick-up coils and superconducting shield, with the magnetization of paleomagnetic sample (e.g., Shibuya and Michikawa 2000; Oda and Xuan 2014). Deconvolution is needed to reconstruct higher-resolution magnetization time series with less distortion (e.g., Dodson et al. 1974; Constable and Parker 1991; Oda and Shibuya 1996; Guyodo et al. 2002; Jackson et al. 2010). Oda and Xuan (2014) recently developed an improved deconvolution algorithm based on ABIC minimization method considering realistic errors in sample length and measurement position. Based on the algorithm, stand-alone graphical software UDECON was developed to directly read pass-through measurement data and perform fast and reliable deconvolution optimization (Xuan and Oda 2015).

Citation: Hirokuni Oda, Chuang Xuan, and Yuhji Yamamoto, paleomagnetic measurements, DOI: 10.1186/s40623-016-0493-2.

Accurate measurements of sensor response including cross-terms are required to conduct deconvolution (e.g., Parker and Gee 2002; Jackson et al. 2010; Oda and Xuan 2014). Oda and Xuan (2014) developed a practical tool to measure sensor response systematically on 5-mm-spacing grids in three orthogonal directions. However, it is possible to improve the efficiency, repeatability and accuracy of the sensor response measurements by better constraining the point source position relative to individual grids as the authors used a double-sided tape to fix the Point Source Cube on the guide block.

Based on repeated measurements of a u-channel sample, Oda and Xuan (2014) demonstrated that accuracy of the deconvolved magnetizations is significantly controlled by SRM measurement noises. However, we currently have little knowledge on the characteristics of noises on the SRMs. A main source of SRM measurement noise is the positioning error of SRM tracking system (e.g., Jackson et al. 2010; Oda and Xuan 2014), which could contaminate both sensor response and sample measurements.

In order to facilitate implementation of the UDECON software (Xuan and Oda 2015) and to further understand uncertainties associated with pass-through measurements and deconvolution, we present new tool and procedure for accurate and more efficient measurements of SRM sensor response and use laser interferometry to evaluate the positioning error of the SRM tracking system. The new tools and procedures were used to measure the sensor response of SRM at Kochi Core Center (KCC), Japan, one of the core repositories of International Ocean Discovery Program (IODP), to ensure better reproducibility and accuracy of point source positions on the grid with improved orthogonality of three-axes orientation of the magnetic point source. The laser interferometry-acquired sample tray positions were compared with stepping motor counts. We also discuss the characteristics of positions measured with laser interferometry and the positioning errors associated with the SRM track system as well as their potential influence on deconvolution.

RESPONSE FUNCTION DETERMINATION

SRM at KCC

Figure 1 shows a schematic view of the u-channel pass-through SRM system (2G Enterprises model 755R) at KCC. The flat sample tray is made of fiber-reinforced plastics and is placed on a track. Both ends of the tray are connected to the original thread provided by the SRM manufacturer (2G Enterprises). The thread is tied to a laddered plastic rope that is connected to a gear attached to a stepping motor. At each end of the track there are clockwise (CW) and counterclockwise (CCW) limit switches. The SRM system at KCC uses the 2G Long Core software (version 3.4) installed on a Windows PC for its operation. Sample Handler Utility included in the software allowed us to control the position of the sample tray on the track by driving the stepping motor at two different speeds (slow or fast).

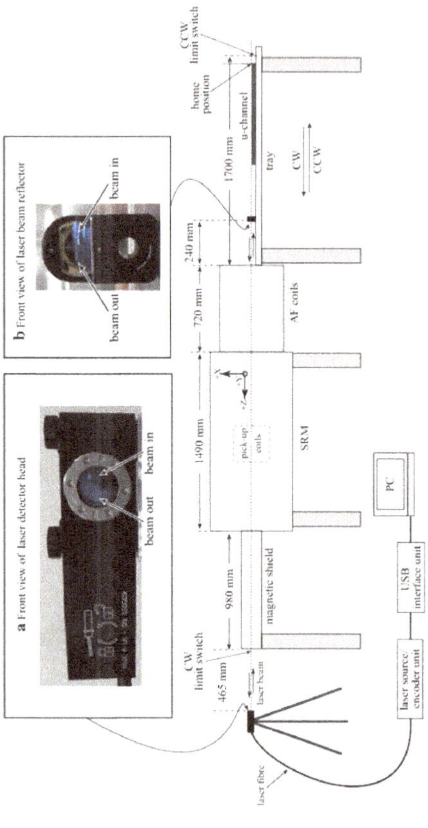

Figure 1: Schematic diagram illustrating the setup for laser interferometry measurement. Laser beam is emitted from **a** the laser detector head, enters the bore of SRM from the side of CW limit switch, and is then reflected by **b** the laser beam reflector attached on the sample tray. Reflected laser beam goes back to the laser detector head through the bore of SRM, which is used for laser interferometry together with the source beam

Sensor Response Measurements

Oda and Xuan (2014) used a 5-mm plastic cube with a point source in the center (Point Source Cube hereafter) and a larger cube with 25 mm edge length and one surface having 5 × 5 = 25 marked grid points. The Point Source Cube was attached onto the grid point on the surface of the larger cube using double-sided tape (Fig. 1b in Oda and Xuan 2014). The orientation of Point Source Cube on the X- and Y-axes plane was controlled with a plastic rod, and orientation along the Z-axis depends on the flatness of the double-sided tape. Here, we use new tool and procedure to facilitate the precise measurements of sensor response for the SRM at KCC. We modified one surface of the larger cube to include 16 grid positions evenly distributed with 6 mm × 6 mm spacing, each of which has four surrounding plastic walls to accurately hold the Point Source Cube (Fig. 2a, b). Point Source Cube was placed to each of these grid positions to measure sensor response with magnetization orientation parallel to +X-, +Y- or +Z-axes. The measurement data collected at every 1-mm interval for 300 mm along the track were then interpolated (or extrapolated for the bottom grid points) into 1-mm grids from 6-mm measurement grids over the cross-sectional area of the larger cube surface. With the above grid data, integration was conducted over the cross-sectional area of a u-channel sample (i.e., horizontal = 18 mm; vertical = 19 mm; bottom height from the tray = 2 mm; centered along track).

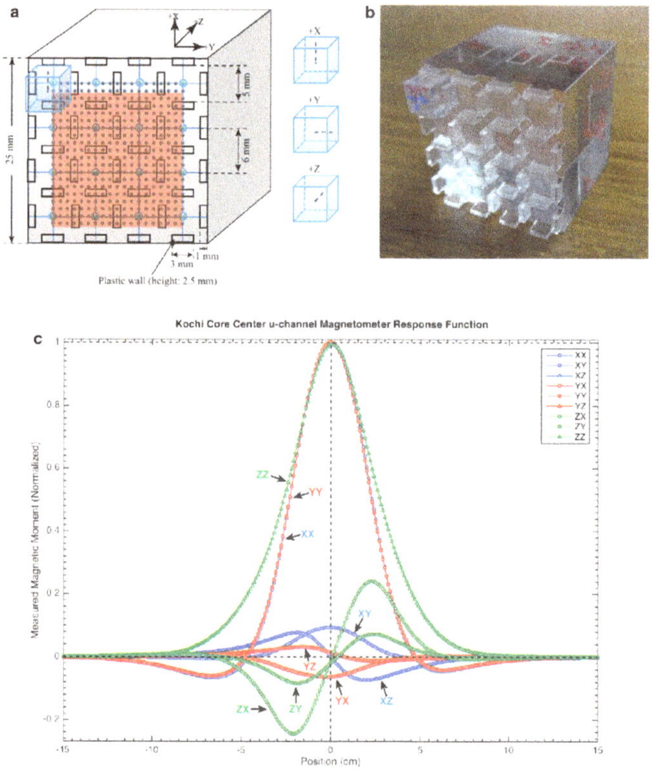

Figure 2: **a** Schematic diagram of sensor response measurement tools. A plastic cube with edge length of 25 mm was used to hold the Point Source Cube during measurement. One surface of the cube has 4 × 4 grid positions, each of which has 5 mm × 5 mm *rectangular space* enclosed by four plastic walls with 3 mm width, 2.5 mm height, and 1 mm thickness. Point Source Cube can tightly fit in each of the 16 grid positions (numbered from 1 to 16). Each *small dot* corresponds to 1-mm-spacing grids. *Shading with orange color* shows the area corresponding to the cross section of a u-channel, which is used for integration. **b** A picture showing the actual plastic cube and Point Source Cube during a measurement on the KCC SRM. **c** Sensor response of SRM at KCC integrated over u-channel cross-sectional area at 1-mm interval along Z-axis. "AB" in legend means sensor response of "B"-axis pick-up coil to magnetization oriented along "A"-axis, where "A" and "B" are either "X," "Y," or "Z." Note relatively large ZX cross-term with anti-symmetry

Magnetic moment of the Point Source Cube was measured on a spinner magnetometer (Natsuhara Giken Co. Ltd.) at the Geological Survey of Japan. It has declination, inclination, and magnetic moment of 0.2°, 0.3°, and 5.30×10^{-7} Am^2, respectively.

KCC Sensor Response

Complete tensor components of sensor response curve for the u-channel SRM at KCC (KCC SRM sensor response) are shown in Fig. 2c. The three diagonal terms (i.e., "XX," "YY," and "ZZ" in Fig. 2c) of the KCC SRM sensor response tensor show single peaks with their full widths at half maxima (FWHM) of 46 mm (XX), 46 mm (YY), and 54 mm (ZZ). The positions of central maximum of XX, YY, and ZZ are −0.02, −0.04, and 0.07 mm, respectively. The XX and YY terms have negative lobes on both sides. There are also cross-terms between X- or Y-axis and Z-axis (i.e., XY, YX, XZ, and ZX). The prominent feature is that the ZX cross-term shows negative and positive peak values of ~25 % relative to the central peak value of the main ZZ term. This could be explained by the relatively large offset of the center of X-, Y-, and Z-coils relative to the center of u-channel about 8-mm downward (-X direction) for KCC sensor response, which is calculated by the X–Y position of minimum of peak values along Z-axis. The XY or YX terms are about 10 or 6 % of peak values in the XX or YY main terms, which could be minimized by rotating X-axis magnetization and Y-axis magnetization 4.5° counterclockwise along Z-axis (looking at SRM from the home position). This observation suggests that X–Y coil system is rotated 4.5° counterclockwise relative to the plane of tray and track.

COMPARISON AND INTERPRETATION OF SENSOR RESPONSES

The KCC SRM sensor response shows different shapes from that of SRM at Oregon State University (OSU SRM sensor response). Oda and Xuan (2014) reported that OSU SRM sensor response has FWHM of 80 mm (XX), 76 mm (YY), and 73 mm (ZZ) for the main terms, which are apparently 1.4–1.7 times wider than those of the KCC SRM sensor response. FWHM of XX or YY components are narrower than ZZ component for the KCC SRM sensor response, whereas opposite is true for the OSU SRM sensor response. Other features to note are that there is a single peak in the ZZ term for the KCC SRM sensor response, while the OSU SRM sensor response shows double peak separated ~20 mm from each other. In addition, XX and YY terms of the KCC SRM sensor response have negative lobes on both sides, whereas the OSU SRM sensor response has almost no such negative lobes.

These differences in the sensor responses could be caused by different designs between SRMs at KCC (cooled with liquid helium) and OSU (cooled with pulse tube). For instance, the KCC SRM pick-up coils may have narrower widths along the SRM track. Jackson et al. (2010) noted that the geometry of the superconducting shield in the "dry" SRM (cooled with pulse tube) is different from that in the conventional liquid He-cooled SRM, and the sensor response function is different accordingly. It is therefore critical to estimate the sensor response accurately for both types of SRM in order to successfully utilize the deconvolution. The narrow width of pick-up coils with narrow FWHM of KCC SRM sensor response is suitable for reconstruction of higher-resolution paleomagnetic signals with deconvolution.

Among the same type of SRMs, there are also some difference in sensor response functions. For example, SRM at OSU is a similar type to SRM at the Institute of Rock Magnetism (IRM), University of Minnesota. However, SRM at IRM has asymmetry with negative lobe on one side of sensor responses of XX and YY (Jackson et al. 2010). The asymmetry of SRM at IRM might have originated from the offset of center of the superconducting shield relative to the center of pick-up coils of X- and Y-axes.

Rotation of ~5° for X–Y pick-up coils relative to the plane of the tray seems to be a common feature of the 2G SRMs (e.g., Parker and Gee 2002; Jackson et al. 2010; Oda and Xuan 2014). This might be related to the construction of SRMs and/or the installation at the laboratories. 2G Enterprises documented the following in their Long Core v1.0 manual in 1996: "This misalignment angle can be determined by measuring a core once, then measuring it again after rotating the sample 180° about its +X axis." This description is not included in later versions of manual for the 2G Long Core software (e.g., v3.0 and v3.2).

Considering the fact that SRM users are not provided with the information on the geometry of pick-up coils and superconducting shield, rotation angle of X–Y pick-up coils, and sensor responses in real shape, it is necessary to estimate sensor responses accurately with a point source for each laboratory. An integrated sensor response can then be constructed for cross-sectional area corresponding to measured samples such as the u-channels. The new tool and procedure provide an effective approach in improving the efficiency and accuracy of sensor response curve estimates.

LASER INTERFEROMETRY OF SRM TRAY POSITION

Experimental setup of Laser Interferometry

Accurate positions of the SRM sample tray were measured with laser interferometry relative to the home position (-Z direction; see Fig. 1). The measurements were conducted using a laser encoder unit RLU10 (Renishaw plc) together with a laser detector head (RLD 90° RRI), a laser beam reflector (RLR10-A3-XF), and a USB interface unit (RSU10). The laser encoder unit was connected to a PC with the USB interface unit, and the interferometry was conducted with a positioning accuracy measurement software provided by Renishaw plc.

The laser detector head (Fig. 1a) was attached on an adjustable support, fixed on a stable tripod, and placed ~465 mm behind the CW limit switch of the SRM system. The laser beam reflector (Fig. 1b) was mounted on a small adjustable support and firmly attached to the sample tray using double-sided tape. Laser beam generated from the laser encoder unit is transmitted through a laser fiber to the laser detector unit. It was then emitted from the laser detector passing through SRM and reflected with the laser beam reflector attached to the sample tray. The reflected laser beam was received with the detector head, which was used for the interference measurements together with the source laser beam. Approximate distance between the laser detector head and the laser beam reflector when the tray is at the home position was measured as ~3895 mm.

Alignment of laser beam was conducted with special care using the tripod and the adjustable support while observing the beam on a paper guide attached on the laser beam reflector at the home and CW limit switch positions. The final adjustments were made with the software by maximizing output signal of the interferometry. Each of the laser interferometry measurements was conducted relative to the home position at a sampling frequency of 1 kHz. Environmental corrections were conducted for all measurements with temperature of 23.5 °C, humidity of 40 %, and atmospheric pressure of 101.45 kPa. Static measurements without motion of the tray show a standard deviation of 0.7 μm, possibly due to background mechanical vibrations and variability of environmental parameters.

The laser interferometry was conducted in two directions of the tray movements, i.e., toward (CW direction) or away (CCW direction) from the SRM, at two speeds (slow or fast) with or without a sample. The stepping motor was driven incrementally by manual operation with 100 stepping motor counts (corresponding to 4.78 mm for KCC SRM) either in a manner that the tray moved toward or away from the SRM, while the laser interferometry data were being collected. The positions measured with laser interferometry were recorded and compared with the distance expected from the stepping motor counts. Table 1 summarizes laser interferometry measurements conducted, while the tray is moving with and without a sample. A 1020-mm-long u-channel filled with sediment (weight: 770 g) was used for laser interferometry measurements with sample.

Table 1: List of measurements for laser interferometry

Run #	With/without U-channel	CW/CCW	Measured length (mm)
7	Without U-channel	CW	568.8
8	Without U-channel	CCW	616.6
9	Without U-channel	CW	568.8
10	Without U-channel	CCW	607.1
11	Without U-channel	CW	425.4
12	Without U-channel	CCW	358.5
12_2[a]	Without U-channel	CCW	262.9
13	Without U-channel	CW	568.8
14	Without U-channel	CCW	282.0
14_2[a]	Without U-channel	CCW	243.8
15	With U-channel	CW	90.8
16	With U-channel	CW	90.8
17	With U-channel	CW	109.9
18	With U-channel	CW	105.2
19	With U-channel	CCW	138.6
20	With U-channel	CW	105.2
21	With U-channel	CCW	129.1
22	With U-channel	CW	224.7
23	With U-channel	CCW	248.6
25	With U-channel	CW	229.4
26	With U-channel	CCW	253.3

Run #1 through #6 are test runs without appropriate setup of temperature, humidity, and pressure and discarded. Run #24 is a static measurement for the measurement of ambient noise

CW clockwise, *CCW* counterclockwise

[a]Measurements restarted after losing track of laser

SRM Tray Position Measured With Laser Interferometry

When no sample was placed on the tray (a u-channel sample was placed on the tray), continuous measurements of laser interferometry were successful up to about 600 (200) mm from the home position, where signal suddenly became weaker than the threshold for continuous tracking of interference. The difficulty of laser beam tracking beyond a few tens of cm could be related to irregularities (or "topography") of the SRM track.

We started the collection of measurements either from the home position (CW; toward SRM) or from the position where the previous measurements were terminated due to poor interference signal (CCW; away from SRM). The tray movement was controlled by manually clicking the "Move" button in the Sample Handler Utility of the 2G Long Core software. Figure 3 shows the position, speed, and acceleration of the tray for laser interferometry measurements with (right) and without (left) a u-channel sample. In order to visualize the details, horizontal axis was expanded for data collected within the first ~2 s and is plotted in Fig. 4. The peak amplitude of the vibration is ~200 µm without the sample (Fig. 4a) and ~50 µm with the sample (Fig. 4b). Vibration of the tray position settles down after ~0.6 s without the sample and after ~0.4 s with the sample. Average speed of the tray without the sample is ~50 mm/s (Fig. 4a, middle panel) both for slow and for fast modes. The speed of the tray with the sample is also centered about 50 mm/s (Fig. 4b, middle panel). It seems that the speed of the tray has no significant influence on the position difference.

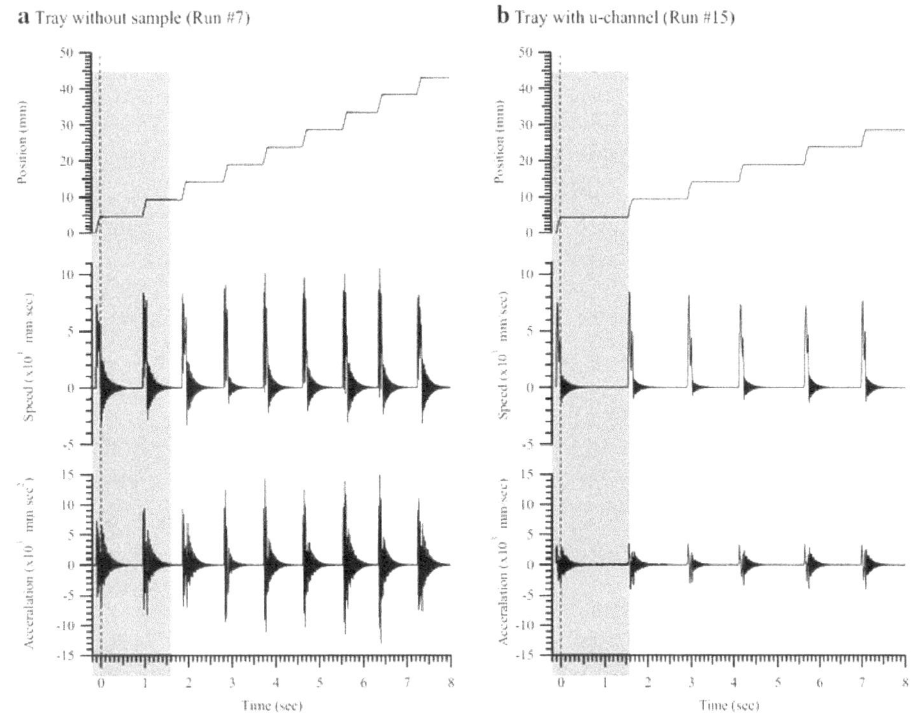

Figure 3: Examples of precise measurement of position (*upper*), speed (*middle*), and acceleration (*bottom*) versus time for the tray of SRM at KCC using laser interferometry **a** without (Run #7) and **b** with (Run #15) a u-channel sample on the tray. *Vertical shading* is the interval shown in Fig. 4. Time is set as zero for the first trigger point (sharp drop in speed). The diagram shows only first nine or six stop positions out of a total of 19–119 positions

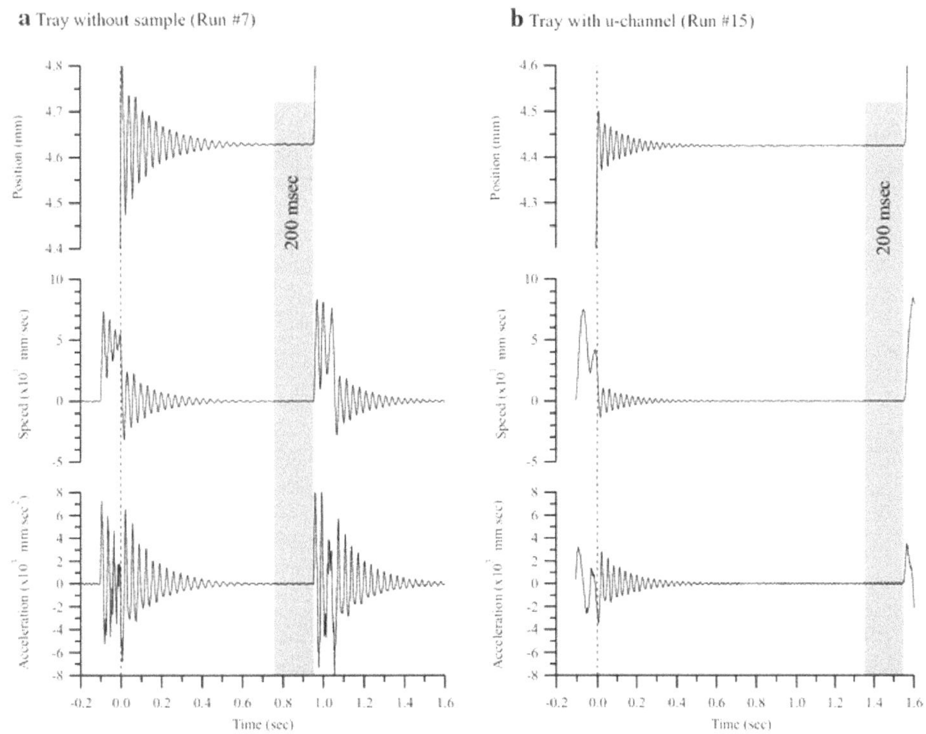

Figure 4: Details of precise position (*upper*), speed (*middle*), and acceleration (*bottom*) versus time for the tray of SRM at KCC measured using laser interferometry **a** without (Run #7) and**b** with (Run #15) u-channel sample on the tray shown in Fig. 3. *Vertical shading* (200 ms before each move) is the interval used to calculate average positions in Fig. 5

Positions of the SRM tray at KCC measured with laser interferometry were compared with those expected from the corresponding stepping motor counts, and the differences between the two (position difference) are plotted against the latter in Fig. 5. The position acquired by laser interferometry for each "move" was calculated by averaging laser interferometry data for a 200-ms time interval right before the next "move" (shaded area in Fig. 4). The differences without sample in the direction toward SRM (CW direction; Fig. 5a) from stepping motor count are within −0.4 to 0.3 mm for the entire 600-mm measured interval. It is notable that the position differences show comparable patterns in addition to random noises (Fig. 5a). Histograms of the position differences (Fig. 6a) show distributions with means between −0.12 and 0.05 mm and fairly consistent standard deviations between 0.10 and 0.12 mm.

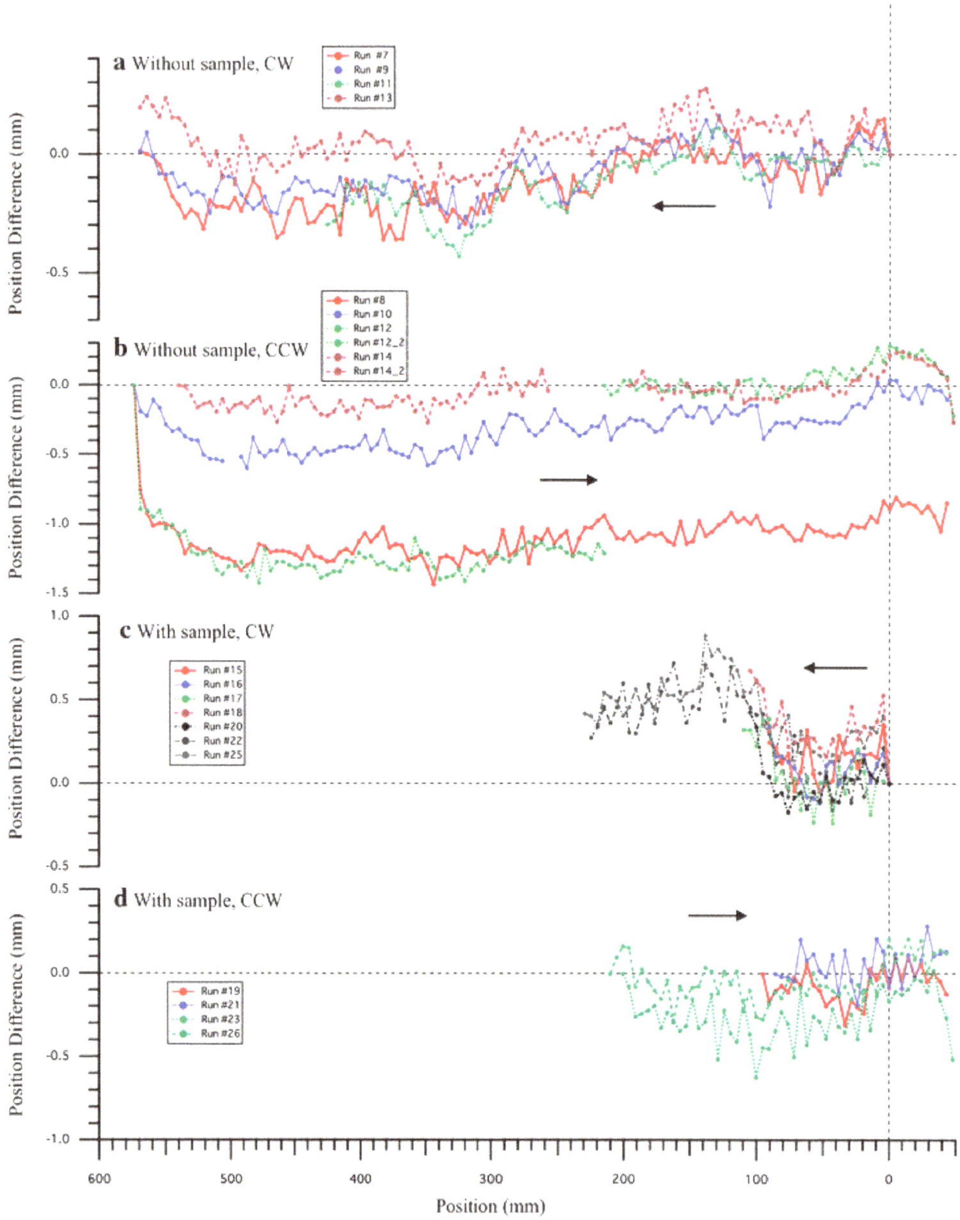

Figure 5: Position difference (expected from stepping motor counts—position measured with laser interferometry) without u-channel sample on the tray for **a** clockwise (CW) and **b**counterclockwise (CCW) moving directions. Similarly, position difference with u-channel sample on the tray is shown in **c**, **d** for CW and CCW moving directions, respectively

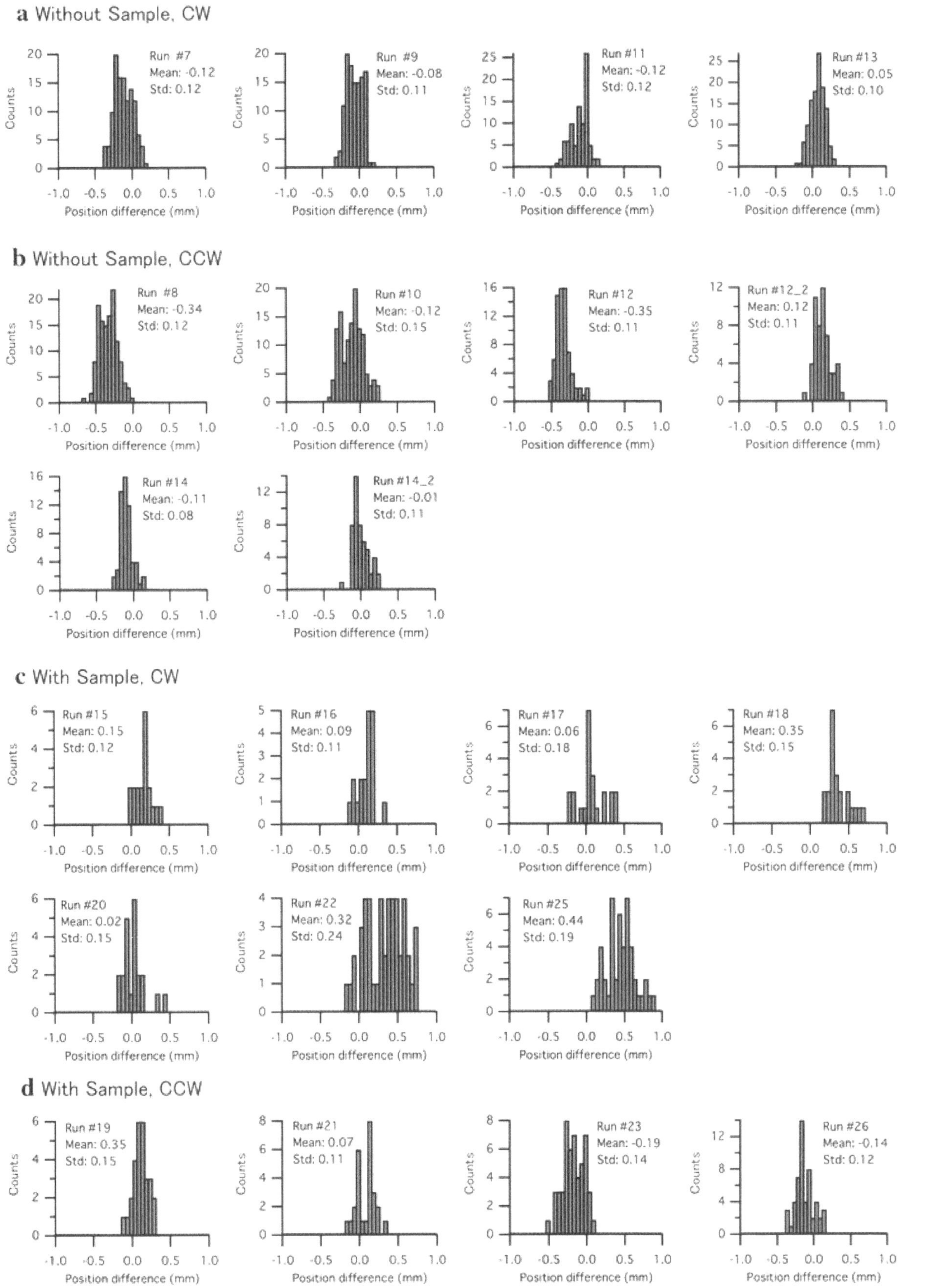

Figure 6: Histograms of the position difference for individual laser interferometry runs listed in Table 1. **a** Runs in CW direction without sample on the tray, **b** runs in CCW direction without sample on the tray, **c** runs in CW direction with sample on the tray, and **d** runs in CCW direction with sample on the tray. *Each diagram* shows frequency versus position difference, which is accompanied by run number, mean, and standard deviation (Std) in mm

The position differences without sample in the direction away from SRM (CCW direction; Fig. 5b) show a gap between the first measurement (reference position for laser interferometry, which is always zero) and the second measurement (the position after the first move) away from the tray. This feature may imply a backlash related to change in the direction of stepping motor motion. Similar to that observed for CW measurements, fluctuation of the position differences shows comparable pattern on position length scales of >a few centimeters for all four measurements (note that Run #12_2 or Run #14_2 is the continuation of Run #12 or Run #14 after recovery of laser tracking). Histograms of the position differences (Fig. 6b) show distributions with means between −0.35 and 0.12 mm and standard deviations between 0.08 and 0.15 mm. The dominating negative means for the runs in CCW directions might be due to backlash mentioned above.

Although the records with sample were obtained for shorter intervals, the calculated position differences are between −0.3 and 0.9 mm for the total move of 250 mm in the direction toward SRM (CW direction; Fig. 5c) and are between −0.7 and 0.3 mm away from SRM (CCW direction; Fig. 5d). A significant increase of ~0.5 mm in the position differences was observed at position ~100 mm for the two runs toward SRM (CW direction; Fig. 5c). This might be related to the irregularity of the stepping motor belt or the roughness (or "topography") of the track surface. The histograms for the runs in CW direction (Fig. 6c) show distributions with means between 0.02 and 0.44 mm and standard deviations between 0.11 and 0.24 mm. The position differences calculated for the runs in the CCW direction (Fig. 6d) show distributions with means between −0.19 and 0.35 mm and standard deviations between 0.11 and 0.15 mm.

POSITIONING ERROR OF SRM TRACK

The gap observed in multiple laser interferometry measurements conducted while moving the tray away from the SRM (CCW direction; Fig. 5b), possibly due to a backlash of the stepping motor, could be accounted for by the new parameter "position shift" introduced to the optimized deconvolution by Oda and Xuan (2014). The reason why such gap is absent in the CW direction measurements (e.g., Figure 5a) might be because the measurement in this direction starts from the home position, which is always calibrated through an initial move from the CCW limit switch (Fig. 1).

The positioning error of measurements on a SRM depends on the configuration in each laboratory. The source of stochastic errors observed in Fig. 5 might be random errors originating from the combination of multiple components of the tracking system (i.e., stepping motor, handler rope, pulley, and tray). Based on repeated measurements of a 1.5-m-long u-channel sample, Oda and Xuan (2014) estimated >0.06-mm positioning error for the SRM at OSU. It should be noted that the positioning error estimated from the standard deviation of repeated pass-through measurement is not absolute error but deviation from the average measurement at each position.

The first direct precise measurement of tray positions with laser interferometry conducted in this study revealed repeatable fluctuations (on length scales of >a few centimeters) in positioning of the SRM track system, which might be related to the condition of the various components of the track system (e.g., track or laddered rope). The apparently repeatable large position error of ~0.5 mm observed at ~100 mm for the measurements with u-channel samples (in CW direction, Fig. 5c) may need to be accounted for to achieve better deconvolution results. It should be noted that the repeatable stepwise positioning error could influence all the demagnetization steps and better to be avoided. Figure 7a shows a histogram of position difference distribution calculated with all laser interferometry measurements in both CW and CCW directions, while no sample was placed on the tray. The average of the distribution is −0.12 mm with a standard deviation of 0.18 mm. The distribution for all laser interferometry measurements with a u-channel sample has average of 0.11 mm with a standard deviation of 0.27 mm (Fig. 7b). It appears that standard deviation of the position difference is larger for the measurements with a u-channel sample on the tray.

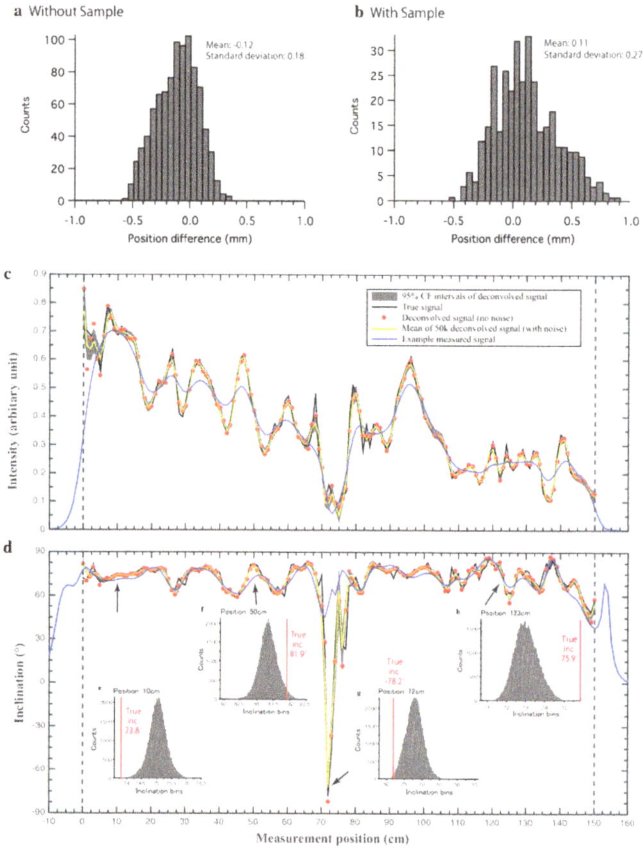

Figure 7: **a** Histogram of all position differences from measurements in CW and CCW directions without u-channel sample on the tray. **b** Histogram of all position differences from measurements in CW and CCW directions with u-channel sample on the tray. Simulation results for **c** intensity and **d** inclination data. *Solid black curve* is true signal used to produce synthetic measurement, and *blue curve* is example of a measured signal with position errors added. *Red dots* are deconvolved signal using measurement and response function data without adding positioning errors. *Yellow curves* are mean of 50,000 deconvolved signals using measurement and response data with realistic positioning errors added. The *gray shading* represents 95 % confidence intervals of the deconvolved data. *Insets* in **d** are histograms of inclination for 50,000 deconvolved signals at *e* 15 cm, *f* 50 cm, *g* 72 cm, and *h* 123 cm positions, respectively. *Red vertical lines* represent true inclination values for the four positions

DECONVOLUTION EXPERIMENTS USING SIMULATED DATA WITH POSITIONING ERROR

In order to investigate the effect of the observed positioning error on deconvolution, we conducted optimized deconvolution experiments 50,000 times using synthetic data with realistic position errors. Synthetic "ideal" measurements were first produced by convolving a synthetic magnetization signal that contains an "excursion" event used by Oda and Xuan (2014) with the KCC SRM sensor response. We then introduced random position errors to the KCC SRM sensor response data according to the histogram of position difference distribution acquired from the laser interferometry measurements, while no sample was placed on the tray (Fig. 7a). Similarly, at each sample measurement position, a random position error was introduced based on the histogram of position differences acquired from the laser interferometry measurements with a u-channel sample on the tray (Fig. 7b). For both sensor response and synthetic sample measurement data, signal with positions at regular interval of 1 cm was used for the simulations. For each simulation the "ideal" measurement and response function data were resampled by spline interpolation based on positions displaced by adding random positioning errors. The measurement and response function resampled from the displaced positions were used for deconvolution as if the positions were not displaced.

The intensity and inclination data from the simulations are shown in Fig. 7c, d. Solid black curve is the true signal (artificial magnetization data set) used to produce synthetic measurements. An example of synthetic measurement is shown in blue curves, which was produced by convolving the true signal with the sensor response. Compared with the true signal, the measured signal shows significant smoothing and distortion due to the convolution effect of the pass-through system. SRM at KCC cannot successfully resolve the ~8-cm-long "excursion" event in the original true signal. In Fig. 7c, d, red dots are deconvolved data using measurement and response function data without adding any position errors. Mean (yellow curve) and the 95 % confidence intervals (shaded area) for all simulated deconvolution data (a total of 50,000 times) using the both measurement and response function data with realistic position errors suggest that deconvolution can successfully restore the true signal (both in amplitude and in direction) overcoming majority of the smoothing and distortion caused by the convolution effect. The "excursion" event was clearly recovered in the deconvolved measurements. Examples of histograms of inclination for 50,000 deconvolved signal show Gaussian-like distributions with standard deviations of ~0.3°, ~0.3°,~3°, and ~0.8° for positions at 15, 50, 72, and 123, respectively. A prominent evidence is that the midpoint of "excursion" with true inclination of −78.2° at 72 cm could be recovered successfully with relatively small uncertainty, which demonstrates the effectiveness of ABIC-minimizing deconvolution for measurement and response function data with realistic position errors.

CONCLUSIONS

New practical tool and procedure were developed to facilitate rapid and accurate measurements of SRM sensor response. Systematic measurements of Point Source Cube on SRM at KCC, Japan, were used to produce an accurate estimate for the sensor response, which can be used for deconvolution of u-channel pass-through measurements made on the SRM at KCC. A possible 4.5° counterclockwise rotation (looking at SRM from the home position) of the X–Y coil system relative to the Z-axis and significant ZX cross-term components were observed. As one of the main research facilities associated with the core repositories of the IODP, the SRM at KCC plays an important role in the collection of critical pass-through data for paleomagnetism research. The accurate estimate of sensor response function of the SRM at KCC, therefore, provides a valuable tool with broader impact to restore high-resolution signals for measurements through deconvolution.

Accurate measurements of SRM tray positions were conducted with and without a 1020-mm-long sediment u-channel at intervals of 4.78 mm using laser interferometry. Position vibrations were observed in all measurements following the stop of the tray movement. Measurements without (with) u-channel show vibrations with peak amplitudes of ~200 μm (~50 μm), which diminishes in ~0.6 s (~0.4 s). Comparison with the position expected from the stepping motor counts indicates random discrepancies with standard deviations of ~0.1–0.2 mm. large gaps were observed just after the change in the direction of the stepping motor, which might be related to backlash of the stepping motor. Stepwise change of ~0.5 mm was recognized for measurements with u-channel sample on the tray, presumably due to the condition of the tray. Positioning error is larger with u-channel sample. Reproducible features in the position differences suggest influence of track/rope shapes on individual positions. Positioning error is generally between 0.06 and 1 mm. Although the current positioning system is accurate enough to reproduce 8-cm-long "excursion" event, it is recommended to build tray tracking system with better positioning accuracy to improve the pass-through measurements and performance of optimized deconvolution for higher resolution and accuracy.

We have conducted deconvolution experiments using synthetic data with realistic position errors acquired from the laser interferometry measurements. Synthetic measurement data were produced by convolving synthetic magnetization signal that includes an "excursion" event with the sensor response measured for KCC SRM. For both measurements and sensor response, positioning errors were randomly introduced based on the laser interferometry measurements conducted in this study. Deconvolution using the synthetic measurement data and response function estimates for SRM successfully restored the true signal overcoming majority of the smoothing and distortion caused by the convolution effect. The "excursion" event was clearly recovered in the deconvolved measurements with a maximum error angle of ~3°.

ACKNOWLEDGEMENTS

Hirokuni Oda was supported by JSPS Grant-in-Aid for Scientific Research (A) Funding Nos. 25247082 and 25247073. Chuang Xuan is supported by a startup fund provided by the University of Southampton. JSPS provided a Visiting Fellowship for Foreign Researchers (Award No. PE14034) for Chuang Xuan to visit Geological Survey of Japan (GSJ), National Institute of Advanced Industrial Science and Technology (AIST), to conduct work related to this research. This study was performed under the cooperative research program of Center for Advanced Marine Core Research (CMCR), Kochi University (Accept Nos. 14B040 and 15A048). Part of the measurements was conducted with facilities of *GSJ-Lab*., and a plastic block for the point source was fabricated with the help of Tomomi Kobayashi at AIST. The authors would like to thank Ayako Katayama at AIST who helped the production of figures. The study could not be completed without the construction of SRMs provided by 2G Enterprises, especially the pioneering works by William S. Goree.

COMPETING INTERESTS

The authors declare that they have no competing interests.

REFERENCES

1. Brown MC, Donadini F, Nilsson A, Panovska S, Frank U, Korhonen K, Schuberth M, Korte M, Constable CG (2015) GEOMAGIA50.v3: 2. A new paleomagnetic database for lake and marine sediments. Earth Planets Space 67:70. doi:10.1186/s40623-015-0233-z
2. Channell JET, Xuan C, Hodell DA (2009) Stacking paleointensity and oxygen isotope data for the last 1.5 Myr (PISO-1500). Earth Planet Sci Lett 283:14–23. doi:10.1016/j.epsl.2009.03.012
3. Constable C, Parker R (1991) Deconvolution of longcore palaeomagnetic measurements—spline therapy for the linear problem. Geophys J Int 104:453–468. doi:10.1111/j.1365-246X.1991.tb05693.x
4. Dodson RE, Fuller MD, Pilant W (1974) on the measurement of the remanent magnetism of long cores. Geophys Res Lett 1:185–188. doi:10.1029/GL001i004p00185
5. Goree WS, Fuller MD (1976) Magnetometers using RF-driven SQUIDs and their applications in rock magnetism and paleomagnetism. Rev Geophys Space Phys 14:591–608
6. Guyodo Y, Valet JP (1999) Global changes in intensity of the Earth's magnetic field during the past 800 kyr. Nature 399:249–252. doi:10.1038/20420
7. Guyodo Y, Channell JET, Thomas RG (2002) Deconvolution of u-channel paleomagnetic data near geomagnetic reversals and short events. Geophys Res Lett 29:1845. doi:10.1029/2002GL014927
8. Jackson M, Bowles JA, Lascu I, Solheid P (2010) Deconvolution of u channel magnetometer data: experimental study of accuracy, resolution, and stability of different inversion methods. Geochem Geophys Geosyst 11:Q07Y10. doi:10.1029/2009GC002991
9. Oda H, Shibuya H (1996) Deconvolution of long-core paleomagnetic data of Ocean Drilling Program by Akaike's Bayesian Information Criterion minimization. J Geophys Res 101:2815–2834. doi:10.1029/95JB02811
10. Oda H, Xuan C (2014) Deconvolution of continuous paleomagnetic data from pass-through magnetometer: a new algorithm to restore geomagnetic and environmental information based on realistic optimization. Geochem Geophys Geosyst 15:3907–3924. doi:10.1002/2014GC005513
11. Ohno M, Murakami F, Komatsu F, Guyodo Y, Acton G, Kanamatsu T, Evans HF, Nanayama F (2008) Paleomagnetic directions of the Gauss–Matuyama polarity transition recorded in drift sediments (IODP Site U1314) in the North Atlantic. Earth Planets Space 60:e13–e16
12. Parker RL, Gee JS (2002) Calibration of the pass-through magnetometer-II. Application. Geophys J Int 150:140–152. doi:10.1046/j.1365-246X.2002.01692.x
13. Roberts AP (2006) High-resolution magnetic analysis of sediment cores: strengths, limitations and strategies for maximizing the value of long-core magnetic data. Phys Earth Planet Inter 156:162–178. doi:10.1016/j.pepi.2005.03.021
14. Roberts AP, Tauxe L, Heslop D (2013) Magnetic paleointensity stratigraphy and high-resolution Quaternary geochronology: successes and future challenges. Quat Sci Rev 61:1–16
15. Shibuya H, Michikawa T (2000) Calculation of superconducting rock magnetometer response. Kumamoto J Sci Earth Sci 16:1–16
16. Tauxe L, Labrecque JL, Dodson D, Fuller M (1983) "U" channels—a new technique for paleomagnetic analysis of hydraulic piston cores. EOS Trans Am Geophys Union 64:219
17. Valet JP, Meynadier L (1993) Geomagnetic field intensity and reversals during the past four million years. Nature 366:234–238. doi:10.1038/366234a0
18. Valet JP, Meynadier L, Guyodo Y (2005) Geomagnetic dipole strength and reversal rate over the past two million years. Nature 435:802–805. doi:10.1038/nature03674
19. Weeks R, Laj C, Endignoux L, Fuller M, Roberts A, Manganne R, Blanchard E, Goree W (1993) Improvements in long-core measurement techniques: applications in palaeomagnetism and palaeoceanography. Geophys J Int 114:651–662
20. Xuan C, Channell JET (2009) UPmag: MATLAB software for viewing and processing u-channel or other pass-through paleomagnetic data. Geochem Geophys Geosyst 10:Q10Y07. doi:10.1029/2009GC002584
21. Xuan C, Oda H (2015) UDECON: deconvolution optimization software for restoring high-resolution records from pass-through paleomagnetic measurements. Earth Planets Space 67:183. doi:10.1186/s40623-015-0332-x

INDEX